Advanced groundwater remediation

Active and passive technologies

Advanced groundwater remediation
Active and passive technologies

Edited by

F.-G. Simon

Federal Institute for Materials Research and Testing (BAM),
Berlin, Germany

T. Meggyes

Federal Institute for Materials Research and Testing (BAM),
Berlin, Germany

C. McDonald

School of Civil Engineering, University of Leeds,
Leeds, UK

Published by Thomas Telford Publishing, Thomas Telford Ltd, 1 Heron Quay, London E14 4JD
URL: http://www.thomastelford.com

Distributors for Thomas Telford books are
USA: ASCE Press, 1801 Alexander Bell Drive, Reston, VA 20191-4400
Japan: Maruzen Co. Ltd, Book Department, 3–10 Nihonbashi 2-chome, Chuo-ku, Tokyo 103
Australia: DA Books and Journals, 648 Whitehorse Road, Mitcham 3132, Victoria

First published 2002

A catalogue record for this book is available from the British Library
ISBN: 0 7277 3121 1

Typeset by Helius, Brighton and Rochester
Printed and bound in Great Britain by MPG Books, Bodmin

Contents

Contributors   xiii

Preface   xi

Part I. Basics of the pump-and-treat and permeable reactive barrier systems   **1**

1.   Groundwater remediation using active and passive processes   **3**
      F.-G. Simon, T. Meggyes and T. Tünnermeier

      1.1.   Introduction   3
      1.2.   The basics of pump-and-treat systems   3
      1.3.   The basics of PRB systems   6
             1.3.1.   Organic pollutants   6
             1.3.2.   Heavy metals   10
      1.4.   Cost comparison between pump-and-treat and PRB systems   15
      1.5.   Engineering of permeable reactive barriers   16
             1.5.1.   Construction of cut-off walls   18
             1.5.2.   Construction of reactive barriers   21
      1.6.   Outlook   28
      1.7.   Acknowledgements   29
      1.8.   References   29

Part II. Groundwater remediation engineering   **35**

**2.   Remediation of chromium-contaminated groundwater in subsurface
      Fe^0 reactor systems**   **37**
      M. Schneider

      2.1.   Introduction   37
      2.2.   Characterization of the field site and hydrogeological setting   37
      2.3.   Groundwater remediation concept   38
      2.4.   Results and discussion   41
      2.5.   References   43

3. **Current R&D needs and tailored projects for solving technical, economic, administrative and other issues concerning permeable reactive barrier implementation in Germany**   **45**
H. Burmeier, V. Birke and D. Rosenau

3.1. Introduction   45
3.2. Current status of PRB technologies worldwide   46
3.3. Development in Germany   47
3.4. Mission, goals and structure of RUBIN   48
3.5. Appendix   52
    1. Bernau   52
    2. Denkendorf   55
    3. Dresden   56
    4. Edenkoben   57
    5. Nordhorn   61
    6. Offenbach   62
    7. Rheine   65
    8. Wiesbaden   71
    9. University of Applied Sciences of North-East Lower Saxony, Suderburg (general project)   71
    10. Christian Albrechts University, Kiel (general project)   72
    11. Eberhard Karls University, Tübingen (general project)   72
3.6. References   73

4. **Engineering design of reactive treatment zones and potential monitoring problems**   **75**
S. Jefferis

4.1. Introduction   75
4.2. The location of the contamination   75
4.3. Natural RTZs   76
4.4. Reaction time   77
4.5. Reaction mechanisms   78
    4.5.1. First-order reactions   78
    4.5.2. Second-order reactions   78
4.6. Types of reactor   78
    4.6.1. Batch reactors   78
    4.6.2. Plug-flow reactors   79
    4.6.3. Stirred-tank reactors   79
    4.6.4. Reactor types used in RTZs   79
4.7. Degree of reaction in the reactor   79
4.8. Use of a reactor recycle (recycling)   80
4.9. Dispersion and diffusion   82
4.10. Short circuiting/by-passing   82
4.11. Monitoring   85
4.12. Conclusions   85
4.13. References   86

5. **Performance monitoring of a permeable reactive barrier at the Somersworth Landfill Superfund Site**   **87**
T. Sivavec, T. Krug, K. Berry-Spark and R. Focht

    5.1.   Introduction   87
    5.2.   Site description and characteristics   88
    5.3.   PRB installation and development   89
    5.4.   Cored material testing   90
    5.5.   Groundwater wells and monitoring   92
    5.6.   Hydraulic testing   97
    5.7.   Acknowledgements   99
    5.8.   References   99

Part III.   **Sorptive removal and natural processes**   **101**

6. **Metals loading on sorbents and their separation**   **103**
K. A. Matis, A. I. Zouboulis, G. P. Gallios and N. K. Lazaridis

    6.1.   Introduction   103
    6.2.   Flotation   105
    6.3.   Biosorption   106
        6.3.1.   Sorptive flotation   108
        6.3.2.   Electroflotation   110
    6.4.   Conclusion   112
    6.5.   References   112

7. **Sorption mechanisms of heavy metal ions on inorganic solids**   **115**
M. Fedoroff

    7.1.   Introduction   115
    7.2.   Distribution coefficients   116
    7.3.   Sorption isotherms   117
    7.4.   Sorption models   118
        7.4.1.   Ion exchange model   119
        7.4.2.   Surface complexation models   120
    7.5.   Speciation in solution   123
    7.6.   Sorption kinetics   123
    7.7.   Sorption mechanisms   124
    7.8.   Influence of other factors   125
    7.9.   Conclusion   126
    7.10.   References   126

8. **Experience with monitored natural attenuation at BTEX-contaminated sites**   **129**
W. Püttmann, P. Martus and R. Schmitt

    8.1.   Introduction   129

8.2.	Analytical monitoring of natural attenuation	130
8.3.	Biodegradation of saturated hydrocarbons under aerobic conditions	131
8.4.	Biodegradation of saturated hydrocarbons under anaerobic conditions	132
8.5.	Biodegradation of aromatic hydrocarbons under aerobic conditions	133
8.6.	Biodegradation of aromatic hydrocarbons under anaerobic conditions	133
8.7.	References	136

9. Heavy metal speciation and phytoextraction   **141**
D. Leštan

9.1.	Introduction	141
9.2.	Materials and methods	142
	9.2.1. Soil samples and analysis	142
	9.2.2. Sequential extraction	144
	9.2.3. Column phytoextraction experiments – disturbed soil profile	144
	9.2.4. Column phytoextraction experiments – undisturbed soil profile	144
	9.2.5. HM analysis	144
	9.2.6. Phospholipid analysis	145
9.3.	Results and discussion	145
	9.3.1. Sequential extractions	145
	9.3.2. Chelate-induced phytoextraction of lead, zinc and cadmium	148
9.4.	Conclusions	153
9.5.	Acknowledgement	154
9.6.	References	155
9.7.	Further reading	156

Part IV. Enhancing the efficiency of remediation processes   **157**

10. Development of novel reactive barrier technologies at the SAFIRA test site, Bitterfeld   **159**
H. Weiss, M. Schirmer and P. Merkel

10.1.	Introduction	159
10.2.	The SAFIRA project	159
	10.2.1. The structure of SAFIRA	160
10.3.	The SAFIRA site	160
	10.3.1. Geology and hydrogeology	160
	10.3.2. Groundwater contamination	162
10.4.	Pilot facility	163
10.5.	Reactor technologies	165
	10.5.1. Some preliminary results	165

10.6.   Conclusions   170
10.7.   Appendix   171
10.8.   References   171

11.   Electrokinetic techniques and new materials for reactive barriers   173
K. Czurda, P. Huttenloch, G. Gregolec and K. E. Roehl

11.1.   Introduction   173
11.2.   Electrokinetic techniques   173
         11.2.1.   Electrokinetic soil remediation   174
         11.2.2.   Fundamental transport processes   174
         11.2.3.   Electrode reactions   176
         11.2.4.   Application   177
11.3.   Innovative sorbents for PRBs   180
         11.3.1.   Zeolites   180
         11.3.2.   Chlorosilane surface-modified natural minerals   185
11.4.   References   190

Part V.   Groundwater remediation following mining activities   193

12.   Kinetics of uranium removal from water   195
B. J. Merkel

12.1.   Sources of uranium   195
12.2.   Toxicity of uranium   196
12.3.   Water chemistry   197
12.4.   Natural attenuation processes   198
12.5.   Case study   200
         12.5.1.   Königstein   200
         12.5.2.   Experiments with zero-valent iron (ZVI)   205
12.6.   Conclusions   206
12.7.   Appendix   207
12.8.   References   208

13.   Flooding strategies for decommissioning of uranium mines –
a systems approach   211
R. Gatzweiler, A. Jakubick, G. Kiessig, M. Paul and J. Schreyer

13.1.   Introduction   211
13.2.   Mine remediation and flooding   212
13.3.   Pump and treat versus collect and treat   214
13.4.   Flooding strategies at Wismut mines   214
13.5.   Summary and conclusions   220
13.6.   References   221

14. **Investigation into calcium oxide-based reactive barriers to attenuate uranium migration** 223
M. Csővári, J. Csicsák and G. Földing

14.1. Introduction 223
14.2. Leaching of uranium and other heavy metals from the wastes 223
14.3. Results of laboratory experiments 224
    14.3.1. Main steps of the process 225
    14.3.2. Open-air experiments 225
    14.3.3. Building reactive barriers in practice 234
14.4. Conclusions 235
14.5 References 235

Part VI. **Groundwater flow modelling** 237

15. **Observed and modelled hydraulic aquifer response to slurry wall installation at the former gasworks site, Portadown, Northern Ireland, UK** 239
R. Doherty, U. S. Ofterdinger, Y. Yang, K. Dickson and R. M. Kalin

15.1. Introduction 239
15.2. Geology and topographic setting 240
15.3. Installation phases 242
15.4. Hydrogeology 242
15.5. Numerical modelling 245
15.6. Discussion 246
15.7. Conclusion 249
15.8. Acknowledgements 249
15.9. References 249

16. **A finite-volume model for the hydrodynamics of flow in combined groundwater zone and permeable reactive barriers** 251
D. B. Das and V. Nassehi

16.1. Introduction 251
    16.1.1. Modelling approaches 252
    16.1.2. The finite-volume method 253
16.2. Formulation of the mathematical model 254
16.3. Numerical results and discussion 256
16.4. Conclusions 262
16.5. Acknowledgement 262
16.6. References 262

Part VII. **Active and passive methods – a comparison** 265

17. Technical and economic comparison between funnel-and-gate and
pump-and-treat systems: an example for contaminant removal
through sorption   267
P. Bayer, C. Bürger, M. Finkel and G. Teutsch

  17.1.   Introduction and background   267
  17.2.   Design optimization framework   268
  17.3.   Case study   268
  17.4.   Hydraulic performance evaluation   270
      17.4.1.   Conventional pump-and-treat systems   270
      17.4.2.   Innovative pump-and-treat systems   270
      17.4.3.   Funnel-and-gate systems   271
  17.5.   Evaluation of remediation costs   272
      17.5.1.   Conceptual model of sorptive removal – assumptions
              and input parameters   272
      17.5.2.   Results for pump-and-treat systems   273
      17.5.3.   Results for funnel-and-gate systems   275
      17.5.4.   Comparison of the three remediation options   278
  17.6.   Summary and conclusions   281
  17.7.   Acknowledgement   281
  17.8.   References   281

18. Engineering and operation of groundwater treatment systems:
pump and treat versus permeable reactive barriers   283
E. Beitinger

  18.1.   Evaluation of best available techniques (BATs)   283
  18.2.   Selection criteria and evaluation process   284
  18.3.   Data gathering and data gaps   292
  18.4.   Conceptual design/dimensioning criteria   295
  18.5.   Treatability studies/final design   296
  18.6.   Operation/maintenance/monitoring/risks   300
  18.7.   Costs   300
  18.8.   References   302

19. Discussion: status, directions and R&D issues   303
C. McDonald

  19.1.   Introduction   303
  19.2.   Setting the scene   304
      19.2.1.   Is there a level playing field?   304
      19.2.2.   Where do our findings come from?   304
      19.2.3.   Treating what with what?   305
      19.2.4.   Do they work and how do they compare?   306
      19.2.5.   How fundamental are the distinctions anyway?   307
  19.3.   Setting about remediation   309
      19.3.1.   What contaminants have we got here?   309
      19.3.2.   How should remediation approach be selected?   309

19.4.   Combined approaches   311
    19.4.1.   How can technologies vary over place and time?   311
    19.4.2.   What other 'treatment trains' are there?   312
19.5.   Targets and agents   313
    19.5.1.   How is the active agent selected or improved?   313
    19.5.2.   Are the chemical processes understood?   314
19.6.   Outcomes   315
    19.6.1.   How well do the reactors work?   315
    19.6.2.   How well do the system hydraulics work?   316
19.7.   Monitoring and durability   317
    19.7.1.   How are reactor systems monitored?   317
    19.7.2.   How often and for how long are they monitored?   318
    19.7.3.   For how long are the reactor systems effective?   318
    19.7.4.   Can contaminants be remobilized?   319
    19.7.5.   Can contaminants be recovered?   320
19.8.   Comparative economics   320
    19.8.1.   Have the economics of the various systems been
          investigated?   320
    19.8.2.   How is a proper basis for comparison made?   321
    19.8.3.   How do their (modelled) costs compare?   321
19.9.   Investment, risk and acceptability   322
    19.9.1.   How much remediation is affordable?   322
    19.9.2.   How can risk be reduced?   323
    19.9.3.   How can acceptance be gained for solutions?   324
19.10. Conclusion   324
19.11. References   325

Index   **327**

Contributors

Peter Bayer studied Geology at the Universities of Würzburg and Tübingen, Germany. He graduated in 1999 at the Centre for Applied Geosciences, University of Tübingen. He is now a PhD student at the University of Tübingen in the D-SITE work group 'Techno-economical Assessment of Remediation Systems'.

Eberhard Beitinger is an engineer and works as a manager at the headquarters of URS Germany.

Karen Berry-Spark received a BSc in earth sciences and an MSc in contaminant hydrogeology from the University of Waterloo, Ontario, Canada. She has over 14 years experience in soil and groundwater contamination and remediation of industrial sites. Her specific technical expertise is in the development of innovative and cost-effective strategies for remediation, risk management, and redevelopment of complex industrial sites, including permeable reactive barriers for treatment of chlorinated solvents.

Volker Birke is an environmental chemist who received his PhD in organic chemistry from the University of Hanover, Germany, in 1996. He has been working for over 10 years on several R&D and commercial projects concerning the remediation of hazardous wastes and contaminated sites, in particular on the removal of poly-halogenated pollutants through the use of innovative dehalogenation techniques.

Claudius Bürger studied geology at the University of Würzburg, Germany, the University of Texas at Austin, USA and the University of Tübingen, Germany. He graduated in 2001 at the University of Tübingen. At present he is a PhD student at the University of Tübingen in the D-SITE work group 'Techno-economical Assessment of Remediation Systems'.

Harald Burmeier is a civil engineer, and received his diploma from the University of Hannover, Germany. He has gained expertise in the remediation of contaminated sites over more than two decades. He was managing director of WCI Germany (Woodward Clyde International) before becoming, in 1997, a professor of civil engineering, water and environmental management at the University of Applied Sciences–NE Lower Saxony, Lüneburg/Suderburg, Germany, where he teaches site remediation and project management. He is currently chairman of ITVA ('Ingenieur-technischer Verband Altlasten'), the German Association of Remedial Engineers.

József Csicsák received his MSc in geology from the Eötvös Loránd University of Science, Budapest, Hungary. He has been employed by Mecsekérc Rt, the former uranium mining company of Hungary, since 1988, working in the field of hydrogeology, especially on water research and groundwater monitoring. From 2000 he has been Head of Environmental Monitoring, dealing with the consequences of uranium mining. His current field of interest is the *in situ* remediation of contaminated groundwater.

Mihály Csővári graduated in nuclear engineering from the Leningrad Technical University, Russia, and has been employed by Mecsekérc Rt, the former uranium mining company of Hungary. He was awarded his PhD in 1970. He has specialized in hydrometallurgy, primarily ion exchange processes, separation of rare earth elements and water treatment, and has 40 years experience in this field. His current focus is on the remediation of former uranium mining and milling sites, and in particular the *in situ* remediation of contaminated groundwater. From 1996 he has been lecturing at the Technical Department of Pécs University on environmental restoration in the nuclear industry.

Kurt Czurda earned his PhDs at Innsbruck University, Austria, and Budapest Technical University, Hungary. He is a professor at Karlsruhe University, and Head of the Institute for Applied Geology.

Diganta Bhusan Das obtained his bachelor of engineering (BE) degree in chemical engineering at Rajasthan University, India, in 1996. He received his PhD in chemical engineering from Loughborough University, UK, in 2001. At present he is a senior research fellow at Delft University of Technology, The Netherlands, and a visiting fellow at Princeton University, USA.

Keith Dickson gained his BSc in geology from Collingwood College, University of Durham, UK, in 1997 and his MSc in engineering geology from the University of Leeds, UK, in 1998. He was a geotechnical engineer with Arcadis Geraghty & Miller International Inc., Leeds, from 1998 to 2000. Since 2000 he has been a research assistant at the Questor Centre EERC, Queen's University Belfast.

Rory Doherty is currently a research fellow with the Environmental Engineering Research Centre (EERC) at Queen's University Belfast. Previously, he worked as an environmental engineer specializing in *ex situ* bioremediation for Bilfinger & Berger UK. His qualifications include an MSc in environmental biogeochemistry from the University of Newcastle upon Tyne, and a BSc in geology from Queen's University Belfast.

Michel Fedoroff is the scientific coordinator of the solid–liquid and solid–gaseous interface activities at the Centre National de la Recherche Scientifique, Centre d'Etudes de Chimie Métallurgique (CECM), Vitry sur Seine, France.

Michael Finkel graduated in civil engineering at the University of Stuttgart, Germany, in 1990. From 1990 to 1994 he was employed at an environmental consulting company, working on water resources and infrastructure planning, where he led the groundwater modelling group. He received his PhD in 1998. Since 1998 he has been undertaking postdoctoral research at the Centre for Applied Geoscience in Tübingen, and is head of the D-SITE working group 'Techno-economical Assessment of Remediation Technologies'.

Robert Focht received a BASc in chemical engineering and an MSc in hydrogeology from the University of Waterloo, Ontario, Canada. He has been involved with granular iron technology since the start of his MSc in 1992. After completion of his MSc he remained at the University of Waterloo, focusing on enhancements to the technology. During this time he became involved with EnviroMetal Technologies Inc., joining the company in 1995. His responsibilities include evaluating innovative construction methods for permeable reactive barrier installation, which involves research on the interaction between biopolymers and granular iron technology.

Gábor Földing obtained his MSc in geology from the Eötvös Loránd University of Science, Budapest, Hungary. He has been employed by Mecsekérc Rt, the former uranium mining company of Hungary, since 1988. He has 5 years of experience in environmental geology and hydrogeology, focusing on groundwater monitoring and hydraulic testing. He has been Head of Data Processing since 2000, working on databases and GIS applications. His current field of interest is the *in situ* remediation of contaminated groundwater.

George P. Gallios earned his BSc in chemistry in 1980 and his PhD in 1987 at the University of Thessaloniki, Greece. Since 1994 he has been an assistant professor at the same institute.

Rimbert Gatzweiler was Head of the Department of Conceptual Planning, Engineering and Permitting in Wismuth GmbH before his retirement in September 2001. From 1992 he has held an Honorary Professorship in the Faculty of Mining, Metallurgy and Geosciences of RWTH (Rhineland–Westphalia Technical University) Aachen.

Gabi Gregolec gained her MSc in geology from Karlsruhe University, Germany. She is currently a PhD student in Professor Czurda's research group at Karlsruhe University.

Petra Huttenloch gained a master's degree in geology from Karlsruhe University, Germany. She finished her PhD thesis in 2002 in Professor Czurda's research group at Karlsruhe University.

Alex Jakubick received his BSc in geotechnical engineering from the Charles University of Prague, Czech Republic, in 1968 and his PhD in environmental physics from the Ruprecht-Karl University of Heidelberg, Germany, in 1972. After working at the Institute of Nuclear Waste Management at the Karlsruhe National Research Centre, Germany, and at the Canadian electric utility Ontario Hydro, Toronto, he joined Wismut GmbH in Chemnitz, Germany, in 1994. Since 2001 he has been Divisonal Head of Planning and Engineering at Wismut GmbH.

Stephan Jefferis graduated in natural sciences and chemical engineering from the University of Cambridge. He went on to King's College London, where he was awarded a PhD in soil mechanics in 1971, and in 1992 an MSc in construction law and arbitration. From 1971 to 1987 he was a lecturer and then a reader in geotechnical processes at King's College London. He was briefly at Queen Mary and Westfield College before joining Golder Associates, a major international geo-environmental consulting company in 1992. He moved to the University of Surrey in February 2000, where he is Professor in Civil Engineering and Director of Research for the School of Engineering in the Environment. He leads their geo-environmental research group.

Bob Kalin is Professor and Head of the Environmental Engineering Research Centre, School of Civil Engineering, Queen's University Belfast, UK.

Gunther Kiessig has held leading technological positions at Wismut GmbH, Chemnitz, Germany.

Thomas Krug has a BSc and MSc in chemical engineering from Queen's University, Kingston, Ontario, Canada. He has over 15 years of private consulting experience in water and wastewater treatment and in innovative and cost-effective approaches to remediation, management and redevelopment of contaminated properties. He is currently managing several projects which involve the design and installation of permeable reactive barriers for the treatment of chlorinated solvents in groundwater and innovative source removal technologies for DNAPL sites.

Nicholaos K. Lazaridis earned his BSc in chemistry in 1979 and his PhD in chemical technology in 1990 at the University of Thessaloniki, Greece. Since 2000 he has been a lecturer at the same institute.

Domen Leštan obtained his degree in chemical technology from the University of Ljubljana, Slovenia, in 1986, and his DSc in chemistry in 1992, and finished his postdoctoral study at the US Department of Agriculture, Forest Product Laboratory, Madison, Wisconsin, USA, in 1996. In 1998 he was elected an assistant professor at the Biotechnical Faculty, University of Ljubljana, where he has been employed since 1996.

Peter Martus obtained his diploma in geology from the University of Karlsruhe, Germany, in 1996. From 1996 to 1998 he worked as an environmental geologist in industry. Since 1999 he has been a PhD student at the University of Frankfurt, taking his final examination in October 2002.

Kostas A. Matis gained his degree in chemistry in 1973 from Aristotle University in Thessaloniki, Greece. He received an MSc in 1975 and a PhD in 1977 in chemical engineering from the University of Newcastle, UK. He is now a professor at Aristotle University, and is the Deputy Head of the Department and Director of the Division of Chemical Technology.

Tamás Meggyes graduated in petroleum engineering from the Technical University of Heavy Industry, Miskolc, Hungary. He has a doctor's title in fluid mechanics and a PhD in landfill engineering from Miskolc University. He has been a research coordinator at the Department of Environmental Compatibility of Materials, Federal Institute for Materials Research and Testing (BAM), Berlin, since 1990, and has participated in major national and European environmental research programmes. His main fields of interest include fluid mechanics, hydraulic transport, landfill engineering, groundwater remediation and tailings facilities.

Broder J. Merkel received his diploma in geology and hydrogeology in 1978, and his PhD from the Technical University Munich in 1983. He became a professor of hydrogeology at the University of Kiel in 1992, and in 1993 became a full professor in hydrogeology and environmental geology at the Technical University Bergakademie Freiberg. He is currently Head of the Department of Geology and Dean of the Faculty for Geoscience, Geotechnics and Mining.

Peter Merkel is presently scientific coordinator of the SAFIRA research programme. He holds a PhD from the University of Tübingen, awarded in 1986.

Chris McDonald earned his PhD in city and regional planning at the University of Pennsylvania, USA, in 1972. He has worked in transportation, river basin and minerals planning, as an engineer and project manager on major land reclamation and waste disposal schemes in the Yorkshire coalfield, and at senior levels for UK local and central government. He currently holds research and teaching posts in waste management at the School of Civil Engineering at the University of Leeds, UK.

Vahid Nassehi obtained his first MSc, in chemical engineering, at the University of Teheran, Iran, in 1974 and was awarded a second MSc degree at the University of Wales, UK, in 1978. He also has a PhD in chemical engineering from the University of Wales, gained in 1981. He is Professor of Computational Modelling at Loughborough University, UK.

Ulrich Ofterdinger obtained his PhD from ETH Zurich, Switzerland, on groundwater systems in fractured crystalline rocks of the Swiss Alps. Prior to his PhD he studied at Kiel and Ghent University, UK, working on the field and laboratory determination of the hydraulic conductivity of natural and artificial liner materials. He is presently working at Queen's University Belfast, UK, with Professor Kalin, on contaminant hydrogeology and the application of reactive barrier technology for the remediation of contaminated groundwater.

Michael Paul has a diploma in geology and a PhD in mineralogy, both gained at the University of Greifswald, Germany. He joined Wismut GmbH in Chemnitz, Germany, in 1991, and has more than 10 years of experience in hydrogeology, geochemistry, exposure pathway analysis and risk assessment for tailings ponds and mining sites, including conceptual work for tailings pond remediation, mine flooding and waste rock relocation projects. Since 2001 he has been Head of Engineering at Wismut GmbH.

Wilhelm Püttmann received his diploma in chemistry at the Institute for Organic Chemistry of the Universität Cologne, Germany, in 1977, and his PhD in 1980 and his postdoctorate in 1994 at the same institute. Since 1996 he has been a professor at the Johann Wolfgang Goethe University, Frankfurt, and Director of their Centre for Environmental Research.

Karl Ernst Roehl gained his PhD at Karlsruhe University, Germany. He is a scientific assistant at the Institute for Applied Geology in Karlsruhe.

Diana Rosenau is a civil engineer and received her diploma in 2000 from the University of Applied Sciences-NE Lower Saxony, Germany, with work on the use of reactive walls for the remediation of contaminated sites.

Mario Schirmer has been with the UFZ Centre for Environmental Research Leipzig-Halle, Germany, since 1999, and is presently a postdoctoral researcher. He received an MSc in geophysics from Freiberg University of Mining and Technology, Germany, 1991 and a PhD in earth sciences from the University of Waterloo, Canada, in 1999.

Reinhard Schmitt gained his diploma in geology at Rhineland-Westphalia Technical University, Aachen, Germany, in 1994. He completed his PhD thesis at the same institute in 2000 on the anaerobic *in situ* metabolism of aromatic hydrocarbons. Since 2000 he has been working as a consultant.

Michael Schneider received his diploma in geology and his PhD from the Berlin Technical University, Germany. He has been a consultant in hydrogeology and environmental and engineering geology since 1987, and a lecturer on hydrogeology at the Free University of Berlin since 1999.

Jochen Schreyer received his diploma in chemistry from TH Merseburg, Germany, in 1979, after which he was employed by Wismut GmbH in Chemnitz, Germany. Since 1991 he has been a project manager for mine rehabilitation in Saxonia, including water treatment and engineering for the flooding of the Königstein mine.

Franz-Georg Simon graduated in chemistry from the University of Frankfurt/Main, Germany, in 1985, and gained his PhD in atmospheric chemistry at the Max Planck Institute for Chemistry, Mainz, Germany, in 1989. He is currently a professor and Director and Head of the Division of Waste Treatment and Remedial Engineering at the Federal Institute for Materials Research and Testing (BAM), Berlin. Before joining BAM he was group leader of the Waste and Residues Treatment Division at the ABB Corporate Research Centre, Baden, Switzerland.

Timothy Sivavec is a systems integration leader in the Information and Decision Technology Laboratory at the General Electric Global Research Center, Niskayuna, New York, USA. He has been at General Electric since 1986. He received a BSc in chemistry at William and Mary College, London, in 1982 and a PhD in chemistry from Columbia University, USA, in 1986. His areas of research at General Electric have included the development and application of a variety of remediation and pollution prevention technologies. Recently, his research has focused on groundwater restoration technologies with an emphasis on permeable reactive barriers. He has also been developing new site characterization and monitoring technologies.

Georg Teutsch completed his study in geology and hydrogeology at the University of Tübingen, Germany, and University of Birmingham, UK, in 1980, and was awarded an MSc in hydrogeology. In 1988 he gained his PhD in the field of groundwater modelling in karst aquifers, and in 1991 his postdoctorate in the study of heterogeneous subsurface environments. He then became a professor of geohydrology at the University of Stuttgart, Germany. Since 1993 he has been Director of the Chair of Applied Geology at the Geological Institute of Tübingen, Germany, where he heads the new Centre for Applied Geoscience, which comprises more than 50 academics from different fields. He is the chairman of the recently established EC ANCORE network, which comprises more than 60 research institutions throughout Europe.

Torge Tünnermeier received a diploma in geology from the Freiberg University of Mining and Technology, Germany. He was a scientist at the Laboratory of Remedial Engineering, Federal Institute for Materials Research and Testing (BAM), Germany.

Holger Weiss has been with the UFZ Centre for Environmental Research Leipzig-Halle, Germany, since 1992, and is presently Head of the Department of Industrial and Mining Landscapes. He has been an associate professor at the University of Quebec, Montreal, Canada, since 1997, and a lecturer in environmental geology at the University of Tübingen, Germany, since 2000.

Yuesuo Yang obtained his PhD from Changchun University of Science and Technology, China, in 1995 on hydrogeology and water resources. He is currently working at Queen's University Belfast, UK, as a research officer on hydrogeology, numerical modelling and design for the remediation of contaminated land and groundwater. Previously he was employed by Changchun University of Science and Technology and Jilin University, China, as an associate professor and professor. He has experience in regional water resources management, GIS applications for water resources and multi-variant analysis of water quality data. He is interested in validation of numerical models by various methods such as field tests, and isotopic and geochemical interpretation.

Anastasios Zouboulis received his PhD in 1986 from Aristotle University in Thessaloniki, Greece, where he has held several academic positions (scientific researcher, lecturer and assistant professor). He has been an associate professor at Aristotle University since 1996 .

Preface

Life was created in water, and water has been our inseparable fellow traveller through history. Life is impossible without water; our bodies and those of animals and plants consist overwhelmingly of water. Half of our blood is water; the water content is above 90% in new-born babies and only drops below 70% during old age. We can survive without food for several weeks, but without water the chances of survival drop drastically after 3 days.

Seventy-one per cent of the earth's surface is covered with water: this is why the weather is mainly determined by meteorological phenomena generated by the oceans. Changes among the three phases of water – ice, liquid water and water vapour – not only contribute to shaping the terrain of the continents but influence our everyday lives. Major shifts in the phase balance of the earth's water reserves during glacial and interglacial periods have resulted in tremendous alterations in the extent of dry land, due to rises and falls in sea level.

The continents' supplies of fresh water stem from the oceans, where 97% of the earth's water reserves are stored. The ongoing circulation of water from the oceans – through clouds and rain, followed by water infiltration into the uppermost layers of the earth's crust, and then through the rivers back to the oceans – is maintained by the sun, the source of our energy. This water circulation is of fundamental importance to all living organisms, since some of the main sources of drinking water are the water bodies in various soil and rock layers. In Germany, for instance, about three-quarters of drinking water supplies are extracted from groundwater.

Due to the inseparability of human and animal life from some form of fresh water, human settlements started near spring lines, rivers, lakes or, more rarely, other sources of water (snow or ice). Our life and activities developed around the ancient centres of such water-side settlements over historical eras. Using the earth's mineral and vegetable resources, gradually we became able to produce objects useful in our everyday lives (tools, clothes, buildings, etc.), plus those for helping to hunt and kill each other (weapons). As time passed, we began to produce more sophisticated materials and objects (metals, chemicals, etc.).

Technological developments, and increasing demand for better products, created more intensive large-scale agriculture and industry, which were not only able to generate value and wealth but introduced a new and disadvantageous type of activity: environmental pollution. One example is the Bitterfeld Region in Germany, where in the course of a century, open-pit lignite mining and chemical industries have contaminated the groundwater over an area of 25 km² with a total volume of approximately 200 million m³. Similar examples of extensive, polluted areas can be found in, for example, Silesia, the British Midlands, the Ruhr District, Erzgebirge, and former uranium mining areas (e.g. Southern Hungary).

Sadly, polluting the environment has not been a privilege of our generation alone: even the 'clever' ancient Greeks managed to pollute land and sea with the waste products of their lead mining, metallurgy and processing, some 2000 years ago. However, the increasing speed of technological advances, especially after the Industrial Revolution, multiplied the amounts and types of emissions, and have added more complex types of contaminant to the established ones: chemicals, toxic heavy metals, dense non-aqueous phase liquids (DNAPLs), radioactive wastes, etc. In doing so, they have generated a steadily increasing threat to our environment, and our drinking water in particular.

This situation has called for remedies to prevent mankind from poisoning its own water resources. In the second half of the twentieth century, increasing efforts were made towards containing and cleaning contaminated groundwater. A number of techniques can be used to achieve such goals, traditionally either by treating or isolating the soil *in situ*, or by removing it for washing or disposal. However, such technologies tend to be accompanied by high energy consumption, and can lead to new environmental problems. More efficient and economical techniques are therefore needed to remediate contaminated soil and groundwater.

The easiest approach appeared to be to apply well-established methods of waste water treatment. This gave rise to the so-called 'pump-and-treat' method, in which contaminated groundwater is removed from the ground by pumping and treated in a treatment plant on the surface. It has the advantage of using proven techniques, is easy to control and the treated groundwater can be re-injected into the ground or discharged into rivers or lakes. The main disadvantages are that it disturbs the groundwater flow regime and requires steady energy and other inputs. Also, certain slowly emitted contaminants, e.g. polyaromatic hydrocarbons (PAHs) with low bio-availability, and contaminants in heterogeneous sediments, are not easily accessible by this technique.

In contrast, permeable reactive barriers (PRBs) are a relatively new technology for groundwater remediation: a trench arranged downstream of the contaminant source, and filled with reactive material, allows the treatment of contaminated groundwater passing slowly through. Chemical, physical and biological methods have so far been used successfully, with reduction, precipitation, adsorption and oxygen release to promote biodegradation as the most common reactions. The targeted chemicals to be removed from the groundwater by the reactive agents are either decomposed to other, less dangerous, compounds or fixed to the reactive material. Their mobility, availability and toxicity may be significantly reduced in both cases.

Groundwater remediation using PRBs is an *in situ* method with low energy demand and offers the prospect of a more cost-effective remediation technique, significantly enhancing natural groundwater protection. However, there are still many questions to be answered about the installation, cost-effectiveness and long-term behaviour of PRBs.

Many of the reactions utilized in PRBs are similar to or the same as those occurring in natural attenuation, which may be defined as a means by which the concentration of groundwater pollution is reduced to an acceptable level by natural processes. In that context, adsorption, biological degradation, cation and anion exchange, dilution, filtration and precipitation reactions are the dominating attenuation mechanisms.

The main advantage of natural attenuation is its cost-effectiveness: in most cases costs are only incurred in connection with monitoring. Its disadvantages are the long time necessary to produce results, and the limited possibilities for influencing the processes. Attempts to provide more favourable conditions to accelerate attenuation (enhanced natural attenuation) tend to involve some of the techniques and expense of other remediation approaches. Research and development is underway, in Europe and world-wide, on contrasting approaches to these remediation techniques.

This volume contains a selection of the most up-to-date research results on reactive barrier and pump-and-treat techniques, with the main focus on heavy metal removal. After dealing with general issues of groundwater remediation using active and passive processes, remediation engineering practice is discussed, with the emphasis on chromium remediation, innovative applications, engineering design and performance monitoring. There is particular focus on sorptive removal and on natural processes, including metals loading on sorbents, sorption mechanisms, monitored natural attenuation and phytoextraction. Novel approaches aim at enhancing the efficiency of remediation processes and involve new barrier technologies and electrokinetic techniques.

Uranium mining is one of the most infamous sources of heavy metal pollution: its mitigation is the objective of investigations into the kinetics of uranium removal, flooding strategies in uranium mine decommissioning and calcium oxide-based reactive barriers. Groundwater flow modelling is looked into, both in regard to numerical simulation and from an engineering viewpoint. Funnel-and-gate and pump-and-treat methods are compared, using sorption as an example and considering engineering aspects.

The research and results presented in this book are mainly based on the workshop 'In-situ Reactive Barriers Versus Pump-and-Treat', jointly organized by the Division of Waste Treatment and Remedial Engineering, Department of Environmental Compatibility of Materials, Federal Institute for Materials Research and Testing (BAM), Berlin, Germany, and the School of Civil Engineering of the University of Leeds, UK, on 18–19 October 2001 and held at BAM. The workshop was supported by the Groundwater Pollution (GPoll) Programme of the European Science Foundation (ESF) (Website: http://www.esf.org), and their support is gratefully acknowledged.

The GPoll Programme lasted 4 years, from January 1998 until December 2001. The activities included workshops, exchange grants and summer schools. The ESF supports a limited number of exploratory workshops each year, which allow leading

European scientists to meet and explore novel ideas at the European level, with the aim of spearheading new areas of research. GPoll was a long-term programme that initiated and promoted multinational, multidisciplinary research on groundwater pollution caused by toxic chemicals, metals, pathogenic organisms, radionuclides and excess nutrients. It focused on the fate of pollution in groundwater systems, because of their significance to human and environmental health. The research had an urgency, due to recent sharp increases in the incidence of pollution-related disorders and the increasingly conspicuous damage caused by pollution to natural ecosystems. Since most groundwater participates in the hydrological cycle, the residence time may vary from months to centuries. The processes causing groundwater pollution may, therefore, often become apparent only in the long term. European collaboration provides added value, since pollution is international, and environmental problems require interdisciplinary research.

BAM's Department of Environmental Compatibility of Materials is concerned with the environmental assessment of materials and the investigation of long-term interactions at the material–environment interface, covering technical, ecological and economic aspects. Activities within the Division of Waste Treatment and Remedial Engineering are focused on the minimization of pollutants in soil and groundwater, on processes for their remediation and on the conversion and reuse of contaminated material. The workshop benefited from extensive help by BAM's personnel, which is greatly appreciated.

It is hoped that this book will contribute to improving groundwater quality: firstly, through enhanced efficiencies in research investment, by avoiding needless overlap/competition between approaches; secondly, by achieving a clearer recognition of situations for which a particular approach is best suited, and of the specialist capabilities essential to the informed delivery of each; and, finally, by achieving recognition for the value and maturity of the whole field of groundwater remediation.

F.-G. Simon
T. Meggyes
C. McDonald

Part I
Basics of the pump-and-treat and permeable reactive barrier systems

1. Groundwater remediation using active and passive processes

F.-G. Simon, T. Meggyes and T. Tünnermeier
Federal Institute for Materials Research and Testing (BAM),
Unter den Eichen 87, D-12205 Berlin, Germany

1.1. Introduction

Groundwater quality is of great public interest these days. In Germany the new Soil Protection Act and Ordinance describe quality goals for groundwater and define action, trigger and precautionary values for the soil–groundwater pathway (Federal Ministry for the Environment (D), 1998, 1999). This was not always the case. Thirty years ago it was believed that groundwater was protected by natural filters. However, groundwater contamination can occur in various ways, from the ground surface (by infiltration of polluted surface water, spills, waste dumps, tailings facilities, airborne particles, agricultural activities, etc.), from above the water table (leaks in pipelines or underground storage tanks, leachates from landfills, cemeteries, etc.), or from below the water table (mines, drainage wells, canals, groundwater withdrawal, etc.) (US Office of Water, 1990). Different remediation options exist for contaminated groundwater: containment of the contaminants to prevent migration (e.g. using geotechnical measures), removal of the pollution (e.g. by excavation and subsequent soil washing), groundwater treatment at the point of use, removal of the groundwater for treatment and replacement, or remediation of the groundwater *in situ*. The present work compares pump-and-treat systems and those using *in situ* permeable reactive barriers (PRBs) for groundwater remediation in more detail, because they exhibit similar features in their underlying processes. Information on a wider range of remediation technologies can be found elsewhere (Suthersan, 1996).

1.2. The basics of pump-and-treat systems

Pump-and-treat systems have been used for more than 20 years in the remediation of groundwater contamination. In these systems, contaminated groundwater is extracted from the ground, treated overground and finally discharged or re-injected. Figure 1.1 illustrates their function.

The contaminated groundwater can be treated in a variety of ways. Table 1.1 lists the most common processes, which are described in more detail in the following sections. Passive systems using PRBs apply the same treatment principles *in situ*.

A major drawback of pump-and-treat systems is the rather long operational time needed for proper remediation. The concentration of the contaminants received in the pumping wells decreases only slowly during operation, and residual concentrations can still exceed the clean-up standards. When pumping is stopped because the concentration of contaminants pumped has reached low levels, it is often observed that contaminant concentration rises again. These phenomena are called tailing and rebound, and are illustrated in Fig. 1.2. Among the reasons for this behaviour are contaminant desorption, dissolution of precipitates and variation of the ground-water flow velocity (Eastern Research Group, 1996).

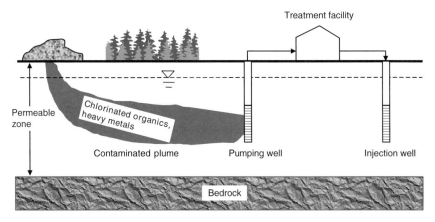

Fig. 1.1. Example of a pump-and-treat system

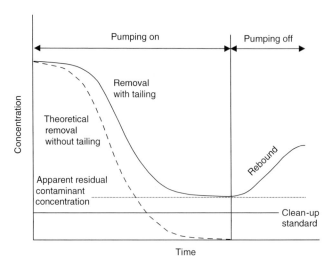

Fig. 1.2. Concentration as a function of pumping on- and off-phases showing tailing and rebound effects (Eastern Research Group, 1996)

Table 1.1. Widespread groundwater contamination and applicable treatment processes (Eastern Research Group, 1996)

Contaminant	Treatment process											
	Precipitation	Oxidation	Reduction	Adsorption to surfaces	Ion exchange	Biological degradation	Electrochemical treatment	Ultraviolet/ozone	Air stripping	Flotation	Gravity separation	Distillation
Heavy metals	+		o	o	+		+			+	+	
Cr^{VI}	+		+	o	+		+					
Mercury	+	o	+	+	+					o	o	
Arsenic		+		o	+			o				
VOCs		+		+	o	o		o	+			+
SVOCs		+		+	+	+		o		o	o	+
Pesticides		+		+	+	o		+		o	o	+
PCBs	+	+			+	o		+		+	+	+
Mineral oil	+				+	o		+		+	+	+

VOCs, volatile organic compounds; SVOCs, semivolatile organic compounds; PCBs, polychlorinated biphenyls
+, applicable; o, potentially applicable

1.3. The basics of PRB systems

PRB technology is a novel groundwater remediation method which enables physical, chemical or biological *in situ* treatment of contaminated groundwater by means of reactive materials. The most commonly used processes/reactions are: redox reactions, precipitation, adsorption, ion exchange and biodegradation.

The reactive materials are placed in underground trenches downstream of the contamination plume, forcing it to flow through them. By doing so, the contaminants are treated without soil excavation or water pumping (Smyth *et al.*, 1997). Generally, this cost-effective clean-up technology impairs the environment much less than other methods. The arrangement of a PRB is illustrated in Fig. 1.3.

Most PRBs are arranged in one of the following two patterns: (1) continuous reactive barriers enabling a flow through their full cross-section, and (2) funnel-and-gate systems in which only special 'gates' are permeable to the contaminated groundwater.

Remediation costs using PRB systems are reported to be up to 50% lower than those of the pump-and-treat methods used so far (Schad and Gratwohl, 1998). Their operational life is expected to be 10–20 years; however, experience over such periods is not yet available.

This chapter is organized with respect to classes of pollutant rather than the various remediation processes. The most extensively studied and applied reactive barrier type uses granulated iron particles. Here, elemental iron acts as a reducing agent and generates a ferrous ion by the redox reaction

$$Fe \rightarrow Fe^{2+} + 2e^- \tag{1}$$

However, other processes (see Table 1.1) are also used in both PRBs and pump-and-treat-systems.

1.3.1. Organic pollutants

1.3.1.1. Halogenated hydrocarbons

A well-studied application of iron-based PRBs is the degradation of halogenated hydrocarbons (R–X) (Gillham and O'Hannessin, 1994). Numerous patents exist for this application (Gillham, 1993; Cherry *et al.*, 1996; Gillham *et al.*, 1999). The pollutants are removed from the groundwater by a redox reaction, which can be written as follows:

$$R-X + Fe + H_2O \rightarrow Fe^{2+} + R-H + OH^- + X^- \tag{2}$$

A redox reaction consists of two half-reactions, i.e. the oxidation of one species and the reduction of another. In iron-based PRBs the elemental iron acts as the reducing agent, and is oxidized to the ferrous ion Fe^{2+} (see reaction (1)). The potential of this half-reaction is independent of the pH, and is given as 440 mV in tables of the electrochemical series (in which the reaction is displayed in the opposite direction, $Fe^{2+} + 2e^- \rightarrow Fe$ with a potential of –440 mV, because the oxidized form is always written on the left-hand side).

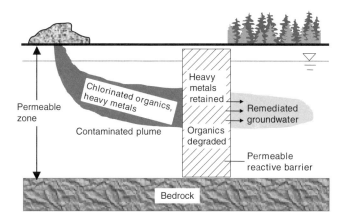

Fig. 1.3. Groundwater remediation using a PRB

The reduction of chlorinated hydrocarbons can proceed via three different mechanisms (Matheson and Tratnyek, 1994). Tetrachloroethene is taken as an example for the following reactions, which represent the three possibilities:

(a) Direct transfer of electrons:

$$C_2Cl_4 + H^+ + 2e^- \rightarrow C_2HCl_3 + Cl^- \tag{3}$$

(b) Reduction by Fe^{2+}:

$$2Fe^{2+} \rightarrow 2Fe^{3+} + 2e^- \tag{4}$$

$$C_2Cl_4 + H^+ + 2e^- \rightarrow C_2HCl_3 + Cl^- \tag{5}$$

(c) Reduction by H_2:

$$2H_2O + 2e^- \rightarrow H_2 + 2OH^- \tag{6}$$

$$C_2Cl_4 + H_2 \rightarrow C_2HCl_3 + H^+ + Cl^- \tag{7}$$

In mechanism (b) the ferrous ion Fe^{2+} is the reducing agent. This reaction has a potential of -770 mV (in standard form $Fe^{3+} + e^- \rightarrow Fe^{2+}$, $E^0 = 770$ mV) and is also independent of pH. In mechanism (c), water is reduced and elemental hydrogen is generated, which acts as the reducing agent in a catalytic reaction. It has been shown that the presence of elemental iron surfaces is necessary for the dehalogenation and that addition of hydrogen and Fe^{2+} has no influence on the degradation. Therefore, it is most likely that the dehalogenation proceeds via sequence (a). More recent investigations postulate surface-bound Fe^{2+} to be involved in the reaction (Sivavec and Horney, 1995). In any case, reaction (6) does take place, and the production of hydrogen in PRBs is significant (Puls, 1998). In aerobic systems, oxygen is a preferred oxidant for iron, leading to rapid corrosion:

$$O_2 + 2H_2O + 4e^- \rightarrow 4OH^- \tag{8}$$

Dissolved oxygen is rapidly consumed at the entrance of an iron-bearing barrier (reaction (8)). The resulting precipitates could have an influence on the hydraulic performance (Mackenzie *et al.*, 1995). The effect of other microbiological and geochemical processes on the long-term performance of barriers with zero-valent iron was studied by Gu *et al.* (1999). According to their results, the function of the walls could be impaired by the accumulation of iron hydroxides, carbonates and sulfides due to decreased permeability and reactivity.

In reactions (6) and (8) OH⁻ is the product, which also leads to an increase in pH in the PRBs (Puls, 1998). The rate of reaction is decreased at elevated pH values, and a linear dependency of the observed rate and pH was measured (Matheson and Tratnyek, 1994). The reason for this behaviour is precipitation of slightly soluble iron carbonates and oxides on the reactive iron surface (Vogan *et al.*, 1995; Sivavec, 1996). Other researchers (Johnson and Tratnyek, 1995) have discussed the precipitation products siderite ($FeCO_3$), goethite ($FeOOH$) and 'green rust' ($Fe_2(OH)_4Cl \cdot nH_2O$), depending on the composition of the groundwater. In column experiments in laboratory studies it was observed that the permeability was decreased by the precipitates (Sivavec, 1996). However, in pilot-scale and full-scale installations the function was not impaired by precipitation reactions.

Redox reactions occur spontaneously if the sum of the potentials of the two half-reactions is positive. The potentials of half-reactions with organic compounds are not readily available in the literature. However, they can be calculated from the free energy ΔG_R^0 of the reaction. The redox reaction takes place in aqueous solutions so that gas phase values have to be transformed to aqueous phase values using the following equation (Vogel *et al.*, 1987):

$$\Delta G_f^0(\text{aq}) = \Delta G_f^0(\text{g}) + RT \ln H \tag{9}$$

where H is the Henry's law constant.

ΔG_R^0 can be transformed to the potential E^0 by dividing by nF, where n is the number of exchanged electrons and F is the Faraday constant:

$$E^0 = -\Delta G_R^0 / nF \tag{10}$$

Reaction (3) is dependent on pH because H^+ is involved, and the potential for a pH of 7 is 580 mV (Vogel *et al.*, 1987). The potentials for various aliphatic hydrocarbons vary between 500 and 1500 mV (Matheson and Tratnyek, 1994). Combining reactions (1) and (3) yields

$$C_2Cl_4 + Fe + H_2O \rightarrow Fe^{2+} + C_2HCl_3 + OH^- + Cl^- \tag{11}$$

Reaction (11) occurs spontaneously because the sum of the potentials of the two half-reactions is positive ($E = 580 + 440 = 1020$ mV). Trichloroethene as product of reaction (11) is further degraded stepwise in hydrogenolysis reactions to ethene:

$$C_2HCl_3 + Fe + H_2O \rightarrow Fe^{2+} + C_2H_2Cl_2 + OH^- + Cl^- \tag{12}$$

$$C_2H_2Cl_2 + Fe + H_2O \rightarrow Fe^{2+} + C_2H_3Cl + OH^- + Cl^- \tag{13}$$

$$C_2H_3Cl + Fe + H_2O \rightarrow Fe^{2+} + C_2H_4 + OH^- + Cl^- \tag{14}$$

1.3.1.2. Sorptive removal of organic pollutants

Sorptive removal is a good solution for the attenuation of non-halogenated organic compounds in groundwater. Organic contaminants are retarded by any natural material with a high organic carbon content (Cohen, 1991). Even more efficient is adsorption on to activated carbon, which is already used for *ex situ* cleaning of drinking water (Jekel, 1979). Sorption rates and permeability must be optimized for the application of sorption in PRBs, and long regeneration cycles (of the order of several years) are required. Sorption is a function of temperature, and is favoured at low temperatures. At elevated temperatures, desorption predominates. The behaviour can be described with isotherms which show the amount of sorbed material plotted against concentration in the solution. In most cases there is no linear dependency. The Freundlich-type adsorption isotherm gives a good approximation of the ability to remove pollutants:

$$K_d = C_s/C_w = K_{Fr}C_w^{1/n-1} \tag{15}$$

where K_d is the distribution coefficient (l/kg), C_s is the sorbed concentration (mg/kg), C_w is the groundwater concentration (mg/l), K_{Fr} is the Freundlich sorption coefficient and $1/n$ the Freundlich exponent (which lies between 0 and 1, $1/n = 1$ yielding a linear behaviour).

K_d decreases with increasing aqueous concentration of the solute. For hydrophobic compounds removed by natural adsorbents, K_d is related to the organic carbon content of the solid. In addition, an empirical correlation exists with the water–octanol partition coefficient (Gratwohl and Peschik, 1997). The Freundlich sorption isotherm is widely applied, but has the fundamental problem that there is theoretically no upper sorption limit. Unlike Freundlich's isotherm, the Langmuir adsorption isotherm has a curved shape and approaches a maximum asymptotically (Fetter, 1993).

The retardation R_d of contaminants, which is the degree of reduction of velocity due to the sorption, can be calculated from K_d according to the following equation (Fetter, 1993):

$$R_d = 1 + K_d\rho/\Theta \tag{16}$$

where ρ is the bulk density and Θ is the porosity.

ρ and Θ for activated carbon are around 0.5, so that the retardation factor R_d is approximately equal to K_d. For benzene, toluene, ethylbenzene and xylene (BTEX) compounds and polyaromatic hydrocarbons (PAHs) the values of K_d on activated carbon (F300) are between 10^3 and 10^7 l/kg (Gratwohl, 1997; Schad and Gratwohl, 1998) at equilibrium. The K_d values obtained from kinetic experiments are lower because equilibrium conditions are reached only after 10^2–10^3 hours. Therefore, in spite of the high retardation factors, a residence time long enough to provide efficient contaminant removal from the groundwater flow has to be ensured by design, permeability, etc. Aspects of the sorption kinetics are discussed elsewhere (Fetter, 1993; Schad and Gratwohl, 1998).

The removal of organic contaminants by sorption in funnel-and-gate systems is enhanced by biological degradation. However, this could lead to a decrease in the permeability of the barrier, known as bio-fouling (Gratwohl, 1997). This effect was also observed in iron PRBs (Powell *et al.*, 1998) with iron-oxidizing microbes. Reduction in permeability is undesirable in the reaction zone, i.e. the gate in a funnel-and-gate system. However, this phenomenon suggests an interesting application for the construction of the funnels themselves. In one such case, flow rates were reduced by 99.9% and above, and it was possible to maintain them for 4 months without nutrient addition (see Section 1.5.2.9 on biobarriers).

1.3.2. Heavy metals

Cationic metals usually have limited mobility in soil and groundwater with a high clay and organic content, high alkalinity and low permeability (Fetter, 1993). However, complexing agents such as carbonates, hydroxides, sulfates, phosphates, fluorides and, possibly, silicates, which are present in natural waters, increase the solubility of metals (Langmuir, 1978). Precipitation is a possible means to lower the concentrations of heavy metal contamination in groundwater. Another possibility is sorption or precipitation subsequent to a chemical reduction, or a combination of the different processes.

Precipitation of heavy metals is a commonly used process in waste water treatment plants. Chemical precipitation is a process by which a soluble substance is converted into an insoluble form by a reaction with the precipitant. Frequently used precipitants are hydroxides, sulfides, phosphates and carbonates. The solubility of heavy metal hydroxides, sulfides and carbonates is pH-dependent. Metal hydroxides exhibit amphoteric behaviour, i.e. their solubility is high both at low pH (removal of hydroxide anions, see reaction (8), where Me represents a two-valent heavy metal) and high pH (formation of soluble hydroxo complexes, see reaction (19)). A minimum solubility for most heavy metals is observed between pH 9 and 11 (Chung, 1989).

$$Me^{2+} + 2OH^- \rightarrow Me(OH)_2(s) \qquad (17)$$

$$Me(OH)_2(s) + 2H^+ \rightarrow Me^{2+} + 2H_2O \qquad (18)$$

$$Me(OH)_2 + OH^- \rightarrow Me(OH)_3^- \qquad (19)$$

A cheap precipitant for forming hydroxides is lime. Metal concentrations lower than 1 mg/l, and sometimes less than 0.1 mg/l, are achievable.

The solubility of metal sulfides is much lower than that of hydroxides, and they are not amphoteric. However, at pH values above 8, hydrogen sulfide (H_2S) is emitted due to hydrolysis of the sulfides, resulting in increasing metal concentrations in the solution.

$$Me^{2+} + S^{2-} \rightarrow MeS(s) \qquad (20)$$

$$MeS(s) + 2H^+ \rightarrow Me^{2+} + 2H_2S \qquad (21)$$

PRBs are systems with low energy consumption, and it is therefore important that the precipitates are not changed back to soluble forms. Unlike in waste water treatment plants, the precipitated material remains in the barrier for the whole period of operation. Spent barrier materials can be replenished by placing the reactive matrix in a double-walled structure with prefabricated elements (Beitinger and Fischer, 1994) or canisters. The stability of the resulting precipitates is therefore an important issue for the application of precipitation reactions in PRBs.

1.3.2.1. Lead

Lead is a major contaminant in solid wastes and soils. Its concentration in drinking water is controlled in most countries to very low limits (e.g. 0.04 mg/l in Germany). Lead forms various compounds with low solubility in water. A good example of the formation of very stable precipitates is the reaction of lead ions with hydroxyapatite $(Ca_{10}(PO_4)_6(OH)_2)$. After dissolution of hydroxyapatite, lead is precipitated as lead hydroxypyromorphite (solubility product log K_{sp} = –62.79 (Allison et al., 1991; Ma et al., 1993).

$$Ca_{10}(PO_4)_6(OH)_2 + 14H^+ \rightleftharpoons 10Ca^{2+} + 6H_2PO_4^- + 2H_2O \qquad (22)$$

$$10Pb^{2+} + 6H_2PO_4^- + 2H_2O \rightarrow Pb_{10}(PO_4)_6(OH)_2(s) + 14H^+ \qquad (23)$$

Another reaction type which may control the immobilization of lead with hydroxyapatite is surface adsorption, but it was found not to be feasible by Ma et al. (1993). In the presence of chloride the precipitation of the even more insoluble chloropyromorphite $(Pb_{10}(PO_4)_6Cl_2)$ is possible (solubility product log K_{sp} = –62.79 (Allison et al., 1991; Ma et al., 1994). The reaction between heavy metals and apatite is rapid (Chen and Wright, 1997), and the resulting precipitates are extremely insoluble. Apatite is therefore suitable as a precipitant for PRBs.

1.3.2.2. Chromium

Chromium exists in natural waters in the oxidation states +3 and +6. Hexavalent chromium occurs as an oxyanion in the form of CrO_4^- or as $Cr_2O_7^{2-}$. Chromates are more soluble and mobile than Cr^{III} compounds. Due to their negative charge these anions are not attracted to negatively charged mineral surfaces (Powell et al., 1998). Hexavalent chromium cannot be precipitated as hydroxide. Prior to precipitation, chromium(VI) has to be reduced to chromium(III). The removal of chromium(VI) is performed with sulfide in a single step combining reduction and precipitation (Chung, 1989):

$$Cr_2O_7^{2-} + 2FeS + 7H_2O \rightarrow 2Cr(OH)_3 + 2Fe(OH)_3 + 2S + 2OH^- \qquad (24)$$

Stupp (1999) describes in detail the remediation of groundwater contaminated with chromium. Water at a rate of 3 m^3/h was injected through 20 injection wells to mobilize adsorbed chromium. Then, water at 9 m^3/h was pumped downstream through pumping wells to a treatment plant applying precipitation, flocculation and sedimentation. Cleaned water was finally discharged.

Reduction of chromate with elemental iron in PRBs has been studied intensively (Blowes and Ptacek, 1992; Powell et al., 1995, 1998). The overall reaction can be written as

$$CrO_4^{2-} + Fe^0 + 8H^+ \rightarrow Fe^{3+} + Cr^{3+} + 4H_2O \tag{25}$$

Reaction (25) occurs spontaneously, as shown in Fig. 1.4. Couples with lower standard electrode potential reduce couples with higher potential. The stability lines as a function of pH were calculated by the computer program Minteqa2 (Allison *et al.*, 1991).

In a further step, iron and chromium are precipitated as oxyhydroxides (Powell *et al.*, 1998):

$$(1-x)Fe^{3+} + xCr^{3+} + 2H_2O \rightarrow Fe_{(1-x)}Cr_xOOH + 3H^+ \tag{26}$$

Sorption on zeolites

A further possibility for chromate removal from groundwater is sorption on zeolites. Zeolites are naturally occurring alumosilicates with open, cage-like structures. They exhibit a high cation exchange capacity, e.g. for transition metal cations such as Pb^{2+}. By modification of the surface with quaternary amines (e.g. hexadecyltrimethyl-ammonium, HDTMA, $(C_{16}H_{33})((CH_3)_3N^+)$ the zeolites can remove non-ionic organic compounds such as benzene and chlorinated hydrocarbons, due to the organic carbon content of the modified sorbent, without lowering the sorption affinity for metal ions (Haggerty and Bowman, 1994). Negatively charged oxyanions such as chromates are sorbed via ion exchange by the cationic surfactant (Bowman, 1999). The organo-zeolite is also applicable for the removal of other oxyanions such as selenate and sulfate. The modified surface was found to be stable at pH values from 3 to 10, high ionic strengths and on exposure to organic solvents.

The removal of metal ions with different inorganic sorbent materials was examined in a comparative study by Lehmann *et al.* (1999). They found that various modifications of ferric oxyhydroxide are strong sorbents for several heavy metals such as chromium or zinc.

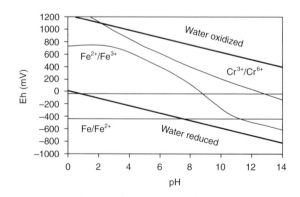

Fig. 1.4. pH–Eh diagram showing the stability fields for water and some iron and chromium species

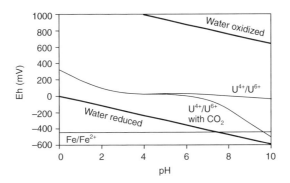

Fig. 1.5. pH–Eh diagram showing the boundary line between U^{IV} and U^{VI} without and in the presence of 0.01 bar CO_2

1.3.2.3. Uranium

Uranium is the heaviest naturally occurring element. All isotopes are radioactive; the half-lives of the two most relevant isotopes ^{238}U and ^{235}U are 4.5×10^9 and 7.0×10^8 years, respectively (Seelmann-Eggebert *et al.*, 1981). If uranium is present in groundwater, e.g. from mine tailings, it is dangerous not because of its radioactivity but because of its toxicity as a heavy metal. From the German Radiation Protection Act, the radiation limit can be calculated to be 0.3 mg/l (7.0 Bq/l of the natural mixture of isotopes). However, with regard to its toxicity a limit between 0.02 and 0.002 mg/l is under discussion (Merkel *et al.*, 1998).

Uranium occurs mainly in the oxidation states +4 and +6. The hexavalent uranium, i.e. the uranyl ion UO_2^{2+}, is more mobile than U^{IV} compounds, similarly to Cr^{VI} and Cr^{III}. Uranyl ions can be precipitated like lead ions, with phosphate forming $(UO_2)_3(PO_4)_2$ (log K_{sp} = –49.09 (Brown *et al.*, 1981)). With hydroxyapatite or bone char (which is hydroxyapatite with a small amount of carbon) the precipitation of autunite $(Ca(UO_2)_2(PO_4)_2)$ occurs (log K_{sp} = –47.28 (Brown *et al.*, 1981)). The reaction is similar to ion exchange processes. It is likely that ion exchange materials can also be used in reactive barriers.

Other ion exchange processes for the adsorption of uranium (and other fission product solutions) use silicates, manganese dioxide, zirconium phosphate and titanium oxide. A detailed overview on inorganic ion exchangers for uranium is given by Löwenschuss (1979).

Good removal results have been obtained by the reduction of U^{VI} to U^{IV} by elemental iron:

$$Fe + UO_2^{2+}(aq) \rightarrow Fe^{2+} + UO_2(s) \tag{27}$$

The reduction of UO_2^{2+} occurs spontaneously, as illustrated in Fig. 1.5. The solubility of uraninite (UO_2) is of the order of 10^{-8} mol/l in a pH range between 4 and 14. Below pH 4 uranium becomes soluble. The behaviour of uraninite has been intensively studied because of its application as a nuclear fuel and because it is a major constituent of uranium pitchblende (Bickel *et al.*, 1996). Measurement of the

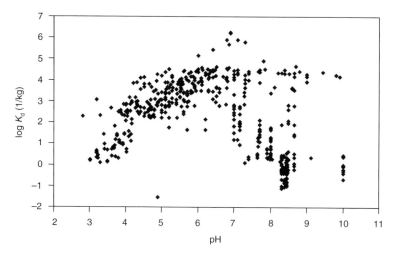

Fig. 1.6. Distribution coefficient K_d *of uranium as a function of pH*

dissolution of spent fuel in deionized water under non-oxidizing conditions gave results in the range of 10^{-9}–10^{-5} mol/l. Under oxidizing conditions UO_2 can be transformed into the uranyl ion, and, if complexation reactions can occur, the solubility is enhanced. It is therefore important that E_h is below the boundary of the U^{IV}/U^{VI} curve in the U^{IV} stability field. With CO_2 present, the boundary line is shifted towards lower E_h at high pH. This behaviour is shown in Fig. 1.5.

At elevated pH values in the presence of CO_2 (0.01 bar which is typical for groundwater conditions) it is more difficult to transform uranium to the oxidation state +4. An excellent review on the complex uranium solution equilibria is given by Langmuir (1978).

Adsorption of uranium on to surfaces
Uranium can also be removed by adsorption on surfaces. Morrison and Spangler (1992) have evaluated a range of uranium and molybdenum adsorption tests using a variety of materials. Good removal results have been obtained for lime, haematite, peat, ferric oxyhydroxide, phosphate and TiO_2, while clays exhibited low sorption potential. Precipitation and adsorption on to a surface are processes which can occur simultaneously in a chemical barrier. Sorption of uranium from groundwater was studied in a series of publications (Morrison and Spangler, 1992, 1993; Morrison *et al.*, 1995). Surface site complexation can be described with different models with or without an electrostatic influence on charged surfaces (Allison *et al.*, 1991). Using SOH to denote a surface site, the adsorption reaction of uranium on to ferric oxyhydroxide as the sorbent can be written as follows:

$$SOH + UO_2^{2+} + 3H_2O \rightleftharpoons SOH–UO_2(OH)_3 + 2H^+ \tag{28}$$

Sorption, rather than precipitation, depends strongly on pH. If several contaminants are to be removed from groundwater, an optimum pH for operation of the barrier is needed. This is difficult to achieve for uranium in the presence of molybdenum, because the latter is mobile at pH values above 8 while uranium exhibits low mobility (Morrison and Spangler, 1993). Sorption on to surfaces and reduction are both possible removal reactions in iron-bearing walls. To which degree and at what efficiency the distinct processes take place has been addressed in several articles (Fiedor et al., 1998; Gu et al., 1998). It was found that Fe^0 filings are much more efficient than adsorbents.

The geochemistry of an aqueous solution and the sorbent properties which are most important in controlling the adsorption/retardation behaviour of uranium and other environmentally relevant elements were reviewed in a technical report (US Office of Radiation and Indoor Air and Office of Environmental Restoration, 1999). Values for the distribution coefficient K_d (see Section 1.3.1.2) of uranium at different pH values in various soils were obtained from the literature and listed in the report. The data exhibit large scatter but show a trend as a function of pH. The values of K_d listed in the report are shown in Fig. 1.6. High K_d values were derived from adsorption experiments with ferric oxyhydroxide and kaolinite, and low values from those with quartz, which has low adsorptive properties. The pH dependence arises from the surface charge properties of the soil and from the complex aqueous speciation of U^{VI}. The dissolved carbonate concentration also has a significant influence, due to the formation of strong uranyl-carbonato complexes. Retardation can then be calculated from K_d according to equation (16).

Uranium removal from groundwater was studied using phosphate, zero-valent iron and amorphous ferric oxide barriers at Fry Canyon in Utah, USA (Naftz et al., 1996; Morrison, 1998). Precipitation as phosphate, reduction to immobile U^{IV} compounds and adsorption on to sorbents were studied simultaneously in this field test. At least 90% removal was found in each application, and the Fe^0 barrier was found to be the most effective.

Biosorption of uranium was studied by Yang and Volesky (1999). The uptake of uranium was more than 500 mg/g. Biosorption is an ion exchange process between the uranium ions and the protons introduced to the binding sites of certain types of inactive, dead, microbial biomass (Volesky, 1990). The process is also suitable for other heavy metals (Pb, Cd, Cu, Zn and even Cr).

The possibility of the reduction of U^{VI} using micro-organisms seems to have some potential for field applications. Initial laboratory results on this method are presented elsewhere (Gorby and Lovley, 1992).

1.4. Cost comparison between pump-and-treat and PRB systems

A variety of papers have been published dealing with the costs of pump-and-treat and PRB systems (Reeter et al., 1999; Stupp, 2000; Edel and Voigt, 2001; Federal Remediation Technologies Roundtable (FRTR), 2001). In a report of the US Environmental Protection Agency, capital and operating costs were estimated for 32 pump-and-treat sites (Federal Remediation Technologies Roundtable

(FRTR), 2001). The volume of groundwater treated was between some thousands and a few million cubic metres per year. The operating cost and capital cost in US dollars per cubic metre of water treated per year was highest for the smaller remediation projects (US $40 and US $400/m³ per year, respectively), and decreased to a few US dollars per cubic metre per year for large projects. Of course, these costs depended on the site characteristics and the kind of contaminant. Figure 1.7 shows the data as a function of the groundwater volume treated per year in the plant.

The capital costs for 16 PRB sites were specified in the same report. They ranged from US $43 000 to US $1 600 000. The volume of treated groundwater could not be stated reliably, so unit operating costs were not estimated.

Remediation costs for chlorinated hydrocarbons in groundwater using PRBs were compared with the pump-and-treat method in a report by the US Department of Defence. The cost saving was around 50–70% over the long term for the example discussed in the report, as can be seen in Fig. 1.8 (Reeter et al., 1999).

In this comparison, barrier maintenance costs of US $268 000 every 10 years were adopted. This was equivalent to 25% of the iron medium costs, and was little more than a good guess because no long-term experience with PRBs exists. It is thus uncertain whether or not such cost savings can be achieved. Further research is urgently needed into the long-term behaviour of PRBs using accelerated testing methods.

1.5. Engineering of permeable reactive barriers

Although the chemical, physical and biological principles used in reactive barriers have been extensively investigated during the last two decades, interest has focused on construction methods only recently. One of the reasons is that very simple construction techniques may be used if only shallow depths are encountered (R. W.

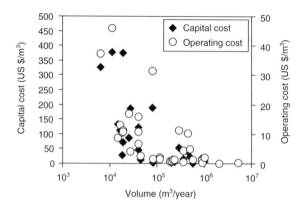

Fig. 1.7. Capital and operating costs of 32 pump-and-treat sites adjusted for treated volume per year

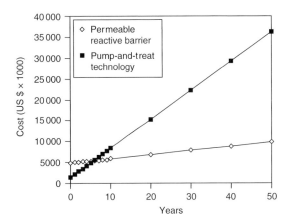

Fig. 1.8. Cost comparison of groundwater remediation between pump and treat technology and PRBs (Reeter et al., 1999)

Puls, personal communication). In addition, cut-off wall construction methods are easy to adapt to reactive barriers due to their similarities, provided the impervious layer does not lie deeper than about 20 m (Schad and Teutsch, 1998; Debreczeni and Meggyes, 1999). At greater depths, however, more efficient construction methods are needed in order to achieve the economic advantages that reactive barriers may offer. In addition to single- and twin-phase diaphragm walls, bored-pile walls, jet grouting, thin walls, sheet pile walls, driven cut-off walls, injection and frozen walls, an increasing number of innovative techniques have recently been introduced utilizing, for example, drilling, deep-soil mixing, high-pressure jet technology, injected systems, column and well arrays, hydraulic fracturing and biobarriers.

The main configurations of permeable reactive barriers are:

- continuous reactive barriers;
- funnel-and-gate systems;
- arrays of wells filled with reactive materials;
- injected systems.

Figure 1.9 shows the most frequently used configurations, i.e. the continuous reactive barrier and the funnel-and-gate system.

In addition to their low operational costs, reactive barriers exhibit further cost-saving features: they are often installed in a semi-permanent fashion, built within the contaminated area and can start operation immediately after installation (Day et al., 1999). The two main types of reactive barrier – continuous permeable barriers and funnel-and-gate systems (Rochmes, 2001) – allow the application of geotechnical construction methods which, as a rule, include slurry trenching, grouting and deep soil mixing.

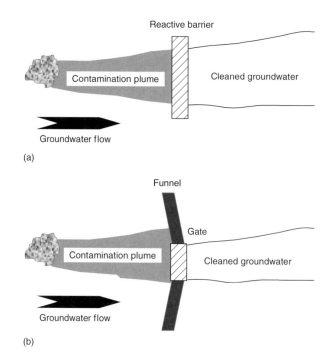

Fig. 1.9. The most frequently used arrangements for PRBs: (a) continuous reactive barrier and (b) funnel-and-gate system

1.5.1. Construction of cut-off walls

The most commonly used cut-off wall construction methods (Meggyes and Pye, 1995) apply one of the following alternatives:

- excavating trenches using supporting fluids capable of solidifying and forming a diaphragm wall (single-phase diaphragm wall) or which – after excavation of the trench – are displaced by another material which, in turn, is capable of solidifying (twin-phase diaphragm wall);
- forming a thin slot by driving a beam into the ground, then consecutively retracting the beam and filling the space formed with a thick slurry (e.g. thin walls);
- driving strong elements into the ground (e.g. steel sheet piles);
- constructing interlocking boreholes and using drill pipes or supporting slurries to stabilize the boreholes until they are backfilled with concrete to form a wall of interlocking columns;
- injecting or placing reactive materials into the ground in a discontinuous fashion.

1.5.1.1. Single-phase diaphragm wall

Single-phase diaphragm walls consist of 0.4–1 m thick panels which are excavated from the soil using grab buckets, clamshells or vertical trench cutters. A self-hardening slurry is pumped in to stabilize the trench walls and form the final fill.

Usually the 'pilgrim's pace' method is applied: primary panels 1, 3, 5, etc., are excavated and filled with slurry first. After a period of 36–48 hours, as soon as the slurry in the primary panels has hardened to a cuttable state, the secondary panels 2, 4, 6, etc., are constructed with the cutter cutting into the primary ones. This ensures an intimate contact between the primary and secondary panels. Verticality is a significant issue (Ghezzi *et al.*, 1999) to prevent gaps being left between adjacent panels. Even minor imperfections impair barrier performance: a 1 m^2 hole lets as much groundwater escape as a 100 000 m^2 high-quality cut-off wall does (Düllmann, 1999).

1.5.1.2. Twin-phase diaphragm wall
Twin-phase diaphragm walls are constructed in two sequences. In the first phase, soil is excavated while a bentonite suspension stabilizes the trench walls. In the second phase, the bentonite suspension is replaced by the cut-off slurry using tremie pipes (contractor method). The individual panels are confined by stop-end tubes. The density of the cut-off slurry must exceed that of the bentonite slurry by 500 kg/m^3 to achieve proper displacement.

1.5.1.3. Composite cut-off wall
In both single- and twin-phase systems, additional elements can be inserted into the cut-off wall to improve strength and/or watertightness. Sheet piles (Jessberger and Geil, 1992), geomembranes (Boyes, 1986) and glass walls/tiles are a few examples – with geomembranes applied most widely. These elements are inserted into the fresh bentonite cement mix immediately after placement. Special construction apparatus is required to insert them into the cut-off wall, and special locks provide water-tight joints between the elements (Meseck, 1987; Stroh and Sasse, 1987; Ghezzi *et al.*, 1999).

1.5.1.4. Thin walls
First sheet piles, then heavier steel beams, are vibrated into the ground. They are then simultaneously retracted while injecting a clay–cement and water mix into the void. The panels cut into the adjacent ones, so providing a given overlap and watertightness for the wall (Arz, 1988). In coarse layers (sand, gravel) the vibration compacts the surrounding medium, thus reducing permeability. The clay–cement mix usually penetrates into the pores of the surrounding soil; thus the final wall thickness achieved can be twice the nominal thickness.

1.5.1.5. Sheet pile walls
Sheet pile walls are constructed using steel, precast concrete, aluminium or wood piles, driven into the ground individually or in groups, and connected to each other by locks (Roth, 1988; Weber *et al.*, 1990; Jessberger and Geil,1992; Rodatz, 1994), with steel sheet piles being the most common option. Steel sheet pile walls are easy to construct and can carry heavy loads, construction time is short and there is no

need for contaminated soil to be disposed of. Watertightness of the locks and corrosion are issues to be addressed.

1.5.1.6. Bored-pile cut-off walls and jet grouting

Bored-pile cut-off walls are constructed of contiguous or secant piles, with or without casing. The primary piles 1, 3, 5, etc., are sunk first, followed by the secondary piles 2, 4, 6, etc. By cutting into the primary piles while sinking the secondary ones an intimate contact is established between them. Cut-off walls can also be constructed of soilcrete columns, also called jet grouting, roddingjet or HPI (high-pressure injection) (Tóth, 1989; Kutzner, 1991). A rotary drilling technique is used to construct the columns, in conjunction with a high-density mud which serves both as a cutting and filling fluid. When the final depth is reached, the drilling mud is diverted to the nozzles and the pump pressure simultaneously increased. The drill rod is retracted while being turned, and in so doing the soil is cut and removed, the resulting void being filled with a cement or clay–cement mix. A novel dual-phase jet grout system (Dwyer, 1998) treats and contains contaminated underground bodies in such a way that contaminated spoils (drill cuttings) remain in the subsurface. In the first step, air and/or water is injected at high pressure to form a hollow underground storage cavern above the contaminated body, while clean soil from the cavern is removed to the surface. In the second step, the contaminated body is treated and stabilized while contaminated drill cuttings are deposited in the hollow cavern formed in the first step.

1.5.1.7. Injection walls

Injection of a solidifying liquid material into the ground fills the pores and fissures, thus reducing permeability. Cement suspensions, artificial resins or water glass-based materials whose compatibility with the groundwater is proven are injected through boreholes. The distance between the injection holes is determined by the rock permeability, the viscosity of the injected material and the highest permissible injection pressure (Kutzner, 1991; Schulze, 1992; German Geotechnical Society, 1993).

1.5.1.8. Frozen walls

Frozen walls are constructed by inserting pipes into the ground in which a coolant or liquid nitrogen circulates. By freezing the soil, a closed, watertight body is produced. Frozen walls were previously considered mainly a temporary measure. Recent developments indicate that frozen soil permeability can be as low as 10^{-12} m/s, and diffusivity around 10^{-9} cm^2/s (Dash $et\ al.$, 1997; Mageau, 1998). Installation of frozen walls requires little or no soil excavation, and they can be removed by stopping the cooling (although they can last months without power), and there is no pollution to deal with and no wastes to dispose of. Feasible depths are a few hundred metres, and frozen walls are effective in both clay and sand/gravel-type soils. Frozen walls may prove to be both an economically and environmentally sound technology. Advanced applications use monitoring by thermistors, electro-potential, etc.

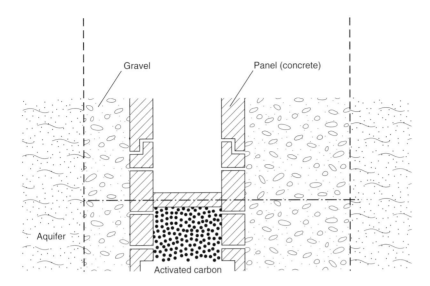

Fig. 1.10. Structure of a reactive barrier (Beitinger and Bütow, 1997)

1.5.2. Construction of reactive barriers

The basic requirements that the structure of a reactive barrier has to fulfil are (Beitinger and Bütow, 1997; Smyth *et al.*, 1997; Beitinger, 1998):

- replaceability of the reactive materials (where appropriate);
- higher permeability than the surrounding reservoir;
- stability against fines washed into barriers from the surrounding soil;
- long life-span.

The selection of construction technique to be used depends mainly on site characteristics (Gavaskar, 1999):

- Depth: probably the most important factor – greater depth involves more specialized equipment, a longer construction time and higher cost.
- Geotechnical considerations: strength of the layers and the presence of cobbles, subsurface utilities, buildings, etc., influence the selection of the construction technique.
- Soil excavation: handling and disposal of (contaminated) soil require additional room on the site and increase costs.
- Health and safety: an important issue, especially when entry of personnel into excavations is required.

In simple cases, not deeper than 8 m, a trench is excavated and simultaneously filled with the granulated reactive material (R. W. Puls, personal communication). In more complex cases (Fig. 1.10) the heart of the reactive barrier consists of the part filled with reactive material. Filter gravel prevents fines from the soil from entering the reactive zone. To exclude any contact with oxygen from the air, the reactive

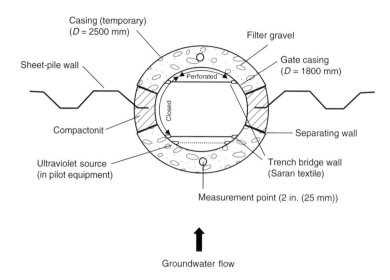

Fig. 1.11. Cross-section of a gate (Schultze and Mussotter, 2001)

barrier is covered with low-permeability material (clay). The overall permeability of the barrier must be higher than that of the surrounding soil to encourage groundwater to flow through the barrier instead of by-passing it.

The design principles of reactive barriers are discussed in greater detail by Beitinger *et al.* in Chapter 18.

1.5.2.1. Cut-off wall technology

The methods used in cut-off wall excavation, described in Section 1.5.1., can be applied to reactive barriers. Reactive materials can be placed into the trench using common earth-moving machines. The funnel elements in funnel-and-gate systems are cut-off walls anyway, and are constructed using the cut-off wall construction methods. Gravelding (1998) reports on the construction of a 400 m funnel-and-gate system in which sheet pile walls were used to construct the funnels, with vibratory installation. Four cells, each 12 m wide, served as gates, containing iron in view of the contaminant occurring. The residence time was determined from the influent chemistry, reaction rates and treatment goals. The residence time and flow velocity, in turn, yielded a 0.6–1.8 m cell thickness, requiring 580 tonnes of iron. The gates were constructed as sheet pile boxes, the bottoms lined with bentonite and geo-textile, on which iron filings were placed surrounded with pea gravel to provide access for the groundwater. The gates were capped by geotextile and bentonite, and native soil was backfilled to grade. Barrier installation costs amounted to US $200–250 per square metre, and gate installation costs were US $850–1000 per square metre.

The Karlsruhe East gasworks site remediation is another example of the use of sheet piles: a funnel-and-gate system was constructed here to clean up PAH and benzene contamination (Schultze and Mussotter, 2001). The funnel is a 240 m long and 19 m deep steel sheet pile wall, which was pressed into the ground using a Silent Piler to avoid damage to nearby historic buildings. The locks connecting the sheet piles were checked both electrically and by measuring the pressing force, and finally filled with Hoesch plastic sealant. The gates were constructed as large-diameter (2.5 m) drillings, and filled with activated carbon (Fig. 1.11).

1.5.2.2. Reactive thin walls

These use the thin-wall technique developed for cut-off walls and unite the advantages of slurry trenching and sheet piling (Jansen and Grooterhorst, 1999). Hollow steel beams, 0.15–0.2 m wide and 0.4–1.0 m long, are vibrated into the ground. They are then retracted while placing reactive material in the void. A number of 'panels' constructed in this way next to each other form a continuous reactive wall. The main advantages of this method are low demand on space, no soil or water extraction, minimum impact on groundwater flow, a maximum depth of around 25 m and recovery of the reactive material using the same hollow steel beams.

1.5.2.3. Drilling and deep soil mixing

Contiguous circular columns containing reactive material can be installed using drilling and deep soil mixing (Day et al., 1999), which then form longer barriers. Caisson drills can be used for the drilling; the column diameters are 0.5–2.5 m. Usually a large circular casing is lowered into the ground, the soil is removed by augers and the hole filled with reactive material. Instead of casing, biodegradable polymer slurries (Hubble et al., 1997) or shear-thinning fluids (Cantrell et al., 1997) may be used as supporting fluids, with significant cost benefit. Deep soil mixing mixes the soil with a slurry in situ without excavating the soil (Gavaskar et al., 2000). A caisson is lowered into the ground, a set of multiple augers penetrates the ground and the reactive material is injected through the hollow kelly bar of the mixing tools. The reactive material is injected in slurry form, which then is mixed with the soil by the augers. The soil remaining in situ 'dilutes' the reactive material to 40–60%, which must be taken into account in calculating residence time.

1.5.2.4. Jet technology

This technique proposes using high-pressure jets and jet pumps (Debreczeni and Meggyes, 1999). The jet cutting head used in the trench excavation is joined to the drilling pipe by a hinged connector and a flexible hose (Fig. 1.12). This allows the drilling pipe to be swung from side to side within the trench by means of a hydraulic mechanism so as to achieve the appropriate length of excavation. Soil excavation in low-strength strata can be performed by slurry jets and in high-strength strata by a milling head which can be attached to the drilling pipe and operated by a hydraulic motor. The drillings can be transported to the surface through the drilling pipe by means of a jet pump. Where slurry jets are used in the trench excavation, both the jet cutting head and the jet pump can be operated by the same slurry pump. Depending on construction process and the final barrier requirements, the slurry can either remain in the trench after excavation is complete and form the final barrier

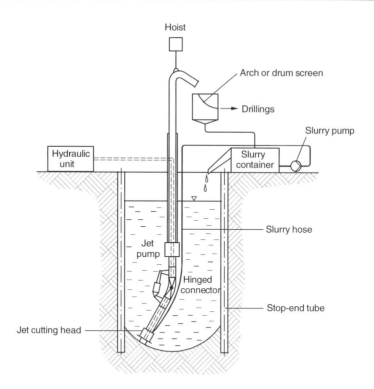

Fig. 1.12. High-pressure jet equipment for cut-off wall or reactive barrier construction

(single-phase technique) or be replaced by the final barrier material (twin-phase technique). After being pumped to the surface the drillings must be separated from the slurry, which can then be recirculated. Solid particles greater than 1 mm can be removed by arch or drum screens, but for finer materials a hydrocyclone must be used. The hydrocyclone can be operated by a second slurry pump. The cutting head and the drilling pipe can be carried by a hoist or drilling mast or supported by stop-end tubes or sheet piles, which can be removed and reused once the panels have been completed. Depending on local conditions, especially where mechanical cutting is used, air lift can be considered instead of jet pumping.

A combined jet method uses iron particles both as the reactive material and the abrasive medium in trench excavation. The granulated iron particles are embedded in high-velocity water jets and erode the soil or rock (O'Hannesin, 2001). This technique enables trench construction and placement of the reactive material simultaneously. The barrier constructed is a hybrid of the reactive wall and injected systems.

1.5.2.5. Applying reactor barrier technology to existing containment systems

Although cut-off technology is highly sophisticated and in most cases provides efficient containment, a groundwater reservoir separated by a cut-off wall from the surrounding geology very often requires treatment which results in additional costs. By applying reactive barrier technology to an existing containment system by opening up the cut-off wall at distinct points and filling the openings with a reactive material (Bradl and Bartl, 1999), thus transforming containment into a funnel-and-gate system, the groundwater can be treated and need not be pumped, which helps reduce costs.

1.5.2.6. Well-based systems

An alternative to continuous barriers is the use of arrays of wells (Wilson and Mackay, 1997; Golder Associates, 1998). This may be helpful at sites where installation of permeable barriers may be impractical for technical or financial reasons. An array of wells can form the gate within a funnel-and-gate system or serve as a set of *in situ* reactors or can release substances that encourage biological or other processes. An array of wells is essentially a non-continuous reactive barrier with the advantage that a drilling technique can be used to construct them, allowing lower installation costs and greater depths. The wells filled with reactive materials have lower hydraulic resistance than the surrounding soil; thus, the groundwater flow converges towards them. The amount of reactive material required for the wells is much less than for a continuous reactive barrier. The residence time must provide for the reactions between the pollutant and the reactive material. The whole contaminant plume must flow through the array of wells, in which regard optimum well spacing is of great importance. If the residence time is high, more than one array of wells may be required. As a rule, well spacing should be twice the well diameter; thus the effective length of an array of wells is about half of that of a reactive barrier. Wells with a fill that discharges substances may have an even greater spacing.

Wells can also be arranged horizontally (HTI, 1995) to deliver treatment agents such as vapour for air stripping, and oxygen and nutrients for bioremediation purposes. Technologies used include rotary rigs, 'mole' systems and trenching methods (Flowmole, 1995; Golder Associates, 1998).

1.5.2.7. Injected systems

Injecting reactive materials into the ground at high pressure without strict geometrical boundaries in terms of a 'wall' is an attractive idea which provides high flexibility. Nevertheless, it is necessary to ensure that the contaminant plume is efficiently taken care of and no by-passing or fingering occurs which may impair the remediation effect. An injected system can control large and deep plumes even if their extension is irregular (Golder Associates, 1998). Researchers and end-users may benefit from the extensive knowledge of, and experience in certain areas in petroleum engineering, where techniques of strata fracturing, secondary recovery, etc., have been developed over many decades. Recovery of reactive material is practically impossible in injected systems. Therefore, degradation is the preferred approach, otherwise reinjection of the reactive material may be necessary.

Two main types of injected systems have been developed so far:

- injection into existing pores in ground;
- ground fracturing.

The range of injection largely depends on the pores available, and injection wells may require a very close spacing in finely grained soils. Some authors suggest that 10^{-5} m/s is the permeability limit below which only pure liquids may be injected. Injecting bacteria, air microbubbles as the oxygen supplier (Duba *et al.*, 1996; Koenigsberg, 1998) or cationic surfactants (Burris and Antworth, 1990, 1992) has also been suggested.

1.5.2.8. Hydraulic fracturing

Ground fracturing, as used in enhancing permeability around oil wells in petroleum engineering, aims at creating an improved zone by cracking the ground with water at high pressures, and pumping sand or similar granular material into the cavity created. Using this method within the concept of barrier technology (Murdoch *et al.*, 1997; Gavaskar, 1999), a part of the material filled in may be reactive material intended for pollutant treatment. As in petroleum engineering, the fractures are initiated from wells and are nearly horizontal structures so that they are capable of intercepting downward-moving contaminant plumes. In addition to introducing reactive materials into the ground, another advantage is that a high-permeability zone is formed which encourages groundwater flow. Fractured zones may also be applied to direct groundwater flow towards the gates in funnel-and-gate systems (Golder Associates, 1998). Murdoch *et al.* (1997) found that the pressure required to fracture the ground increased with the depth, but was surprisingly low: a maximum of 500 kPa was measured during a test at a depth of 2 m. The fractures usually have a preferred direction of propagation, and are therefore asymmetric with respect to the borehole and climb in the preferred direction of propagation. Murdoch *et al.* (1997) report on 7–10 m fracture dimensions in silty clay, with 0.1–1.25 m³ filled-in material and a 5–25 mm fracture thickness. The fracturing procedure can be monitored by recording both pressure and deformation of the ground surface, which will lift to form a gently sloping dome. Applications include materials that alter redox conditions, adsorb contaminants or slowly release useful materials (e.g. oxygen, nutrients, porous ceramic granules, etc.). Using high-energy jets and adjustable outlets, directional fracturing is feasible.

A new application of hydraulic fracturing aims at constructing vertical barriers (Gavaskar, 1999) by creating controlled vertical fractures from a series of wells. The fractures are monitored through, for example, down-hole resistivity sensors to ensure coalescence or overlap of the fractures. A barrier thickness of 8–10 cm can be achieved, and some variability in thickness can be obtained by designing two parallel barriers.

1.5.2.9. Biobarriers

Subsurface biofilm barriers or biobarriers consist of biofilm which in turn comprises deposited cells of micro-organisms and polymers, together with captured organic and inorganic particles (Cunningham *et al.*, 1991, 1997; Sharp and Cunningham, 1998). Biobarriers may selectively plug permeable strata with microbial biomass, thus influencing the hydraulic conductivity. Investigations indicate that reducing the

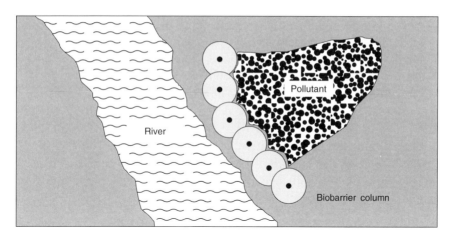

Fig. 1.13. Biobarrier configuration (Hiebert, 1998)

hydraulic conductivity by five orders of magnitude is possible, and in column tests 92–98% reduction in hydraulic conductivity was achieved. Biobarrier technology may also be useful as a means of funnelling contaminated groundwater through subsurface treatment systems (e.g. funnel and gate).

The first step in forming a biobarrier is the isolation and identification of potential barrier-forming bacteria from a field site. Once identified, these bacteria are re-injected to serve as the inoculum for biobarrier formation. Extracellular polymer (EPS)-producing bacterial strains (i.e. the mucoid phenotype) are desirable candidates for barrier formation. Several *Pseudomonas* and *Klebsiella* strains have been isolated from potential field sites. In addition to exhibiting high EPS production, these bacteria also have the ability to biodegrade BTEX compounds – thereby raising the possibility that biobarriers can be constructed which actively biodegrade dissolved contaminants in addition to providing containment. These bacteria were found to grow well on either molasses or distillery waste, which means that a low-cost nutrient source is potentially available for field scale biobarrier construction. The next step is to verify that biobarrier formation and hydraulic conductivity reduction can be achieved in laboratory scale packed-bed reactors. Biobarrier experiments carried out by Cunningham *et al.* (1991, 1997) in 0.9 m long, 15 cm diameter PVC and stainless steel columns under a hydraulic gradient of 1.0 m/m yielded a reduction in the hydraulic conductivity from an initial value of about 4 cm/min down to approximately 0.01 cm/min. Similar experiments were run in $0.3 \times 0.9 \times 0.15$ m^3 stainless steel lysimeters under a head of 0.03 m/m and reductions from 1 cm/min to approximately 10^{-5} cm/min were obtained. Biobarriers can be constructed by using an array of wells to introduce micro-organisms, nutrients and oxygen into the ground. The activity of micro-organisms leads to the development of contiguous columns of low permeability, which in turn form the barrier (Fig. 1.13).

Biobarrier performance was found to be unaffected by exposure to heavy metal contaminants (strontium and caesium) at 1 ppm levels for periods of up to 120 days. Similarly, biobarrier performance and formation were not influenced by the presence of carbon tetrachloride at concentrations of 100 mg/l for extended periods and up to 300 mg/l for short periods.

Preliminary economic estimates by Hiebert (1998) indicated a cost range of US $6.5 million to US $10.7 million for a biobarrier and US $9.8 million to US $13.5 million for a grout curtain for a 30 m deep and 3200 m long barrier. A sheet pile wall of the same length but only 12 m depth would cost US $15 million to US $17 million.

Recent research indicates that reactive barriers combined with biological activity may prove efficient. Phytoremediation can increase clean-up efficacy for PAHs, phenols and heterocyclic aromatics (Rasmussen, 2002). Reactive barriers using compost or wood chips are effective in remediating groundwater contaminated with mine water.

1.5.2.10. Biopolymer trenching

Biopolymers were first used as additives to stabilize trench walls during excavation. Owing to the advantages of this technique, biopolymer trenching has since been developed into a barrier technology (Day et al., 1999; Gavaskar, 1999; Gavaskar et al., 2000; Jayaram, 2001; see also Chapter 5). The procedure exhibits similar features to traditional slurry trenching, the main difference being that a biopolymer (e.g. guar gum) is added to the stabilizing suspension. The reactive material can then be added to the trench using contractors. Sivavec et al. (see Chapter 5) discuss the construction of a 7.6 m long and 10.4 m deep experimental reactive barrier. A 90%/10% iron/sand mixture was placed in the trench using a tremie pipe. Care was taken to minimize segregation, and contact between the iron/sand mixture and the biopolymer. Following placement, a high-pH enzyme breaker was added to the fluid to break down the biopolymer, and clay was placed on the barrier to prevent contact with air.

1.6. Outlook

PRBs provide a wide range of applications for the *in situ* treatment of contaminated groundwater. This includes organic pollutants as well as heavy metals and other inorganic contaminants. The most commonly applied physical and chemical processes are redox reactions, precipitation, adsorption, ion exchange and biodegradation. The types of reactive material used are those changing pH or redox potential (zero-valent iron), those causing precipitation, adsorbents or oxygen-releasing compounds. Considerable cost savings are expected from their use compared with traditional technologies such as pump and treat. Interest in the improvement of construction methods has recently been on the increase. Most of the construction methods used so far rely on techniques developed for cut-off walls. With the development of the reactive barrier technology, deeper and more difficult installations will be undertaken, which will require more advanced construction methods.

Further research into the long-term behaviour of PRBs using accelerated testing methods is urgently needed. This issue is being looked into by the integrated research programme 'Long-term Performance of Permeable Reactive Barriers Used for the Remediation of Contaminated Groundwater' within the Indirect Action

'Energy, Environment and Sustainable Development', Part A, 'Environment and Sustainable Development', Key Action 1, 'Sustainable Management and Quality of Water', in the 5th Framework Programme of the European Union (Perebar, 2000). Another integrated programme, supported by the German Federal Ministry for Education and Research, was started in Germany in 2000, with several technical demonstration projects on the application of reactive barriers for the remediation of contaminated sites (see Chapter 3). The results of these programmes will help extend our knowledge of the long-term behaviour and costs of PRBs.

1.7. Acknowledgements

Support by the European Commission for the project 'Long-term Performance of Permeable Reactive Barriers Used for the Remediation of Contaminated Ground-water' (Perebar, 2000) is appreciated. Permission by the Solid and Hazardous Waste Research Unit of the University of Newcastle Upon Tyne to use parts of the state-of-the-art-report 'Landfill Liner Systems' (Meggyes and Pye, 1995)is acknowledged.

1.8. References

ALLISON, J. D., BROWN, D. S. and NOVO-GRADAC, K. J. (1991). Minteqa2/Prodefa2, A geo-chemical assessment model for environmental systems. *Database of Computer Programs*, Version 3.0. US Environmental Agency, Athens, Georgia.

ARZ, P. (1988). Dichtwandtechnik für seitliche Umschliessungen. *Bauwirtschaft* **110**(B42), 831–835.

BEITINGER, E. (1998). *Permeable Treatment Walls – Design, Construction and Costs, NATO/CCMS Pilot Study. Evaluation of Demonstrated and Emerging Technologies for the Treatment of Contaminated Land and Groundwater (Phase III). 1998 Special Session. Treatment Walls and Permeable Reactive Barriers*, Vol. 229, pp. 6–16. North Atlantic Treaty Organization, Vienna.

BEITINGER, E. and BÜTOW, E. (1997). Konstruktive und herstellungstechnische Anforderungen an unterirdische, durchströmte Reinigungswände zur in-situ Dekontamination. In H. P. Lühr (ed.), *Grundwassersanierung 1997, Berlin*, pp. 342–356. Erich Schmidt, Berlin.

BEITINGER, E. and FISCHER, W. (1994). Permeable treatment bed for use in purifying contaminated ground water streams *in situ*. German Patent P 4425061.4, WCI Umwelttechnik.

BICKEL, M., FEINAUER, D., MAYER, K., MÖBIUS, S. and WEDEMEYER, H. (1996). Uranium, Supplement Vol. C6. In: D. Fischer, W. Huisl and F. Stein (eds), *Gmelin Handbook of Inorganic and Organometallic Chemistry*. Springer-Verlag, Berlin.

BLOWES, D. W. and PTACEK, C. J. (1992). Geochemical remediation of groundwater by permeable reactive walls: removal of chromate by reduction with iron-bearing solids. Subsurface Restoration Conference. Third International Conference on Groundwater Quality Research. Dallas, Texas.

BOWMAN, R. (1999). Pilot-scale testing of surfactant-modified zeolite PRB. *Ground Water Currents* **31**, 4–5.

BOYES, R. G. H. (1986). Wider applications for slurry trench cutoffs. *Civil Engineering* June, 42–57.

BRADL, H. B. and BARTL, U. (1999). Reactive walls – a possible solution to the remediation of old landfills? In: T. H. Christensen, R. Cossu and R. Stegmann (eds), *Sardinia 99, 7th International Waste Management and Landfill Symposium, S. Margherita di Pula, Sardinia. Conference Proceedings*, Vol. IV, pp. 525–540. CISA Environmental Sanitary Engineering Centre, Cagliari.

BROWN, D., POTTER, P. E. and WEDEMEYER, H. (1981). Uranium, Supplement Vol. C14. In: R. Keim (ed.) *Gmelin Handbook of Inorganic Chemistry*. Springer-Verlag, Berlin.

BURRIS, D. R. and ANTWORTH, C. P. (1990). Potential for subsurface *in situ* sorbent systems. *Groundwater Management* **4**, 527–538.

BURRIS, D. R. and ANTWORTH, C. P. (1992). *In situ* modification of an aquifer material by a cationic surfactant to enhance retardation of organic contaminants. *Journal of Contaminant Hydrology* **10**, 325–337.

CANTRELL, K. J., KAPLAN, D. I. and GILMORE, T. J. (1997). Injection of collodial size particles of Fe^0 in porous media with shearthinning fluids as a method to emplace a permeable reactive zone. *Proceedings of the International Containment Technology Conference, St Petersburg, Florida*, pp. 774–780.

CHEN, X. B. and WRIGHT, J. V. (1997). Evaluation of heavy metal remediation on mineral apatite. *Water, Air and Soil Pollution* **98**, 57–78.

CHERRY, J. A., VALES, E. S. and GILHAM, R. W. (1996). System for treating polluted groundwater. US Patent 5487622, University of Waterloo, Canada.

CHUNG, N. K. (1989). Chemical precipitation. In: H. M. Freeman (ed.), *Standard Handbook of Hazardous Waste Treatment and Disposal*, 7.21–7.32. McGraw-Hill, New York.

COHEN, A. D. (1991). Method for *in-situ* removal of hydrocarbon contaminants from groundwater. US Patent 5057227, University of South Carolina.

CUNNINGHAM, A. B., CHARACKLIS, W. G., ABEDEEN, F. and CRAWFORD, D. (1991). Influence of biofilm accumulation on porous media hydrodynamics. *Environmental Science and Technology* **25**(7), 1305–1311.

CUNNINGHAM, A., WARWOOD, B., STURMANN, P., HORRIGAN, K., JAMES, G., COSTERTON, J. W. and HIEBERT, R. (1997). Biofilm processes in porous media – practical applications. In: P. S. Amy and D. L. Haldeman (eds), *The Microbiology of the Terrestrial Deep Surface*, pp. 325–344. Lewis, Boca Raton, Florida.

DASH, J. G., FU, H. Y. and LEGER, R. (1997). Frozen soil barriers for hazardous waste confinement. *Proceedings of the International Containment Technology Conference, St Petersburg, Florida*, pp. 607–613.

DAY, S. R., O'HANNESIN, S. F. and MARSDEN, L. (1999). Geotechnical techniques for the construction of reactive barriers. *Journal of Hazardous Materials* **B67**, 285–297.

DEBRECZENI, E. and MEGGYES, T. (1999). Construction of cut-off walls and reactive barriers using jet technology. In: T. H. Christensen, R. Cossu and R. Stegmann (eds), *Sardinia 99, 7th International Waste Management and Landfill Symposium, S. Margherita di Pula, Sardinia. Conference Proceedings*, Vol. IV, pp. 533–540. CISA Environmental Sanitary Engineering Centre, Cagliari.

DUBA, A. G., JACKSON, K. L., JOVANOVICH, M. C., KNAPP, R. B. and TAYLOR, R. T. (1996). TCE remediation using *in situ* resting state bioaugmentation. *Environmental Science and Technology* **39**(6), 1982–1989.

DÜLLMANN, H. (1999). Statement on cut-off walls. Geotechnical Bureau, Aachen.

DWYER, B. (1998). Remediation of deep soil and groundwater contamination using jet grouting and innovative materials. Subsurface Barrier Technologies, International Business Communications Conference, Tucson, Arizona.

EASTERN RESEARCH GROUP (1996). *Pump-and-Treat Ground-water Remediation, A Guide for Decision Makers and Practitioners*, EPA/625/R-95/005. US Environmental Protection Agency, Washington DC.

EDEL, H. G. and VOIGT, T. (2001). Aktive und passive Grundwassersanierung – ein Verfahrens- und Kostenvergleich. *Terratech* **1**, 40–44.

FEDERAL MINISTRY FOR THE ENVIRONMENT OF GERMANY (1998). Federal Soil Protection Act. *Federal Law Gazette* **I**, 502.

FEDERAL MINISTRY FOR THE ENVIRONMENT OF GERMANY (1999). Federal Soil Protection and Contaminated Sites Ordinance. *Federal Law Gazette* **I**, 1554.

FEDERAL REMEDIATION TECHNOLOGIES ROUNDTABLE (FRTR) (2001). *Cost Analysis for Selected Groundwater Cleanup Projects: Pump and Treat Systems and Permeable Reactive Barriers*, US Environmental Protection Agency, Washington DC.

FETTER, C. W. (1993). *Contaminant Hydrogeology*. Prentice-Hall, Englewood Cliffs.

FIEDOR, J. N., BOSTICK, W. D., JARABEK, R. J. and FARRELL, J. (1998). Understanding the mechanism of uranium removal from groundwater by zero-valent iron using x-ray photo-electron spectroscopy. *Environmental Science and Technology* **32**(10), 1466–1473.

FLOWMOLE (1995). Technical sales literature. Flowmole, Washington DC.

GAVASKAR, A. R. (1999). Design and construction techniques for permeable reactive barriers. *Journal of Hazardous Materials* **68**, 41–71.

GAVASKAR, A., GUPTA, N., SASS, B., JANOSY, R. and HICKS, J. (2000). *Design Guidance for Application of Permeable Reactive Barriers for Groundwater Remediation. Final Report.* Battelle, Columbus, Ohio.

GERMAN GEOTECHNICAL SOCIETY (1993). *Geotechnics of Landfills and Contaminated Land: Technical Recommendations "GLC" for the International Society of Soil Mechanics and Foundation Engineering*, 2. Ernst, Berlin.

GHEZZI, G., GHEZZI, P. and PELLEGRINI, M. (1999). Use of a cement–bentonite–slurry plastic diaphragm with HDPE membrane for MSW landfill. In: T. H. Christensen, R. Cossu and R. Stegmann (eds), *Sardinia 99, 7th International Waste Management and Landfill Symposium, S. Margherita di Pula, Sardinia, Conference Proceedings*, Vol. IV, pp. 549–554. CISA Environmental Sanitary Engineering Centre, Cagliari.

GILLHAM, R. W. (1993). Cleaning halogenated contaminants from groundwater. US Patent 5266213.

GILLHAM, R. W. and O'HANNESSIN, S. F. (1994). Enhanced degradation of halogenated aliphatics by zero-valent iron. *Groundwater* **32**(6), 958–967.

GILLHAM, R. W., FOCHT, R. M., RONIN, P. M. and PRITZKER, M. D. (1999). Water treatment system. US Patent 5868941, University of Waterloo, Canada.

GOLDER ASSOCIATES (1998). *Active Containment: Combined Treatment and Containment Systems.* UK Department of the Environment, Transport and the Regions, London.

GORBY, Y. A. and LOVLEY, D. R. (1992). Enzymatic uranium precipitation. *Environmental Science and Technology* **26**(1), 205–207.

GRATWOHL, P. (1997). Einsatz von Funnel and Gate-Systemen zur *in-situ* Sanierung gaswerks-spezifischer Schadstoffe im Grundwasser. In: V. Franzius (ed.), *Sanierung von Altlasten mittels durchströmter Reinigungswände, Berlin, Conference Proceedings.* Umweltbundesamt, PT AWAS.

GRATWOHL, P. and PESCHIK, G. (1997). Permeable sorptive walls for treatment of hydrophobic organic contaminant plumes in groundwater. *Proceedings of the International Containment Technology Conference, St Petersburg, Florida*, pp. 711–717.

GRAVELDING, D. (1998). Design and construction of a 1200 ft funnel & gate system. Subsurface Barrier Technologies, Conference, Tucson, Arizona.

GU, B., LIANG, L., DICKEY, M. J., YIN, X. and DAI, S. (1998). Reductive precipitation of uranium(VI) by zero-valent iron. *Environmental Science and Technology* **21**(21), 3366–3373.

GU, B., PHELPS, T. J., LIANG, L., DICKEY, M. J., ROH, Y., KINSALL, B. L., Palumbo, A. W. and JACOBS, G. K. (1999). Biochemical dynamics in zero-valent iron columns: Implications for permeable reactive barriers. *Environmental Science and Technology* **33**(13), 2170–2177.

HAGGERTY, G. M. and BOWMAN, R. S. (1994). Sorption of chromate and other inorganic anions by organo-zeolite. *Environmental Science and Technology* **28**(3), 452–458.

HIEBERT, R. (1998). Using biological barriers to control movement of contaminated ground-water. Subsurface Barrier Technologies, Conference, Tucson, Arizona.

HTI (1995). Technical sales literature. HTI, Florida.

HUBBLE, D. W., GILLHAM, R. W. and CHERRY, J. A. (1997). Emplacement of zero-valent metal for remediation of deep contaminant plumes. *Proceedings of the International Containment Technology Conference, St Petersburg, Florida*, pp. 872–878.

JANSEN, T. and GROOTERHORST, A. (1999). Reaktive Schmalwände zur passiven Grund-wasserreinigung. *TerraTech* **3**, 46–48.

JAYARAM, V., MARKS, M. D., SCHINDLER, R. M. and KOHNKE, R. J. (2001). Permeable reactive barrier installation using the biopolymer slurry trench method. Website: http//:www.containment.fsu.edu/cd/content/index.htm.

JEKEL, M. (1979). Biologisch-adsorptive Trinkwasseraufbereitung in Aktivkohlefiltern. *Veröffent-lichungen der Wasserchemie Karlsruhe*, Bd. 11. ZfGW, Frankfurt.

JESSBERGER, H. L. and GEIL, M. (1992). Einsatz von Spundwänden bei Deponien und Altlasten. *Geotechnik* **4**, 237–242.

JOHNSON, T. L. and TRATNYEK, P. G. (1995). Dechlorination of carbon tetrachloride by iron metal: the role of competing corrosion reactions. *209th American Chemical Society National Meeting, Anaheim, California, Division of Environmental Chemistry, Conference Proceedings*, Vol. 35, pp. 699–701.

KOENIGSBERG, S. (1998). The formation of oxygen barriers with ORC, Subsurface Barrier Technologies, Conference, Tucson, Arizona.

KUTZNER, C. (1991). *Injektionen im Baugrund*. Enke, Stuttgart.

LANGMUIR, D. (1978). Uranium solution–mineral equilibra at low temperatures with applications to sedimentary ore deposits. *Geochimica Cosmochimica Acta* **42**, 547–569.

LEHMANN, M., ZOUBOULIS, A. I. and MATIS, K. A. (1999). Removal of metal ions from dilute aqueous solutions: a comparative study of inorganic sorbent materials. *Chemosphere* **39**(6), 881–892.

LÖWENSCHUSS, R. (1979). Inorganic ion exchangers for the selective adsorption of uranium and plutonium from fission product solutions. *Atomkernergie – Kerntechnik* **33**(4), 260–264.

MA, Q. Y., TRAINA, S. J. and LOGAN, T. J. (1993). *In situ* lead immobilization by apatitec. *Environmental Science and Technology* **27**(9), 1803–1810.

MA, Q. Y., LOGAN, T. J. and TRAINA, S. J. (1994). Effects of NO_3^-, Cl^-, F^-, SO_4^{2-} and CO_3^{2-} on Pb^{2+} immobilization by hydroxyapatite. *Environmental Science and Technology* **28**(3), 408–418.

MACKENZIE, P. D., BAGHEL, S. S., EYKHOLT, G. R., HORNEY, D. P., SALVO, J. J. and SIVAVEC, T. M. (1995). Pilot scale demonstration of reductive dechlorination of chlorinated ethenes by iron metal. *209th American Chemical Society National Meeting, Anaheim, California, Division of Environmental Chemistry. Conference Proceedings*, Vol. 35, pp. 796–799.

MAGEAU, D. (1998). Use of a frozen ground barrier to contain groundwater contamination. Subsurface Barrier Technologies, Conference, Tucson, Arizona.

MATHESON, L. J. and TRATNYEK, P. G. (1994). Reductive dehalogenation of chlorinated methanes by iron metal. *Environmental Science and Technology* **28**(12), 2045–2053.

MEGGYES, T. and PYE, N. (1995). Cut-off walls. In: U. Holzlöhner, H. August, T. Meggyes and M. Brune (eds), *Landfill Liner Systems*, N1–28. Federal Institute of Materials Research and Testing (BAM), Berlin, Solid and Hazardous Waste Research Unit, The University of Newcastle, Penshaw Press, Sunderland.

MERKEL, B., HELLING, C., FALCK, E., METZLER, D., FRANCIS, A. J., HURST, S. and KOLITSCH, S. (1998). Zusammenfassung Internationale Konferenz, Uranbergbau und Hydrogeologie II, Freiberg.

MESECK, H. (1987). Dichtwände mit eingestellten Kunststoffbahnen (Kombinationsdichtwände). In: H. Meseck (ed.), *Dichtwände und Dichtsohlen*, Fachseminar 02/03. *Mitteilung des Instituts für Grundbau und Bodenmechanik*, Vol. 23, pp. 155–170. Technische Universität Braunschweig, Braunschweig.

MORRISON, S. (1998). Evaluation of materials for use in reactive barriers for uranium containment. Subsurface Barrier Technologies, Conference, Tucson, Arizona.

MORRISON, S. J. and SPANGLER, R. R. (1992). Extraction of uranium and molybdenum from aqueous solutions: a survey of industrial materials for use in chemical barriers for uranium mill tailings. *Environmental Science and Technology* **26**(10), 1922–1931.

MORRISON, S. J. and SPANGLER, R. R. (1993). Chemical barriers for controlling groundwater contamination. *Environmental Progress* **12**(3), 175.

MORRISON, S. J., SPANGLER, R. R. and TRIPATHI, V. S. (1995). Adsorption of uranium(VI) on amorphous ferric oxyhydroxide at high concentrations of dissolved carbon(IV) and sulfur(VI). *Journal of Contaminant Hydrology* **17**, 333–346.

MURDOCH, L., SLACK, B., SIEGRIST, B., VESPER, S. and MEIGGS, T. (1997). Advanced hydraulic fracturing methods to create *in situ* reactive barriers. *Proceedings of the International Containment Technology Conference, St Petersburg, Florida*, pp. 445–451.

NAFTZ, D. L., FREETHEY, G. W., HOLMES, W. F. and ROWLAND, R. C. (1996). *Field Demonstration of* in situ *Chemical Barriers to Control Uranium Contamination in Ground Water, Fry Canyon, Utah*. Project UT-96–242. US Geological Survey, Water Resources of Utah, Salt Lake City.

O'HANNESIN, S. (2001). Commercial aspects of permeable reactive barriers. Permeable reactive barrier (PRB) technology and its current status. 1st Workshop of the Permeable Reactive Barrier Network, Queens University, Belfast. Website: http://www.prb-net.org.

PEREBAR (2000). *Long-term Performance of Permeable Reactive Barriers Used for the Remediation of Contaminated Groundwater*. European Research Project in the 5th Framework Programme. Website: http://www.perebar.bam.de.

POWELL, R. M., PULS, R. W., HIGHTOWER, S. K. and SABATINI, D. A. (1995). Coupled iron corrosion and chromate reduction: mechanisms for subsurface remediation. *Environmental Science and Technology* **29**(8), 1913–1922.

POWELL, R. M., PULS, R. W., BLOWES, D. W., VOGAN, J. L., GILHAM, R. W., SCHULTZ, D. , POWELL, P. D., SIVAVEC, T. and LANDIS, R. (1998). *Permeable Reactive Barrier Technologies for Contaminant Remediation*. Report EPA/600/R-98/125, US Environmental Protection Agency, Washington DC.

PULS, R. W. (1998). Permeable reactive barrier research at the National Risk Management Research Laboratory, US EPA. In: H. Burmeier (ed.), *Treatment Walls and Permeable Reactive Barriers*. NATO CCMS Series, Vol. 229, pp. 3–5, NATO, Vienna.

RASMUSSEN, G. (2002). Sorption and biodegradation of creosote compounds in permeable barriers. Doctoral scientiarium thesis. Agricultural University of Norway, Ås.

REETER, C., CHAO, S. and GAVASKAR, A. (1999). Permeable reactive wall remediation of chlorinated hydrocarbons in groundwater. *US Department of Defense, Environmental Security Technology Certification Program (ESTCP). Cost and Performance Report*. US Department of Defense, Washington DC.

ROCHMES, M. (2001). Funnel-and-gate-system (Fe^0-Reaktor) in Edenkoben – Bau und Inbetriebnahme einer Grossanlage auf einem Industriegelände. In: G. Burghardt, T. Egloffstein and K. Czurda (eds), *ALTLASTEN 2001 Neue Verfahren zur Sicherung und Sanierung, Karslruhe*. Conference Proceedings, Vol. 4, pp. 59–69. ICP Eigenverlag Bauen und Umwelt.

RODATZ, W. (1994). *Grundbau, Bodenmechanik, Unterirdisches Bauen*. Institut für Grundbau und Bodenmechanik, Braunschweig.

ROTH, S. (1988). Eignung von Stahlspundwänden für Einkapselung von Altlasten. In: V. Franzius, R. Stegmann, K. Wolf and E. Brandt (eds), *Handbuch der Altlastensanierung*. Deckers, Heidelberg.

SCHAD, H. and GRATWOHL, P. (1998). Funnel and gate systems for *in-situ* treatment of contaminated groundwater at former manufactured gas plant sites. In: H. Burmeier (ed.), *Treatment Walls and Permeable Reactive Barriers*, NATO CCMS Series, Vol. 229, pp. 56–65. NATO, Vienna.

SCHAD, H. and TEUTSCH, G. (1998). Reaktive Wände – Aktueller Stand der Praxisanwendung. *Geotechnik* **21**(2), 73–82.

SCHULTZE, B. and MUSSOTTER, T. (2001). Sanierung des ehemaligen Gaswerksgeländes Karlsruhe-Ost mit funnel-and-gate (Aktivkohle). In: G. Burghardt, T. Egloffstein and K. Czurda (eds), *ALTLASTEN 2001 Neue Verfahren zur Sicherung und Sanierung, Karslruhe. Conference Proceedings* Vol. 4, pp. 47–58. ICP Eigenverlag Bauen und Umwelt.

SCHULZE, B. (1992). *Injektionssohlen – Theoretische und experimentelle Untersuchungen zur Erhöhung der Zuverlässigkeit*. Report 126. Institut für Bodenmechanik und Felsmechanik der Universität Karlsruhe, Karlsruhe.

SEELMANN-EGGEBERT, W., PFENNIG, G., MÜNZEL, H. and KLEWE-NEBENIUS, H. (1981). *Chart of Nuclides.* Kernforschungszentrum, Karlsruhe.

SHARP, R. R. and CUNNINGHAM, A. (1998). Fundamentals of biobarriers, design, development and activity. Subsurface Barrier Technology, Conference, Tucson, Arizona.

SIVAVEC, T. M. (1996). Research on passive groundwater remediation technologies at General Electric. Motivation and results. *Passive Systeme zur in-situ-Sanierung von Boden und Grundwasser, Workshop, Dresden, Extended Abstracts*, pp. 24–25.

SIVAVEC, T. M. and HORNEY, D. P. (1995). Reductive dechlorination of chlorinated ethenes by iron metal. *209th American Chemical Society National Meeting, Anaheim, California, Division of Environmental Chemistry, Conference Proceedings*, Vol. 35, pp. 695–698.

SMYTH, D. A., SHIKAZE, S. G. and CHERRY, J. A. (1997). Hydraulic performance of permeable barriers for the *in situ* treatment of contaminated groundwater. *Land Contamination and Reclamation* **5**(3), 131–137.

STROH, T. and SASSE, D. (1987). Beispiele für die Herstellung von Dichtwänden im Schlitzverfahren. In: H. Meseck (ed.), *Dichtwände und Dichtsohlen, Fachseminar 02/03. Mitteilung des Instituts für Grundbau und Bodenmechanik*, Vol. 23, pp. 35–38. Technische Universität Braunschweig, Braunschweig.

STUPP, H. D. (1999). Verfahren zur Reinigung schwermetallhaltiger Grundwässer. *TerraTech* **8**(2), 40–45.

STUPP, H. D. (2000). Grundwassersanierung von LCKW-Schäden durch Pump and Treat oder reaktive Systeme. *TerraTech* **9**(2), 34–38.

SUTHERSAN, S. (1996). *Remediation Engineering: Design Concepts. Environmental Science and Engineering Series.* Lewis, Boca Raton, Florida.

TÓTH, S. (1989). Soilcrete – Dichtwand als vorsorgliche Sicherung zur Verhinderung der Ausbreitung von Kontaminationen. *Geotechnik* **1**, 1–4.

US OFFICE OF RADIATION AND INDOOR AIR AND OFFICE OF ENVIRONMENTAL RESTORATION (1999). Review of geochemistry and available K_d values for cadmium, cesium, chromium, lead, plutonium, radon, strontium, thorium, tritium and uranium. *Understanding Variation in Partition Coefficient, K_d, Values*, Vol. II, EPA 402-R-99-004B. US Environmental Protection Agency and US Department of Energy, Washington DC.

US OFFICE OF WATER (1990). *Citizen's Guide to Ground-water Protection*, EPA 440/6-90-004. US Environmental Protection Agency, Washington DC.

VOGAN, J. L., GILLHAM, R. W. O. and HANNESIN, S. F. (1995). Site specific degradation of VOCs in groundwater using zero valent iron. *209th American Chemical Society National Meeting, Anaheim, California, Division of Environmental Chemistry. Conference Proceedings*, Vol. 35, pp. 800–804.

VOGEL, T. M., CRIDDLE, C. S. and MCCARTY, P. L. (1987). Transformations of halogenated aliphatic compounds. *Environmental Science and Technology* **21**(8), 722–736.

VOLESKY, B. (1990). *Biosorption of Heavy Metals.* CRC Press, Boca Raton.

WEBER, H. H., FRESENIUS, W., MATTHESS, G., MÜLLER-KIRCHBAUER, H., STORP, K. and WESSLING, E. (1990). *Altlasten: Erkennen, Bewerten, Sanieren.* Springer-Verlag, Berlin.

WILSON, R. D. and MACKAY, D. M. (1997). Arrays of unpumped wells: an alternative to permeable walls for *in situ* treatment. *Proceedings of the International Containment Technology Conference, St Petersburg, Florida*, pp. 888–894.

YANG, J. and VOLESKY, B. (1999). Biosorption and elution of uranium with seaweed biomass. In: A. Ballester and R. Amils (eds), *International Biohydrometallurgy Symposium, San Lorenzo de El Escorial, Conference Proceedings*, Vol. B, pp. 483–492. Elsevier, Amsterdam.

Part II
Groundwater remediation engineering

2. Remediation of chromium-contaminated groundwater in subsurface Fe⁰ reactor systems

M. Schneider
Liebermann + Schneider GmbH, 96515 Sonneberg, Germany; Free University of Berlin, Research Field Hydrogeology, Malteser Strasse 74-100, D-12249 Berlin, Germany

2.1. Introduction

The inorganic contaminant chromate (Cr^{VI}), as well as halogenated aliphatics such as trichloroethylene (TCE), *cis-* and *trans*-dichloroethylene (DCE) and vinyl chloride (VC) are not degraded, or only very slowly, in groundwater. It is therefore necessary to eliminate these toxic contaminants, or reduce their further dissemination, through appropriate remedial actions. Passive hydraulic groundwater remediation methods are being discussed intensively for this purpose, as a potential low-cost and efficient remediation alternative to energy- and maintenance-intensive pump-and-treat systems (e.g. McMurty and Elton, 1985; Starr and Cherry, 1994; Teutsch *et al.*, 1997; Schad and Teutsch, 1998). In this approach, subsurface *in situ* reactor systems are installed in the groundwater flow path in order to remove dissolved contaminants.

In redox-sensitive reactor systems, for example, elemental iron can be used to remove chromate and chlorinated solvents by reduction processes. The relevant oxidation, reduction and precipitation reactions are described by Gould (1982), Powell *et al.* (1994), Gillham and O'Hannesin (1994) and Matheson and Tratnyek (1994). At a contaminated site in northern Bavaria (Germany) such an *in situ* treatment unit was installed and tested as part of a full-scale groundwater remediation system, which is operated as a conventional pump-and-treat system (Schneider, 2000).

2.2. Characterization of the field site and hydrogeological setting

The site selected was a galvanizing plant near Coburg in northern Bavaria, which had been in use for over 30 years. While the plant was in operation, chromium wastes and chlorinated solvents contaminated the soil and groundwater. As a first measure, the area was intensively investigated in order to determine the type, source and dissemination of the contaminants. The initial remediation phase comprised demolition of the building, excavation of soil in the area with the highest pollutant concentrations, and disposal of the contaminated material.

The plume of contaminated groundwater had a mean concentration of 2 mg/l of halogenated aliphatics (tetrachloroethylene, PCE, TCE and *cis*-DCE) and 5 mg/l of chromate (maximum value: 80 mg/l). A disused waste pit was identified as the main source of contamination. The estimated total amount of pollutants in the plume was of the order of 100–200 kg for halogenated aliphatics, and 10–20 kg for chromium compounds. For Cr^{VI}, the maximum concentration detected exceeded the target value for remediation by a factor of 400. Figure 2.1 shows the centre part of the contaminant plume, with concentrations above 2 ppm for halogenated aliphatics and 10 ppm for chromate. The field site is located in an area of a groundwater divide, so the flow directions of the shallow groundwater are extremely variable.

The shallow sandstone aquifer (with intercalations of siltstone, Upper Triassic) has a thickness of about 1 m and is covered by low permeability quaternary sediments comprising silt, clay and gravel. The base of the aquifer is 5–6 m below ground level, and consists of shale of Upper Triassic age (Fig. 2.2). Groundwater flow velocity is relatively low, and the hydraulic conductivity lies between 10^{-5} and 10^{-6} m/s.

2.3. Groundwater remediation concept

Within the scope of a research project (financially supported by the Federal Ministry of Economics and Technology, Project No. 826/97), a concept was developed for the pilot scale *in situ* remediation of groundwater contaminated with Cr^{VI} and halogenated aliphatics. This system was to be operated as a part of the pump-

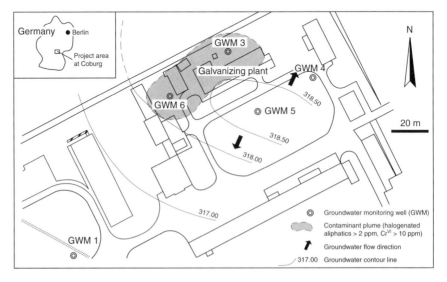

Fig. 2.1. Hydrogeological situation at the field site and position of the contaminant plume

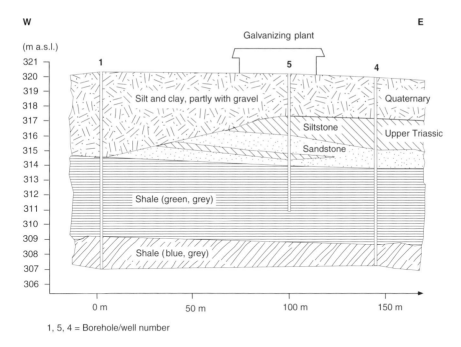

Fig. 2.2. Generalized geological cross-section through the plant area

and-treat facility already installed as a final site remedy and approved by the regulatory agency.

Because of the small aquifer thickness and low groundwater discharge, a sub-surface drainage system was installed approximately 5 m below ground level. This consists of a side drain connected through a reactor shaft to a main drain. The mixed water from both drains finally enters a main shaft and is pumped into an on-site treatment unit where the contaminants are removed by conventional methods (ion exchange and activated carbon). The remediated groundwater from the on-site unit is discharged into a nearby stream. Figure 2.3 shows the installations at the field site and the depression of the water table caused by the drainage system. This concept allows the operation of the experimental *in situ* reactor system under varying conditions, even if the quality of the water discharged from the reactor is not in accordance with the regulations. In a full-scale passive remediation system the remediated water would be infiltrated directly into the aquifer. Figure 2.4 illustrates the installation of the main drain outside of the excavated area.

The structure of the reactor shaft is shown in Fig. 2.5. The shaft has a diameter of 1.5 m, and is subdivided into four compartments in order to optimize the flow path of the contaminated water through the permeable reactive material, which has a volume of approximately 1 m^3. The hydraulic gradient in the reactor is 0.0033.

Iron in various forms (grains, chips and wool) was investigated in laboratory batch and column tests. Owing to its large specific surface, steel wool achieved the best results. For this reason, steel wool was chosen as the reactive material. The content of Fe^0 in the steel wool was 98%.

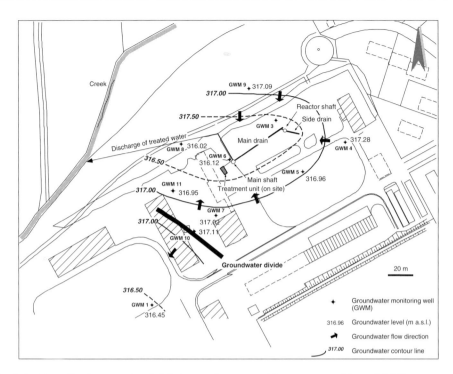

Fig. 2.3. Installations at the field site and water table contours (June 2000)

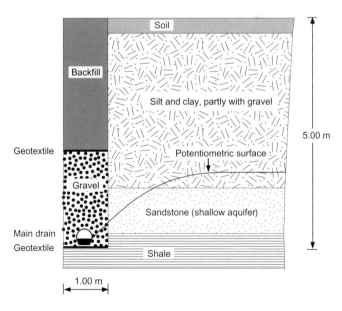

Fig. 2.4. Construction of the main drain

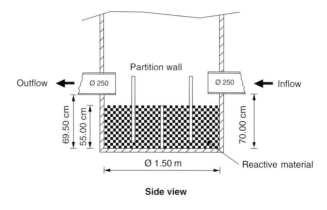

Side view

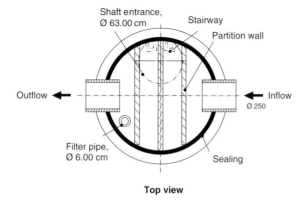

Top view

Fig. 2.5. Construction of the reactor shaft

Figure 2.6 shows the construction work at the field site. Installation of the reactor shaft and the drainage system is shown in Fig. 2.6a. Figure 2.6b shows a general view of the site, with the excavated highly contaminated material and the on-site treatment unit in the background.

2.4. Results and discussion

As a field test, 45 kg of steel wool mixed with calcite in 64 single packs was inserted in the reactor. Over a period of 32 days the mean rate of flow through the reactor was 30 l/h. Water samples were taken at the inflow, outflow and from the various compartments of the reactor system. The mean inflow concentration was 2.1 mg/l for Cr^{VI} and 0.9 mg/l for halogenated aliphatics.

During the test period, Cr^{VI} concentrations in the effluent samples had a mean value of 0.03 mg/l, and were below the target value of 0.2 mg/l at all times (Fig. 2.7). Regarding the halogenated organic compounds, the remediation target (0.04 mg/l in groundwater) was not reached; in most cases the outflow concentration was of the same order as the inflow concentration. It is assumed that passivation of the surface of the steel wool – due to concurrent chromate reduction (Fruth, 1998) – plus insufficient

Fig. 2.6. Installation of (a) the drainage system and (b) the reactor shaft

residence time of the contaminated water in the reactor led to the unsatisfactory degradation performance. The pH values were stabilized between 6.7 and 7.3.

Using steel wool as the reactive material it was shown that a stable remediation performance was achieved throughout the test phase, removing chromate from groundwater even when inflow concentrations and water quantities fluctuated greatly. Reduction of permeability within the reactive material, due to precipitation of mineral phases, was not observed.

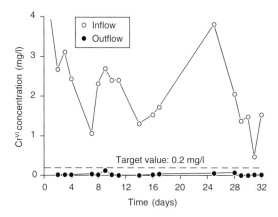

Fig. 2.7. Change in Cr^VI concentration over time

Table 2.1. Essential demands on project planning and structure

Demands on project planning	Demands on structure
Detailed hydrogeological site investigation	Exchangeability and ease of handling the reactive material
Three-dimensional localization of contamination	Reactor permeability must be greater than aquifer permeability.
Modelling of groundwater flow	Stability (if reactive material is exchanged)
Geochemical modelling: the influence of the reactor on groundwater composition	Avoidance of oxygen influx
	Longevity
	Monitoring facilities: supervision of the remediation process

The installation of a passive *in situ* remediation system and its optimal operation require a detailed hydrogeological site investigation. Modelling groundwater flow is advisable in order to optimize the position and dimensions of the subsurface structure. Essential demands on project planning and structure are summarized in Table 2.1.

2.5. References

FRUTH, M. (1998). LCKW-Abbau und Chromatreduktion in Fe⁰-Reaktionswänden: Inhibierungs-prozesse durch Chromat. Diploma thesis, Fachhochschule Weihenstephan, FB Landwirt-schaft und Umweltsicherung, Triesdorf.

GILLHAM, R. W. and O'HANNESIN, S. F. (1994). Enhanced degradation of halogenated aliphates by zero-valent iron. *Ground Water* **32**(6), 958–967.

GOULD, J. P. (1982). The kinetics of hexavalent chromium reduction by metallic iron. *Water Research* **16**, 871–877.

MATHESON, L. J. and TRATNYEK, P. G. (1994). Reductive dehalogenation of chlorinated methanes by iron. *Metals in Environmental Science and Technology* **28**(12), 2045–2053.

MCMURTY, D. C. and ELTON, R. O. (1985). New approach to *in-situ* treatment of contaminated groundwaters. *Environmental Progress* **4**(3), 168–170.

POWELL, R. M., PULS, R. and PAUL, C. J. (1994). *Innovative Solutions for Contaminated Site Management*, pp. 485–496. Water Environment Federation, Miami.

SCHAD, H. and TEUTSCH, G. (1998). Reaktive Wände – Aktueller Stand der Praxisanwendung. *Geotechnik* **21**(2), 73–82.

SCHNEIDER, M. (2000). Passive *in situ* treatment of Cr^{VI}-contaminated groundwater by zero-valent iron: results of a field study. *Proceedings of the XXX IAH Congress, Cape Town*, pp. 823–826.

STARR, E. C. and CHERRY, J. A. (1994). *In-situ* remediation of contaminated ground water: the funnel-and-gate system. *Ground Water* **32**(3), 465–476.

TEUTSCH, G., TOLKSDORFF, J. and SCHAD, H. (1997). The design of *in-situ* reactive wall systems – a combined hydraulic-geochemical-economic simulation study. *Land Contamination and Reclamation* **5**(3), 125–130.

3. Current R&D needs and tailored projects for solving technical, economic, administrative and other issues concerning permeable reactive barrier implementation in Germany

H. Burmeier, V. Birke and D. Rosenau
University of Applied Sciences, North-East Lower Saxony, Department of Civil Engineering, Water and Environmental Management, Office Hanover, Steinweg 4, D-30989 Gehrden, Germany

3.1. Introduction

The interdisciplinary German PRB network RUBIN was initiated and set up by the German Federal Ministry for Education and Research (BMBF) in 2000. RUBIN is an acronym for (in German) 'Reaktionswände und -barrieren im Netzwerkverbund' which translates into English as 'Reactive Wall and Barrier Projects Co-operating in a Network'.

The focus of RUBIN's missions and goals is to meet current R&D needs pertaining to the practical set up and long-term operation of PRBs as a prospective remediation technology through a large-scale, coordinated initiative. In particular, various technical, operational, economic, ecological, toxicological, administrative and legislative issues are addressed and investigated. Therefore, RUBIN will plan, design, implement, monitor and evaluate pilot and full-scale PRB projects in Germany in order to check and assess as thoroughly and precisely as possible the applicability, long-term performance and limits of PRBs in a broad technical scope combined with an intensive, simultaneous, scientific back-up. The network will also investigate novel innovative approaches to the elimination of recalcitrant compounds from contaminated groundwater by means of innovative reactive materials and barrier design and construction methods.

Although a growing number of demonstration sites for PRBs, predominantly involving treatment of chlorinated ethenes by granular elemental iron, have proven

successful in principle in North America, so far PRBs have not been fully accepted
and therefore established as general remediation technologies in Europe. The lack
of general acceptance and absence of incentives to implement PRBs, at full scale
and of diverse type, are due to, among other factors, insufficient reliable information
on long-term effects and performance, and, associated with these, their overall
viability. In Germany, seven pioneering PRB projects (full and pilot scale) have
been implemented since 1999, revealing promising preliminary results. Therefore,
the German BMBF decided to evaluate and assess the performance of PRBs (as well
as other material issues) to a greater extent and in a broader scope by means of
RUBIN. This chapter provides an insight into the mission, tasks and goals as well as
the structure of RUBIN and its individual projects.

3.2. Current status of PRB technologies worldwide

As passive *in situ* groundwater remediation techniques, and therefore avoiding
several drawbacks of active systems *a priori*, PRBs are currently regarded worldwide,
especially across North America and Western Europe, as promising upcoming
alternatives to common active groundwater remediation technologies such as pump
and treat. For instance, PRBs require only very low energy input during operation;
there is no permanent and massive intervention into the aquifer; and the remediation
takes place in the subsurface directly inside the contaminated aquifer, i.e. no costly
installations or specific plant have to be provided, requiring long-term operation
and maintenance.

Since the late 1980s PRBs have been developed and implemented in about 50
projects across North America, with pioneering R&D work undertaken by Robert
Gillham and others at the University of Waterloo, Canada. This research investigated
fundamental construction methods for PRBs, such as the funnel-and-gate principle,
as well as reactive materials for the *in situ* dehalogenation of recalcitrant poly-
halogenated groundwater pollutants.

The first pilot PRB was set up at Borden, Canada, in June 1991, and since the
mid-1990s the number of North American pilot and full-scale projects has increased
notably (US Environmetal Protection Agency, 1999; Birke *et al.*, 2000a; Gavaskar *et
al.*, 2000). These projects predominantly concentrate on chlorinated volatile organic
compounds (cVOCs) such as chlorinated ethenes (mainly tetrachloroethene (also
called perchloroethene, PCE) and trichloroethene (TCE)), which are dehalogenated
via *cis*-dichloroethene (*cis*-DCE) and vinyl chloride (VC) to halogen-free ethene as
the major degradation product using zero-valent (elemental) iron (ZVI) as the
dehalogenation reagent (mainly in the form of iron filings or granules). Some
pilot projects have already been running for more than 5 years and have revealed
consistently high degradation rates for the pollutants, e.g. the funnel-and-gate
system at Mofett Federal Airfield, Mountain View, California, set up in April 1996
(cVOCs dehalogenated with ZVI (US Environmental Protection Agency, 1999;
Birke *et al.*, 2000b)), or the funnel-and-gate PRB at Dover Air Force Base, Delaware,
set up in December 1997 (cVOCs again dehalogenated with ZVI (US Environmental Protection Agency, 1999; Yoon *et al.*, 2000)).

However, PRB technologies have not so far gained general acceptance as establish-ed remediation technologies for two major reasons:

(1) There is a lack of reliable information on long-term performance, longevity and long-term effects, because of the absence of results from long-term tests and lack of reliable long-term modelling.
(2) Insufficient information is currently available on the viability of PRBs, especially their long-term efficacy, if this is assumed to be subject to change.

Detailed knowledge of the long-term performance and effect of PRBs is essential, since a PRB is usually designed to operate for some decades. For example, long-term efficacy and performance are influenced to a high degree by the reactivity of the reactive material towards the contaminants over time. This reactivity depends on the robustness of the material to hydraulic and hydrochemical effects and can potentially be influenced by a range of physical and chemical parameters. Hence, obtaining and predicting long-term performance data regarding engineered PRB installations are of the highest importance (Puls *et al.*, 2000).

Unlike other remediation processes, the viability of PRBs is highly dependent on their stable and predictable long-term performance. Since both investment and construction costs are relatively high at the very beginning of a PRB project – therefore governing the total costs over a longer operational period – it is essential to minimize the overheads during operation: long-term performance must be strictly maintained without the need for additional active measures, such as exchanging or regenerating the reactive material if it loses its reactivity or sorption capacity. Any unforeseen active measure required to improve or re-establish decreased performance can result in potentially high additional costs. Assuming stable performance over time, cost estimates for PRBs can be made by applying suitable cost functions, provided that all influencing parameters are determined precisely (Finkel *et al.*, 1998; Liedl *et al.*, 1999; Bayer *et al.*, 2001). Unfortunately, owing to lack of knowledge of long-term PRB performance, there are uncertainties concerning these cost estimates.

3.3. Development in Germany

In Germany, seven PRB pilot projects, all showing some good initial results regarding efficiency of degradation or removal of contaminants and viability, have been implemented since 1999, namely in Rheine (cVOCs, iron filings and iron sponge, pilot scale, continuous wall; Rochmes, 2000), Tübingen (cVOCs, granular iron, full scale, funnel and gate; Klein and Schad *et al.*, 2000; Rochmes, 2000), Karlsruhe (polyaromatic hydrocarbons (PAHs), activated carbon, full scale, funnel and gate; Rochmes, 2000; Schad *et al.*, 2000), Edenkoben (cVOCs, iron filings, pilot scale, expanded to full scale in 2001, funnel and gate; Rochmes, 2000), Denkendorf (cVOCs, activated carbon, full scale, drain and gate), Bitterfeld (chlorinated hydrocarbons, PAHs, microbiological degradation and palladium and iron plus activated carbon in different reactors, pilot scale with focus on R&D, specific reactor systems; Weiss *et al.*, 1999) and Reichenbach (cVOCs, activated carbon, full scale, specific design; Edel and Voigt, 2001). These projects are described more fully in the appendix to this chapter.

Important R&D work is ongoing at the University of Tübingen, the University of Kiel and at the Umweltforschungszentrum Leipzig.

Owing to these projects, and other initiatives in Germany, the importance of PRBs in groundwater remediation has come to the attention of the German public, and its potential has gained greater recognition. Significant progress has also been made in the understanding and prediction of long-term PRB performance (Wüst et al., 1999; Dahmke et al., 2000). Thus, there is strong support for testing and evaluating this technique, and for developing new concepts and solutions. Hence, in 1999, SAFIRA, the first nationwide PRB initiative, was launched. SAFIRA is an interdisciplinary PRB network for basic R&D work on an engineering scale, using specifically designed in situ reactors for testing different reactive materials at Bitterfeld. It is managed and coordinated by the Umweltforschungszentrum Leipzig in cooperation with research institutions from all over Germany (Weiss et al., 1999). Like RUBIN, SAFIRA is funded by the BMBF.

3.4. Mission, goals and structure of RUBIN

In order to support and promote the technical development of PRBs, the BMBF set up RUBIN in 2000 (Birke et al., 2000b). RUBIN's time span is about 4 years. At least 4 million euros will be provided and spent over that term.

The mission and goals of RUBIN are detailed below:

(1) RUBIN's projects are expected to deliver comprehensive information on and solutions to problems in various areas such as planning and design, construction and operation, monitoring, economics, ecological effects, regulations, and acceptability of the technique.

(2) The RUBIN projects focus on the establishment and operation of pilot and full-scale PRB installations. Experts from research institutions (e.g. universities), developers (universities and companies), planners (consultants, environmental technology and engineering companies), executives (builders and contractors, specialized civil engineering companies) and regulators cooperate in an interdisciplinary way. Data have to be collected from as many different sites and installations as possible.

(3) RUBIN is expected to deliver extensive data for a reliable assessment of the benefits and drawbacks as well as a precise prediction of the applicability and viability of a PRB, in relation to any given single remediation scenario.

(4) Since RUBIN includes existing PRBs such as the Rheine, Tübingen and Edenkoben installations, initial investigations can be made into the long-term aspects of PRBs. Therefore, with the help of RUBIN, experts are getting the opportunity to test thoroughly German PRB installations.

(5) RUBIN is to provide quality standards and a generally applicable quality management scheme for the construction, operation and monitoring of PRBs. Approaches are being developed for improved monitoring and more reliable site investigations.

(6) Both investment and overhead costs of all RUBIN projects will be scrutinized, and will deliver a data set for more precise approaches for calculations of viability, especially compared with the more common pump-and-treat measures.

The 12 RUBIN projects can be classified into two groups: nine projects deal with planning, setting up and/or operating, as well as monitoring and actual PRB construction. Three further projects are scheduled to tackle general issues and problems. An overview is given by Table 3.1. Detailed data for each individual project can be found in the appendix to this chapter. The findings will be available as a single publication consisting principally of a state-of-the-art report and a main model procedure for the planning, design, construction and operation of PRBs in Germany.

On 6 September 2001 the installation of the funnel-and-gate system at the Bernau site was successfully completed, and the facility is now ready for operation. Managed by the 'Brandenburgische Boden Gesellschaft für Grundstücksverwaltung und -verwertung mbH (BBG)', this is the first RUBIN project to erect and operate a new PRB. The initial results from monitoring of the TCE degradation are keenly awaited.

Three other RUBIN projects deal with important general issues and aspects:

(1) At the University of Tübingen (G. Teutsch and M. Finkel) there is ongoing work comprising development of models for estimation and prediction of costs and viability calculations. A comparative economic assessment is being performed for the PRB technique versus 'innovative' pump-and-treat systems.

(2) At the University of Kiel, A. Dahmke, M. Ebert and N. Silva-Send are undertaking, in cooperation with R. Wienberg, Hamburg, a comparative laboratory and site study for the evaluation and further development of site investigation procedures, monitoring and quality management. One focus of this work is on scrutinizing degradation mechanisms and solving issues regarding side reactions, as well as determining mass balances, especially for the dehalogenation of cVOCs such as PCE and TCE by ZVI, using column and field data.

(3) At the University of Applied Sciences, North-East Lower Saxony, Suderburg, the general work and results of the network are coordinated. The university is also responsible for the production of a general publication to include a state-of-the-art report and guidance for implementation of PRBs in the form of a main model procedure, which will have to accord with the existing general regulations and laws for remediation of contaminated sites in Germany. This main model procedure will cover descriptions, advice and instructions for cost calculations, planning, design, administrative regulation and approval, construction and operation, as well as monitoring of PRBs.

More information on the RUBIN network is available on the Internet: http://www.rubin-online.de.

Table 3.1. Overview of the RUBIN projects dealing with an actual PRB installation, arranged by location in alphabetical order (2001)

	Bernau	Denkendorf	Dresden	Edenkoben	Nordhorn	Offenbach	Rheine	Wiesbaden
Topic, pollution	TCE-contaminated groundwater (two aquifers); funnel-and-gate system utilizing regenerable Fe^0 reactors with a horizontal arrangement and single cyclindrical segments	Innovative downgradient remediation of groundwater polluted by TCE, PCE, *cis*-DCE and 1,1,1-TCA at a trading estate in Denkendorf	Investigations for the construction of adsorptive walls at uranium-contaminated sites, especially at Dresden-Coschuetz/Gittersee	Monitoring programme for the full-scale funnel-and-gate system treating TCE, *cis*-DCE, 1,1,1-TCA. Production of operational instructions to be used in a manual	Development of an integrated process for the remediation of groundwater polluted by mixed contaminants (dyes, DNAPL). Former textile factory/dyeing works	Development of and testing a reactor and a treatment wall for the removal of BTEX and PAH. Former tar plant	Evaluation of the long-term behaviour of the pilot Fe^0 reactive wall at Rheine. Case study for the long-term removal of PCE, TCE and *cis*-DCE	Development and evaluation of reactive wall systems applied to arsenic contamination at a site in contact with a River Rhine aquifer
Applicant	Federal State of Brandenburg, BBG mbH, Waldstadt	IMES GmbH, Amtzell	Technical University of Berlin	Peschla und Rochmes GmbH, Kaiserslautern	Administrative District (County) of Bad Bentheim	Hessische Industriemüll GmbH, ASG, Wiesbaden	Mull und Partner Ing. GmbH, Garbsen, and Technical University of Berlin	Hessische Industriemüll GmbH, ASG, Wiesbaden

Wall system	Funnel and gate (horizontal flow). The source is entirely enclosed by a slurry wall	Drain and gate	Still to be designed (will be based on the outcomes of this project)	Funnel and gate (vertical flow)	Still to be designed	Funnel and gate	Continuous trench	Reactive zone at the upgradient, funnel and gate at the downgradient
Reactive material	Fe^0	Palladized zeolites, pelletized and hydrophobic. Addition of hydrogen gas or Fe^0	Alternative adsorbents such as brown coal, brown coke; other adsorbents such as zeolites. Naturally occurring complexing agents (e.g. humic acids), synthetic agents	Fe^0 filings (exception: in the existing gate the upgradient area is merely infilled). Activated carbon in the downgradient area of the still existing gate	Still to be determined: microbiology combined with adsorptive materials and Fe^0?	Microbiology plus activated carbon. Addition of electron acceptors required for the microbiological degradation. Activated carbon	Fe^0 sponge. Granular grey iron/pea gravel mixture	*Plan 1* Readily oxidizable C_{org} phases, solid phases consisting of sulfate (both resulting in sulfides) *Plan 2* Iron oxide: adsorptive fixation
Status	Approved/ running	Approved/ running	Still to be approved	Still to be approved	Still to be approved	Approval expected soon	Approved/ running	Still to be approved

3.5. Appendix

1. Bernau

Topic

In situ decontamination of TCE-contaminated groundwater (two aquifers affected) applying the funnel-and-gate principle and utilizing regenerable ZVI reactors (single cylindrical segments) with a horizontal arrangement.

Applicant

*   The German Federal State of Brandenburg.
*   Representative: Brandenburgische Boden Gesellschaft für Grundstücksverwaltung und -verwertung mbH (BBG), Waldstadt (Dipl.-Geol. Freygang, Dipl.-Geogr. Isenberg).

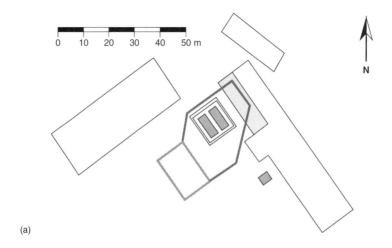

(a)

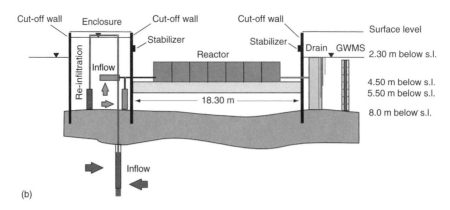

(b)

Fig. 3.1. The Bernau contaminated site. (a) View of the funnel-and-gate system (closed funnel, i.e. surrounding the former TCE tanks, constructed as a slurry wall = hexagon, gate = rectangle attached to the hexagon). (b) Elevation of the gate system, showing the reactor. (Courtesy of BBG mbH (Dipl.-Geol. Freygang), the Technical University of Berlin (Dipl.-Ing. Vigelahn) and INGAAS GmbH (Dipl.-Ing. Hein))

Fig. 3.2. Charging the last gate segments with granular iron in Bernau (September 2001)

Cooperating partners
- Technical University of Berlin (Dipl.-Ing. Vigelahn).
- INGAAS GmbH (Dipl.-Geol. Hein).

Pollutants
- Cause of contamination: dry-cleaning facility (set up in the 1960s); spills of TCE, leakage of large storage tanks, which used to be charged with TCE.
- cVOCs, main contaminant > 90% TCE, average 75–500 mg/l of cVOCs.
- Area where pollutants intrude into the aquifer: 1.1 g/l in the first aquifer.
- Up to a depth of approximately 40 m inside the second aquifer still 60 mg/l of halogenated volatile organic compounds (hVOCs).
- In the unsaturated soil zone ≤ 690 mg/kg, average concentration of cVOCs in soil gas 200 mg/m^3.

Wall system (Figs 3.1–3.3)
- Funnel and gate (horizontal flow through a gate).
- Depth of wall up to 11 m.
- Reactive material: elemental iron (Fe^0).
- Gate structure: several single cylindrical segments with an emergency filter in case of leakage of pollutants.
- Construction of a 120 m long closed sealing wall structured as a funnel in the direction of the groundwater downgradient (combination of a sheet pile and a slurry wall, hexagon in Fig. 3.1) that reaches about 1.5 m into the aquitard (at 10–11 m depth).
- Intensive vertical flow inside the edged area (hexagon in Fig. 3.1) via re-infiltration of water lifted from the lower aquifer by pumping.

Fig. 3.3. (a) Fitting/installation of the single reactor segments and (b) overview of the whole gate construction of the Bernau PRB (September 2001)

2. Denkendorf

Topic
Innovative downgradient remediation on the premises of a trading estate in the town of Denkendorf (near Stuttgart).

Applicant
- IMES GmbH (Dr Schad).

Cooperating partners
- GeoRisk Ingenieur GmbH (Dipl.-Geol. Stiehl).
- University of Tübingen, Geological Institute (Dr Schueth).

Pollutants
- Cause of contamination: six cases of pollution (cHCs) on the premises of the trading estate.
- cVOCs: main contaminants are: TCE, perchloroethylene (PCE), *cis*-DCE and TCA.
- Concentration > 200 mg/l hVOCs within the focus of the contamination (cVOCs are present in phase). Average total sum of hVOCs: ≤ 30 mg/l.

Wall system (Fig. 3.4)
- Drainage system and gate construction (*in situ* reactor, 'drain and gate').
- Depth of wall up to 6 m.
- Reactive material: palladized zeolithes, pelletized and hydrophobic with a palladium loading of 0.5% (w/w).
- Addition of hydrogen gas or ZVI required.
- Collecting/trapping the contaminated groundwater by means of a 20 m long gravel drain, flow through the reactors (hydraulic gradient approximately 2%), direct drawing off of the groundwater into the receiving water.
- Construction of the reactor as a shaft structure.
- Common civil engineering techniques used for the construction of the drainage and the shaft structure up to a depth of 6 m. Experimental set-ups are illustated in Figs 3.5 and 3.6.

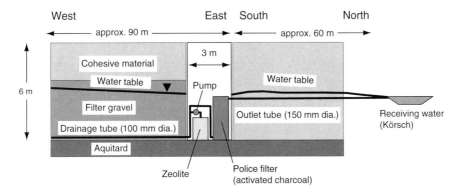

Fig. 3.4. Sketch of the construction of the drain-and-gate-system at the Denkendorf site. (Courtesy of IMES GmbH (Dr Schad))

3. Dresden

Topic

Investigations for the construction of adsorptive walls at uranium-contaminated sites, especially at Dresden-Coschuetz/Gittersee.

Applicant

* Technical University of Berlin (Prof. Rotard, Dipl.-Ing. Borrmann).

Cooperating partners

* Ingenieurbüro Dr Bütow, Berlin (Dr Bütow).
* IGU Institute for Applied Isotopes, Gas and Environmental Investigations.
* City of Dresden.

Pollutants

* Cause of contamination: industrial settling tank; slagheap of the former uranium ore-processing plant.
* Uranium contamination and successive products of uranium ore-processing; arsenic and heavy metals are also present.

Reactive material

* Alternative adsorbents such as brown coal, brown coal coke and zeolites are intended to be tested at laboratory and pilot scale.
* Naturally occurring complexing agents (e.g. humic and fulvic acids).

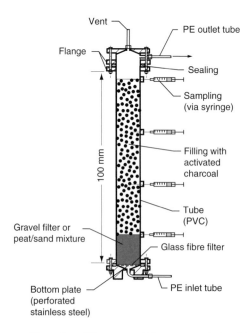

Fig. 3.5. Column set-up, Dresden-Gittersee. (Courtesy of the Technical University of Berlin (Dipl.-Ing. Borrmann))

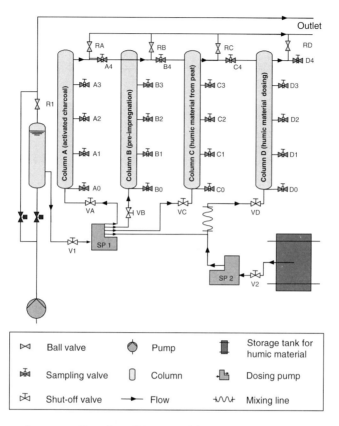

Fig. 3.6. Pilot-scale set-up, Dresden-Gittersee. (Courtesy of the Technical University of Berlin (Dipl.-Ing. Borrmann))

- Synthetic complexing agents such as ethylenediaminetetraacetic acid (EDTA) and nitrilotriacetic acid (NTA).

Wall system
Wall design and construction will be outcomes of this research project.

4. Edenkoben

Topic
Set-up of a monitoring programme for the existing full-scale funnel-and-gate system at the town of Edenkoben.

Applicant
- Peschla and Rochmes GmbH (Dipl.-Geol. Rochmes, Dipl.-Ing. Woll).

Pollutants
- Cause of contamination: solvents used in production processes.
- Average of single cVOCs: 20% TCE, 50% *cis*-DCE, 30% 1,1,1-trichloroethane (1,1,1-TCA).

- At least three single plumes, partly overlapping:
  - plume south: TCE, *cis*-DCE, ≤ 8000 µg/l of cVOCs;
  - plume north: PCE, ~ 2000 µg/l of cVOCs;
  - plume middle: 1,1,1-TCA, TCE, *cis*-DCE, ≤ 20 000 µg/l of cVOCs.

Wall system (Figs 3.7–3.12)

- Funnel and gate (vertical flow).
- Depth of wall: approximately 15 m.
- Reactive material: Fe^0 filings.
- Six gates (each 10 m wide) surrounded by a sheet pile caisson (open towards the bottom) that extends approximately 8 m below ground level.
- Continuous sheet pile wall, 400 m long (14 m below ground level into aquifer base), separating gates into two chambers (each 1.25 m wide); in the area of the gate, the sheet pile wall was buried down to 1 m below the anticipated lowest groundwater level (at 5 m below ground level) serving as an overflow weir between the chambers (vertical flow; flow direction was intentionally lengthened by the reaction zone).
- Outside of the gate, the wall in the middle reaches ground level, forming the funnel.
- Complete connection of the deeper, polluted groundwater areas via vertical drainage systems.

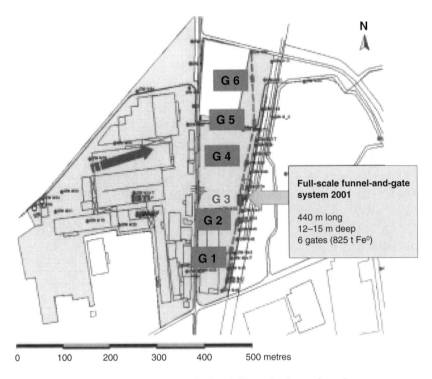

Fig. 3.7. Location and appearance of the full-scale funnel-and-gate system at the Edenkoben site (since 2001) including groundwater flow direction and the position of the pilot installation (denoted G3, originally set up in 1998)

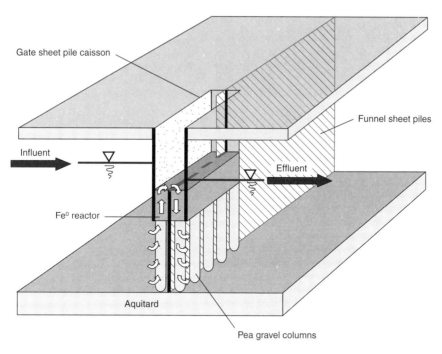

Fig. 3.8. Design and construction of a gate of the funnel-and-gate system in Edenkoben featuring a vertical forced flowthrough

Fig. 3.9. Edenkoben. (a) Set-up of the vertical drainage system using a large-diameter borehole construction method. (b) Filling the gate with Fe⁰. (Courtesy of Peschla & Rochmes GmbH, Dipl.-Geol. Rochmes (Dipl.-Ing. Woll))

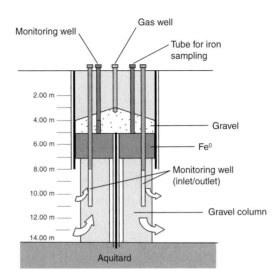

Fig. 3.10. Monitoring system of the Edenkoben wall (elevation). (Courtesy of Peschla & Rochmes GmbH, Dipl.-Geol. Rochmes (Dipl.-Ing. Woll))

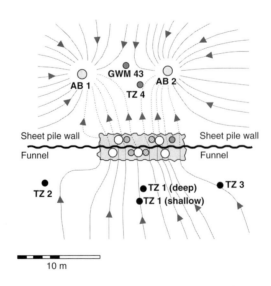

Fig. 3.11. Monitoring wells (top view) of the Edenkoben wall showing gate G3 (see also Fig. 3.7; G3 is identical to the original pilot gate). Using AB 1 and AB 2 for pumping, different constant throughflows can be adjusted and maintained. GWM and TZ denote groundwater monitoring wells. (Courtesy of Peschla & Rochmes GmbH, Dipl.-Geol. Rochmes (Dipl.-Ing. Woll))

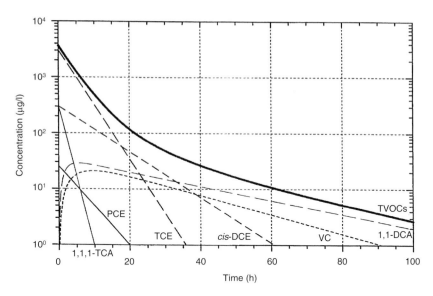

Fig. 3.12. Edenkoben: degradation of VOCs from column experiments. 1,1-DCA, 1,1-dichloroethane; TVOCs, total volatile organic compounds. (Courtesy of Peschla & Rochmes GmbH, Dipl.-Geol. Rochmes (Dipl.-Ing. Woll))

5. Nordhorn

Topic

Development of an integrated process for the remediation of groundwater polluted by mixed contaminants stemming from a former textile factory (Figs 3.13 and 3.14).

Applicant

* NINO Sanierungsgesellschaft mbH, Nordhorn.

Cooperating partners

* Administrative District (County) of Bad Bentheim (Dipl.-Ing. Zwartscholten).
* Dr Wessling Beratende Ingenieure GmbH (Dipl.-Ing. Wortmann).
* EN-PRO-TEC GmbH, Dipl.-Ing. Rongen.
* Umweltforschungszentrum Leipzig-Halle GmbH (Prof. Kopinke, Dr Georgi).

Pollutants

* Cause of contamination: contaminants stemming from a former textile factory/dyeing works.
* Dyes (e.g. azo and alizarin dyes).
* Dense non-aqueous phase liquids (DNAPL).
* cVOCs.
* Chlorinated benzenes and other aromatic compounds.
* Aromatic amines.

Wall system/reactive material(s)

- A permeable reactive barrier will be designed and constructed on the basis of the outcomes of this research project.
- The reactive material has still to be determined (e.g. microbiology combined with adsorptive materials and/or Fe^0).

6. Offenbach

Topic

Developing and testing a reactor and a treatment wall for cleansing a (benzene, toluene, ethylbenzene and xylene (BTEX)- and PAH-polluted aquifer.

Applicant

- Hessische Industriemuell GmbH, ASG, Wiesbaden (Dipl.-Ing. Kayser, Dr Boehmer).

Cooperating partners

- AICON AG (Dr Weiss).
- IMES GmbH (Dr Schad).

Pollutants

- Cause of contamination: area of the former tar plant 'Lang' near the River Main in Offenbach.
- BTEX, PAH (naphthalene, acenaphthene).
- Benzene in the downgradient (primary component of groundwater contamination) ≤ 4000 µg/l.

Fig. 3.13. Aerial photograph of the Nordhorn site (located in the centre of Nordhorn). (Courtesy of NINO Sanierungsgesellschaft mbH (Dipl.-Ing. Zwartscholten))

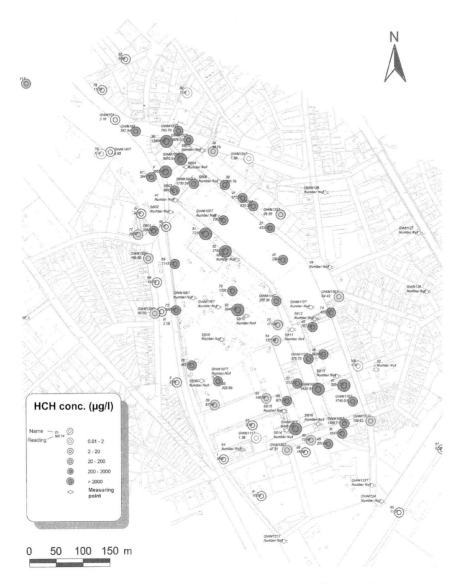

Fig. 3.14. Map showing the contaminated area in Nordhorn. The marked points represent the concentration of cVOCs. (Courtesy of NINO Sanierungsgesellschaft mbH (Dipl.-Ing. Zwartscholten))

- Area of a former impregnation works employing tar oil: 20–80 cm thick tar oil in phase.
- BTEX ≤ 110 000 µg/l, naphthalene ≤ 12 000 µg/l, PAH (excluding naphthalene) ≥ 800 µg/l, phenols ≤ 11 mg/l.
- Highly varying hydrophobicity causing different sorption tendency as regards BTEX and PAH (due to backwaters of the River Main; upgradient there are constant seepage/ infiltration conditions at the site).

Wall system/reactive materials (Fig. 3.15)

- Funnel and gate (microbiology plus activated carbon).
- Reactive material: addition of electron acceptors required for the microbiological degradation.
- Activated carbon.
- 120 m long sealing wall, approximately vertical to groundwater flow downgradient of the contamination source.
- Length and width of the gate (located approximately in the middle of the sealing wall) about 10 m.
- Construction of the gate using sheet pile wall procedure.
- Microbiological step: approximately 10 m long gravel packing, inside (after 2 m) a sheet pile wall intercepts the aquifer; directly in front of and behind the sheet pile wall, filter pipes containing the required electron donors are installed connected to each other by means of a

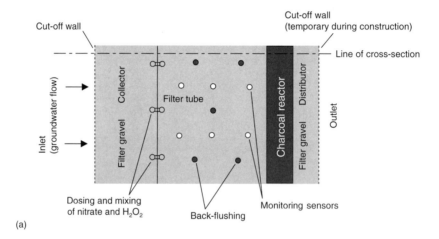

(a)

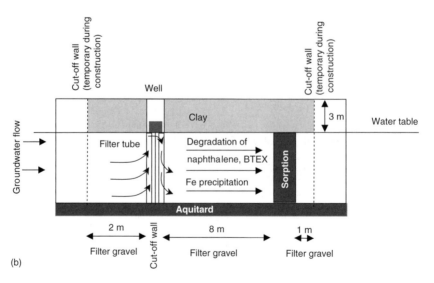

(b)

Fig. 3.15. (a) Top view and (b) elevation of the gate system at the Offenbach site. (Courtesy of HIM ASG (Dipl.-Ing. Kayser))

massive wall pipe; inside the subsequent 8 m gravel filter (approximately 1.5 days residence time) a microbiological BTEX degradation takes place; back-purging devices are provided.
- Filtration by activated carbon follows the biological step (retaining poorly soluble PAH molecules, consisting of 4–6 condensed rings).

7. Rheine

Topic
Evaluation of the long-term behaviour of a ZVI reactive wall using the Rheine site as a model.

Applicant
- Mull und Partner Ingenieur GmbH, Garbsen (Dr Moeller, Dr Wegner (project 1)).
- Technical University of Berlin (Dr -Ing. Steiof (project 2)).
- University of Kiel (Prof. Dahmke, Dr Ebert, Dipl.-Chem. Nilmini Silva-Send (general project)).

Cooperating partner
- Beratende Ingenieure GmbH (Dr Wessling).

Pollutants
- Contamination is based on pollution caused by a former laundry/dry-cleaning facility.
- Concentration: PCE 16 000 µg/l, TCE ≈ 115 µg/l, *cis*-DCE ≈ 320 µg/l.

Wall System (Figs 3.16–3.18)
- Continuous PRB.
- Depth of wall approximately 6 m; 22.5 m long, two-wall segments designed for the comparison between two different infilled materials; a concrete-filled borehole separates the two segments; wall constructed as an overlapping bored pile wall (18% loss of infilled material).
- Active gas drainage above the reactive wall.

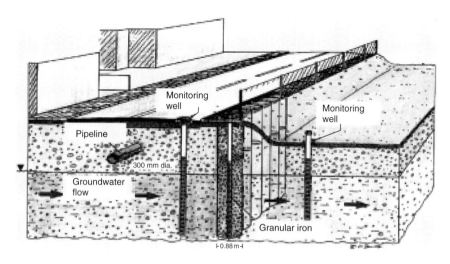

Fig. 3.16. Sketch of the continuous PRB in Rheine. (Courtesy of Mull und Partner Ingenieur GmbH (Dr-Ing. Möller, Dr Wegner))

Fig. 3.17. (a) Set-up of the continuous wall in Rheine using the large-diameter borehole method. (b) Filling the borehole with reactive materials. (Courtesy of Mull und Partner Ingenieur GmbH (Dr-Ing. Möller, Dr Wegner))

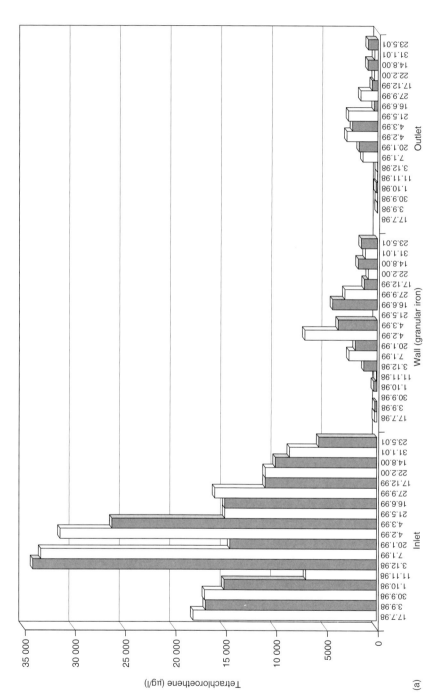

Fig. 3.18. Validation of the PRB performance in Rheine over approximately 3 years. (a) Granular iron. (Courtesy of Mull und Partner Ingenieur GmbH (Dr-Ing. Möller, Dr Wegner))

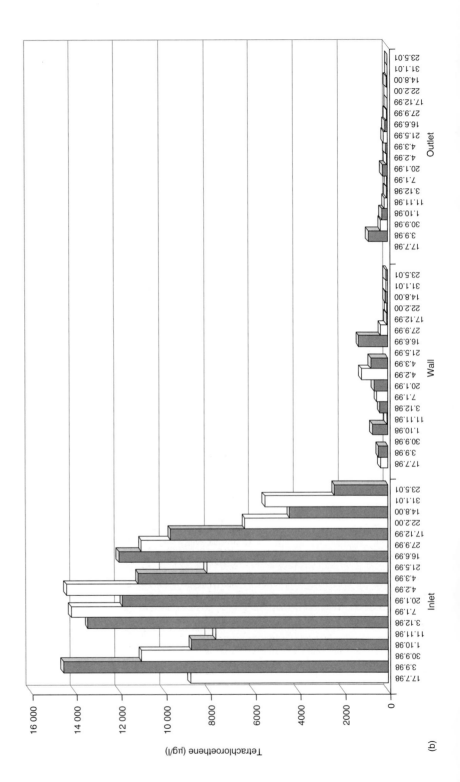

(b)

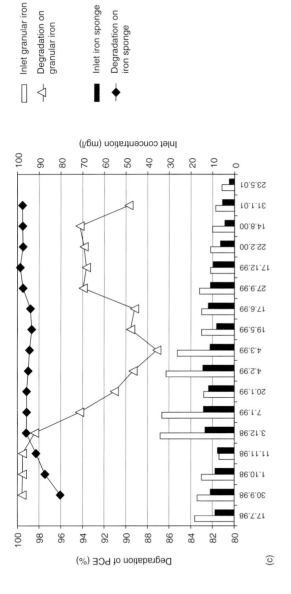

Fig. 3.18 (Contd). Validation of the PRB performance in Rheine over approximately 3 years. (b) Iron 'sponge'. (c) Comparison of PCE degradation. (Courtesy of Mull und Partner Ingenieur GmbH (Dr-Ing. Möller, Dr Wegner))

- Reactive material:
  - iron sponge (85 tonnes);
  - grey iron/pea gravel mixture (69 tonnes of iron), 1:2 volume ratio (34 tonnes each of ZVI and gravel).

8. Wiesbaden

Topic
Development of reactive wall systems to be used for clean up of arsenic contamination at an abandoned site in contact with a River Rhine aquifer.

Applicant
- Hessische Industriemüll GmbH, ASG, Wiesbaden (Dipl.-Ing. Kayser, Dr Boehmer).

Cooperating partners
- Peschla and Rochmes GmbH (Dipl.-Geol. Rochmes, Dipl.-Ing. Woll).
- University of Kiel (Prof. Dahmke).
- Technical University of Hamburg-Harburg (Prof. Foerstner, Dr Gerth).

Pollutants
- Cause of contamination: the former (Lembach und Schleier) chemical plant in Wiesbaden.
- Different arsenic compounds (arsenites, arsenates, partly arsenic trioxide, arsenic pentoxide as well as secondary volatile arsenides).
- Concentrations:
  - soil: total As concentration up to 56 400 mg/kg;
  - 21 samples of eluates: two samples $\approx$ 500 mg/l total As, three samples 10–50 mg/l, rest < 10 mg/l.
- Groundwater: until November 1998, upgradient $\leq$ 27 mg/l of As, downgradient $\approx$ 2 mg/l of As; in April 1999, 10–20 times increased As pollution (cause: long-lasting high water levels releasing As from the groundwater range).

Wall system (Fig. 3.19)
Reactive wall at the upgradient (plan 1), funnel and gate at the downgradient (plan 2).

- Depth of wall approximately 3–5.5 m.
- Reactive material:
  - Plan 1 (upgradient): readily oxidizable C_{org} phases, and, if required, solid phases consisting of sulfate (both emitting sulfide); C_{org}–$CaSO_4 \cdot 2H_2O$.
  - Plan 2 (downgradient): iron oxide for the adsorptive fixation of arsenate; non-redox-sensitive sorbents; modified iron oxides.

Wall-specific data
- *Plan 1 (upgradient)*
  Total width 5 m, consisting of five large-scale drillings (600 mm), extended as a well using filter pipes (300 mm dia.) and filled with the reactive material. The reactive wall is intended to release sulfur compounds (S^{2-}) in order to gain immobilization of solid arsenic phases via conversion to poorly soluble arsenic-sulfides.
- *Plan 2 (downgradient)*
  Funnel-and-gate system: gate 2 m wide, 5 m long; sheet pile walls on both sides, each 5 m long; in addition, passive measures to remove contamination from readily accessible areas in order to reduce the inflow of the pollutants and, therefore, to prolong the life of the adsorbent.

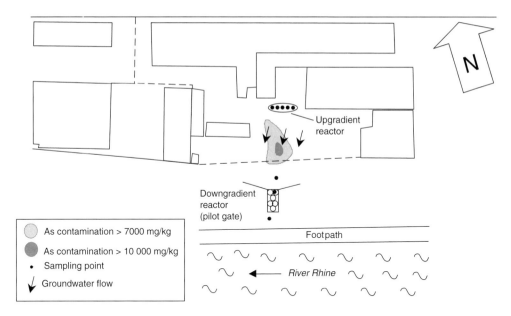

Fig. 3.19. Sketch showing the planned groundwater remediation project at the Wiesbaden-Biebrich site. (Courtesy of HIM ASG (Dipl.-Ing. Kayser))

9. University of Applied Sciences of North-East Lower Saxony, Suderburg (general project)

Topic

Coordination of RUBIN, the German PRB network, and preparing comprehensive document-ation and guidance for PRBs covering a main model procedure for planning, construction, implementation, operation, monitoring and quality standards.

Applicant

• Department of Civil Engineering (Water and Environmental Management) (Prof. Burmeier, Dr Birke, Dipl.-Ing. Rosenau).

Details

• Aim: to gain generalizable and universally applicable knowledge pertaining to applicability and implementation of PRBs:
  – site- and pollutant-specific design;
  – analysis and assessment of the degradation of different pollutants;
  – ecological and economic assessment of PRB performance;
  – operation of PRBs, regulatory acceptance, approval by the regulatory authorities.
• Publication of the results from the RUBIN network studies in a publication to include a state-of-the-art report and to act as the main model procedure for legislators, authorities, problem owners, planners, executives and operators.
• Organization of national conferences and workshops in Germany, publications and public relations work, setting up a website, communication and knowledge exchange with PRB initiatives worldwide, especially in Western Europe and the USA.

10. Christian Albrechts University, Kiel (general project)

Topic

Evaluation and further development of preliminary investigation procedures, monitoring and quality management regarding reactive walls – a comparative laboratory and site study.

Applicant

- Institute for Geological Sciences (Prof. Dahmke, Dr Ebert, Dipl.-Chem. Silva-Send).

Cooperating partner

- Laboratory for Environmental Studies, Hamburg (Dr R. Wienberg).

Details

- Monitoring:
  - conventional groundwater sampling;
  - multilevel sampling;
  - passive sampling;
  - tracer experiments;
  - pumping experiments;
  - indicators for degradation of contaminants.
- Column experiments using site-specific materials:
  - differences in reactivity;
  - reliability of preliminary investigations (reproducibility and precision of prediction);
  - development of tracers;
  - mass balances;
  - indicators for degradation of contaminants.
- Microbiology.

11. Eberhard Karls University, Tübingen (general project)

Topic

Comparative technical and economic assessment of *in situ* PRBs.

Applicant

- Geological Institute, Centre for Applied Geosciences (ZAG) (Prof. Teutsch, Dr Finkel).

Details

- Comprehensive and general technical and economic assessment utilizing data from different RUBIN PRB projects.
- Set up of a uniform database.
- Development of mathematical functions for predicting costs dependent on relevant physical, geological, hydrological and chemical quantities/parameters; capital and operating costs will be covered.
- Comparing predicted costs and actual costs.
- Investigating potential for cost saving.
- PRB versus pump and treat: comparing benefits and drawbacks regarding viability and technical practicability.

3.6. References

BAYER, P., MORIO, M., BÜRGER, C., SEIF, B., FINKEL, M. and TEUTSCH, G. (2001). Funnel-and-gate versus innovative pump-and-treat systems: a comparative assessment. *Groundwater Quality* (in press).

BIRKE, V., BURMEIER, H. and ROSENAU, D. (2000a). Mehr Kosteneffizienz bei Sanierungen gefragt. *TerraTech* **4**, 16–17.

BIRKE, V., BURMEIER, H. and ROSENAU, D. (2000b). Startschuss zum BMBF- Forschungs-vorhaben Anwendung von Reinigungswänden für die Sanierung von Altlasten gefalln. *Altlastenspektrum* **6**, 367–369.

DAHMKE, A., EBERT, M., KÖBER, R., SCHÄFER, D., SCHLICKER, O. and WÜST, W. (2000). *Konstruktion und Optimierug passiver geochemischer Barrieren zur* in-situ *Sanierung und Sicherung CKW-kontaminierter Aquifere*. Abschlussbericht BMBF Vorhaben, Bonn.

EDEL, H. G. and VOIGT, T. (2001). Aktive und passive Grundwassersanierung – ein Verfahrens- und Kostenvergleich. *TerraTech* **1**, 40–44.

FINKEL, M., LIEDL, R. and TEUTSCH, G. (1998). A modelling study on the efficiency of groundwater treatment walls in heterogeneous aquifers. *Proceedings of the Groundwater Quality Conference. IAHS-Publication*, No. 250, pp. 467–474. IAHS, Tübingen.

GAVASKAR, A., GUPTA, N., SASS, B., JANOSY, R. and HICKS, J. (2000). *Design Guidance for Application of Permeable Reactive Barriers for Groundwater Remediation*. Battelle, Columbus, Ohio.

KLEIN, R. and SCHAD, H. (2000). Results from a full scale funnel-and-gate system at the BEKA site in Tübingen (Germany) using zero-valent iron. *Proceedings of the 7th International FZK/TNO Conference on Contaminated Soil, Leipzig*, pp. 917–923.

LIEDL, R., FINKEL, M. and TEUTSCH, G. (1999). Modellierung von Abbaureaktionen in *in-situ*-Sanierungsreaktoren. In: H. Weiss, B. Daus and G. Teutsch (eds), *SAFIRA 2. Statusbericht, Modellstandort, Mobile Testeinheit, Pilotanlage*, UFZ-Bericht, No. 17.

PULS, R. W., KORTE, N., GAVASKAR, A. and REETER, C. (2000). Long-term performance of permeable reactive barriers: an update of an US multi-agency initiative. *Proceedings of the 7th International FZK/TNO Conference on Contaminated Soil, Leipzig*, pp. 591–594.

ROCHMES, M. (2000). Erste Erfahrungen mit Reaktiven Wänden und Adsorberwänden in Deutschland. In: V. Franzius, H. P. Lühr and G. Bachmann (eds), *Boden und Altlasten Symposium, Proceedings*, pp. 225–245.

SCHAD, H., HAIST-GULDE, B., KLEIN, R., MAIER, D., MAIER, M. and SCHULZE, B. (2000). Funnel-and-gate at the former manufactured gas plant site in Karlsruhe: sorption test results, hydraulic and technical design, construction. *Proceedings of the 7th International FZK/TNO Conference on Contaminated Soil, Leipzig*, pp. 951–959.

US ENVIRONMENTAL PROTECTION AGENCY (1999). *Field Applications of* in-situ *Remediation Technologies*, EPA-542-R-99-002. US Environmental Protection Agency, Washington DC.

WEISS, H., DAUS, B. and TEUTSCH, G. (1999). *SAFIRA 2. Statusbericht, Modellstandort, Mobile Testeinheit, Pilotanlage, UFZ Bericht*, No. 17.

WÜST, W., KÖBER, R., SCHLICKER, O. and DAHMKE, A. (1999). Combined zero- and first-order kinetic model of degradation of TCE and *cis*-DCE with commercial iron. *Environmental Science and Technology* **33**, 4304–4309.

YOON, S., GAVASKAR, A., SASS, B., GUPTA, N., JANOSY, R., DRESCHER, E., CUMMING, L. and HICKS, J. (2000). Innovative construction and performance of a permeable reactive barrier at Dover Air Force Base. *Chemical Oxidation and Reactive Barriers. Second International Conference on Remediation of Chlorinated and Recalcitrant Compounds, Monterey, California. Conference Proceedings*, Vol. C2–6, pp. 409–416. Battelle, Columbus, Ohio.

4. Engineering design of reactive treatment zones and potential monitoring problems

S. Jefferis
School of Engineering, University of Surrey, Guildford, Surrey GU2 7XH, UK

4.1. Introduction

This chapter first considers the nature of reactive treatment zones (RTZs, also known as permeable reactive barriers, PRBs) and shows that there are analogies with natural processes and also some repository designs. The effects of flow recycling, dispersion and heterogeneities in the flow and concentration in the input to the reactor on the performance of an RTZ are then investigated. The chapter concludes with a brief discussion on the need to consider heterogeneities when designing monitoring systems for RTZs.

4.2. The location of the contamination

There are many definitions of contamination and pollution, but it is outside the scope of this chapter to review them. However, a point that is sometimes missed is the location. A simple definition could be that contamination is a chemical, a living organism or energy in the wrong place. To take an analogy, a weed is a plant in the wrong place. It follows from this that part of the purpose of a reactive barrier can be to put the contamination in a safe place without the requirement to store large quantities of soil or groundwater.

Figure 4.1a shows a schematic diagram of a simple reactive barrier. A plume of contamination is carried into the RTZ by advection and diffusion. Within the barrier the contamination is stored (e.g. by sorption), destroyed or converted to a safer species. Figure 4.1b shows a repository which could be for the storage of domestic, hazardous or nuclear waste. The aim of the repository is to prevent the generation of a plume of contamination downstream. In both figures the central issue is the relative movement of contaminants with respect to the repository or RTZ. Figures 4.1a and 4.1b show that an RTZ is functionally the inverse of a repository. For a repository the contaminant will be placed in it and tend to leak out whereas an RTZ which stores the contamination or daughter products will, over time, become a repository. It follows that RTZs which become repositories must be subject to very different regulatory controls from those which destroy contaminants or convert them to a harmless form.

It should also be noted that a repository may be designed to achieve chemical as well as physical containment. An example of this is a repository design proposed for low- and intermediate-level radioactive waste storage in the UK. For this it was accepted that there would be some groundwater flow through the repository. To prevent the leaching of radionuclides the waste was to be embedded in and surrounded by a high-pH grout, which would render insoluble some of the more significant radionuclides. An important design parameter for the repository was the time to flush out the high pH material. The high pH buffer grout was designed so that pH reduction would be homogeneous throughout the mass (at the timescale of the leaching) rather than progressing inwards from leached sections of the periphery – a low pH at the periphery would have allowed early release of radionuclides.

4.3. Natural RTZs

The concept of an engineered RTZ is now well established, but it is important to recognize that there are natural analogues, and that even in man-made situations RTZs can self-generate, as considered below.

Consideration of the geology of many mineral deposits shows that changes in, for example, temperature or redox conditions can lead to the deposition of minerals from flowing water. As a rather unusual example of this (and an example of self-generation), consider a situation in which groundwater is leaking into an inner-city rail tunnel. Further assume that this city area in common with most others has some leaking drains so that there is organic contamination of the groundwater. There is also some natural and anthropogenic phosphate and some natural arsenic at a not unusual concentration of 10 μg/l. Let us further assume that the flow rate is 20 m³/h (a significant flow, but nothing special for an 'older' rail tunnel) and that this flow is continuous throughout the year. The situation is summarized in Table 4.1.

Could our example situation cause a problem? For an inner-city area where the groundwater is contaminated the phosphate would not be a concern, and at 10 μg/l

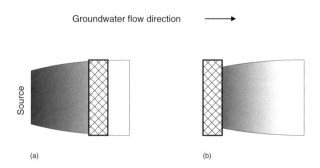

Groundwater flow direction ⟶

Source

(a)   (b)

Fig. 4.1. Comparison of (a) RTZ and (b) repository designs

Table 4.1. Concentrations of arsenic and phosphate in groundwater, and annual fluxes, in the example situation discussed in the text

Species	Concentration (μg/l)	Mass flux (kg/year)
As	10	1.8
PO_4^{3-}	5000	880

the arsenic would be at drinking water control levels. However, could some natural RTZ concentrate the arsenic and the phosphate? Consider now the organic matter in the groundwater. This will degrade and produce reducing conditions. Under such conditions there will be a tendency for iron species to dissolve from the soil minerals, and if the iron were to be present at 0.5 mg/l then the mass of iron entering the tunnel would be 88 kg/year. What is the fate of this iron? Water trickling through track ballast in the invert of the tunnel will be in contact with air and tend to oxidize. Oxidation will lead to precipitation and, as is often the case with iron, accumulation as a voluminous bio-slime with a volume several hundred times that of the precipitated iron hydroxide species alone.

Unfortunately, freshly precipitated iron slimes can be very active and are likely to sorb species such as heavy metals, arsenic and phosphate. Given the fluxes shown in Table 4.1 it is likely that the iron could 'trap' all the arsenic but not all the phosphate. Furthermore, once an iron slime has developed, further arsenic and phosphate in the leak water would tend to be sorbed in the region of first contact with the iron slime, i.e. near the leaks, and to concentrate there as the leaking waters filtered through the slime. From consideration of natural iron–arsenic minerals and sorption processes (Pierce and Moore, 1982) it seems possible that arsenic could accumulate in the iron slimes at concentrations approaching a few per cent by dry weight near the leaks – a surprising result since in the groundwater leaks the arsenic is present at no more than drinking water levels, well below, for example, the Dutch intervention value for contaminated groundwater.

The lessons we can learn from our thought experiment are that:

- RTZs can self-generate in both natural and man-made situations;
- iron bio-slimes can sorb heavy metals;
- the concentration of sorbed species will not be uniform within the slime;
- concentrations may be highest near the source of the water feeding the slime.

4.4. Reaction time

A major difference between the behaviour of an RTZ and a repository is the time-scale each operates on. A repository must be effective for time-scales of perhaps centuries for domestic and industrial waste and hundreds of thousands to millions of years for nuclear waste. In contrast, an RTZ must treat the groundwater within a single passage through the repository. This is considered in more detail in Sections 4.7 and 4.10.

4.5. Reaction mechanisms

When designing a treatment zone, key parameters will be the fluid residence time in the reactor τ and the order of the reaction which is to occur within the reactor. The residence time in a reactor of volume V, and flow porosity e, is given by

$$\tau = Ve/Q \tag{1}$$

where Q is the volumetric flow rate through the reactor (assuming no change in fluid density on reaction, which is reasonable for low-concentration aqueous reactions).

The order of the reaction depends on the nature of the reaction and, for example, whether the reaction is elementary (the rate following the overall stoichiometry of the reaction) or whether it is more complex, perhaps involving a series of activation steps.

4.5.1. First-order reactions

For a simple reversible reaction of the form

$$A \to B \tag{2}$$

if it follows the stoichiometry, the rate of change in the concentration of A, r_a, will be given by

$$-r_a = -dC_A/dt = k_1 C_A - k_2 C_B \tag{3}$$

where C_A and C_B are the concentrations of the reactant A and product B, respectively, and k_1 and k_2 are rate constants. If the reaction is irreversible, then equation (3) becomes

$$-dC_A/dt = k_1 C_A \tag{4}$$

4.5.2. Second-order reactions

The simplest second order reaction is given by

$$2A \to 2B \tag{5}$$

If the reaction is irreversible, the rate of change of concentration of species A is given by

$$-dC_A/dt = k_3 C_A^2 \tag{6}$$

Note that equations (2) and (5) can be reduced to an identical stoichiometry but have different orders of reaction and hence different dependencies on the concentration C_A.

4.6. Types of reactor

There are many different types of reactor. The simplest general types are described below.

4.6.1. Batch reactors

The reactants are put in a reaction chamber, well mixed and left to react for a certain time.

4.6.2. Plug-flow reactors

Reactants enter at one end of a tubular reactor or a packed bed, reaction processes occur within the bed and reaction products are discharged at the downstream end. For analysis the flow is generally assumed to be piston flow, though dispersion, diffusion and short circuiting will occur. These are considered in more detail later.

4.6.3. Stirred-tank reactors

The reactants continuously flow into and out of a well-stirred tank of defined volume. The exit concentration from this reactor will be the same as that within the stirred tank. As a result the reaction processes within the tank will depend on the exit concentration rather than on the inlet concentration. At first sight this may seem undesirable, and indeed in many situations it will be. However, a stirred-tank reactor can resist shock input concentrations, and can thus be useful for microbiological systems where a shock input could be lethal to the system.

4.6.4. Reactor types used in RTZs

So far as the author is aware, to date RTZs have all been designed as plug-flow reactors. The use of a batch reactor will require the use of pumps unless a significant head drop can be arranged across the reactor and some form of dosing system (e.g. a dosing siphon) is used. The use of a stirred-tank reactor will require the input of stirring energy, but could be useful for some biochemical reactors. When considering other types of reactor it is useful to keep in mind the trickling filter and the activated sludge processes for sewage treatment – though both are effectively derived from plug-flow reactors. If biological treatment of groundwater is required, they may offer a useful starting point for design – though the biochemical oxygen demand (BOD) of even the most contaminated groundwater is likely to be much lower than that of sewage.

4.7. Degree of reaction in the reactor

The generalized performance equation for a perfect plug-flow reactor (no dispersion, diffusion or short circuiting and constant fluid density), or for a constant-volume batch reactor, is given by Levenspiel (1972):

$$\tau = \int_{C_i}^{C_o} \frac{dC}{r_A} \tag{7}$$

where τ is the residence time, r_A is the instantaneous rate of reaction of species A (e.g. a contaminant of concern in the inflow) at concentration C, and C_i and C_o are the concentrations of this species entering and leaving the reactor, respectively.

If the reaction rate r_A is known analytically, equation (7) can be integrated directly (otherwise it may be integrated graphically). For example, for the first-order irreversible reaction defined by equation (2), direct integration of equation (7) is possible, to give the ratio of outflow to inflow concentrations as

$$C_o/C_i = \exp(-k\tau)$$ (8)

For orders of reaction other than $n = 1$, which are irreversible and of the type given in equation (5) (where $n = 2$), equation (7) can be integrated to give

$$(C_o/C_i)^{1-n} = 1 + k\tau(n - 1)C_i^{n-1}$$ (9)

where n is the order of the reaction (for $n \neq 1$).

From inspection of equations (8) and (9) it can be seen that reactions of order greater than or equal to 1 cannot go to completion in finite time, i.e. C_o cannot go to 0 in finite time – though fortunately it may go to less than regulatory control levels or analytical detection limits.

For reactions of order less than 1, equation (9) suggests that concentrations can go not only to 0 but to negative values. Clearly, negative concentrations are not possible, and in practice the observed fractional orders will move up towards unity as C_o reduces – an important point to note if high design reductions are required for these types of reaction.

For reversible reactions the integration is significantly more complex. However, it should be noted that reaction systems which show significant reversibility may not be convenient for use in RTZs as the final equilibrium concentration will be sensitive to the molar ratio of the reactants and it may be necessary to add reactants at high concentrations to achieve a reasonable degree of conversion.

The form of equations (8) and (9) is not the standard textbook form, because an RTZ is not a standard chemical reactor. For an RTZ the design aim is likely to be to minimize the output concentration C_o, whereas at a chemical plant there may be other constraints/design issues. Each equation has been written to present the concentration ratio (C_o/C_i) – the ratio of the output concentration to the input concentration. For an RTZ this ratio is an important design parameter, and in this chapter it will be referred to as the design concentration reduction or design reduction. For a design reduction of 0.01, C_o would be $0.01C_i$, i.e. C_i would be reduced by 100 times.

4.8. Use of a reactor recycle (recycling)

Occasionally, use of a reactor recycle has been suggested as a method of increasing the effectiveness of an RTZ. However, it must be used with great care, and generally it will reduce the effectiveness of the treatment zone. For example, if a recycle R is imposed on a plug-flow reactor through which the base flow is Q, then

flow through reactor $= (1 + R)Q$ (10)

and the concentration of the reactant entering the reactor is C_r:

$$C_r = (C_i + RC_o)/(1 + R)$$ (11)

For a first-order irreversible reaction it can be shown (Levenspiel, 1972) that

$$k\tau = (1 + R)\ln\left(\frac{C_i + RC_o}{(1+R)C_o}\right)$$ (12)

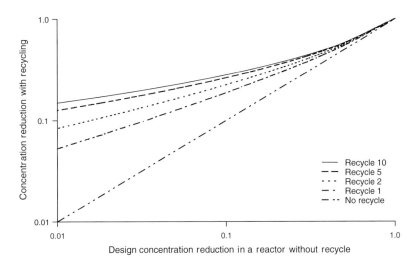

Fig. 4.2. Effect of recycling on the design concentration reduction

where C_i and C_o are the concentrations of the reactant of interest (reaction rate k) entering and leaving the reactor system, respectively (i.e. not the reactor alone, but the reactor plus recycle).

Figure 4.2 shows the effect of the recycle on the design reduction (C_o/C_i) of the system. It can be seen that if high design concentration reductions are required (as will be the norm) the recycle significantly damages the degree of conversion for a first-order irreversible reaction. Similar though potentially more limited effects will be seen for higher-order reactions. Prediction of the effects for higher-order reactions is more complex, because the design reduction in the reactor becomes a function of the inlet concentration C_i. For the low concentrations typical of many groundwater contamination situations, careful analysis will be required to predict the effects.

Reactor recycles are only of use for rather special types of reaction, for example autocatalytic reactions where the reaction is catalysed by the reaction products. This analysis of recycling, however, serves to highlight two very important features of first (and higher) order reaction processes:

- The degree of conversion in a reactor is very sensitive to the residence time in the reactor, τ, and hence the flow rate to through the reactor. For example if, for a first-order irreversible reaction, the flow rate is doubled, so reducing $k\tau$ from 4.6 (which will give a design reduction of 0.01, i.e. 100-fold reduction across the reactor) to 2.3, the reduction in concentration across the reactor will be only 10-fold.
- The mass reduction rate in the reactor is a function of the inlet concentration of the reactant (except for zero-order reactions). If high mass reductions are required, inlet concentrations must be high.

A recycle will reduce both the residence time and inlet concentration.

Although a direct recycle on an RTZ will seldom be beneficial for the reactions currently typical of RTZs, it can be used to exploit any unused reaction capacity in an RTZ – some designs may need to be very conservative. For example, for many RTZ configurations the reactor may have to be upstream of the trailing end of the contaminant plume, i.e. at the time of installation of the reactor there may be a pre-existing plume downstream of it (perhaps because of land ownership boundaries). If a well is sunk into the plume, then contamination may be slowly recovered and treated. But the rate of this 'recycle' must be very carefully controlled, and may have to be only a small fraction of the normal flow through the reactor. An example of a suitable site for recycling of a downstream plume is given by Jefferis *et al.* (1997): this is a case where a system was later installed and is now operating successfully.

4.9. Dispersion and diffusion

The reactants moving through a plug-flow reactor may be subject to axial diffusion and dispersion, which will reduce their residence time in the reactor. The effect will be most pronounced for long residence times, i.e. when high reductions in concentration across the reactor are required. For a first-order irreversible reaction and small deviations from plug flow, the effect of dispersion can be approximated by (Levenspiel, 1972)

$$C_o/C_i = \exp(-k\tau + (k\tau)^2 D/uL) \tag{13}$$

where D is a combined coefficient for diffusion and dispersion, u is the interstitial velocity in the reactor and L is the thickness of the reactor. If diffusion/dispersion is to increase the output concentration C_o by less than a factor of $(1 + b)$ – where b is small, e.g. 10% – then for a first-order irreversible reaction it can be shown that

$$(k\tau)^2 D/uL < b - b^2/2 \tag{14}$$

Thus, if $k\tau$ is 4.6 (which will give a design reduction of 0.01 without dispersion), L is 1 m and D is 10^{-8} m^2/s, then dispersion will increase the output concentration by more than 10% if u is less than 2.3×10^{-6} m/s – a situation that could occur in moderate- to low-permeability soils. A case in point would be a continuous wall reactor in a soil of permeability 1.1×10^{-4} m/s with a regional gradient of 0.01 and reactor porosity of 50%. For a reactor thickness of 1 m, an interstitial velocity of 2.3×10^{-6} m/s corresponds to a residence time of 5.2 days, which is relatively long for an RTZ. However, the situation is complex, since mechanical dispersion will be a function of the velocity in the reactor, increasing as the velocity increases; and dispersion could be significant at shorter residence times.

4.10. Short circuiting/by-passing

In a chemical plant care will be taken to ensure that the inflow reaction stream is uniformly distributed over the cross-sectional area of the reactor. However, in a funnel-and-gate reactor there may be preferential flow, for example near the boundaries of the gate. In a continuous permeable wall the adjacent strata may vary quite significantly in permeability and hydraulic gradient.

To analyse the effect of heterogeneous flow in the reactor, consider a first-order irreversible reaction for which the concentration ratio is given by equation (8) (ignoring diffusion/dispersion).

If chemical conditions within the reactor are constant (reaction rate k is constant), then the design reduction depends only on the residence time, τ, in the reactor. If there is a distribution of residence times across the reactor there will also be a distribution of exit concentrations (even for a uniform input concentration, though this can also vary, e.g. from the centre to the edge of a plume).

The effect of variation in residence time in a reactor can be seen by considering two reactors in parallel with the flow divided between them (a real situation could be many parallel reactors). The performance of these two reactors can then be examined as a function of the split in the flow between them. The average concentration downstream of the two reactors is

$$C_o = [Q_1 C_1 \exp(-kV_1 e/Q_1) + Q_2 C_2 \exp(-kV_2 e/Q_2)]/(Q_1 + Q_2) \tag{15}$$

where C_1 and C_2 are the influent concentrations to reactors 1 and 2, V_1 and V_2 are the volumes of the reactors, Q_1 and Q_2 the flows, and e their porosity (assumed constant).

If the same input flux were passed through a single reactor, of volume equal to the sum of the volumes of the parallel reactors, then the concentration in the discharge from this reactor would be

$$C_o = (Q_1 C_1 + Q_2 C_2)\exp[-k(V_1 + V_2)e/(Q_1 + Q_2)]/(Q_1 + Q_2) \tag{16}$$

From equations (15) and (16) it follows that if the residence time in the two reactors and in the combined reactor are all equal (i.e. the exponential terms are identical), then variations in concentrations of the influent to the reactors, C_1 and C_2, will not affect the overall performance of the reactors in comparison with the combined reactor. However, variation in the flow proportion to the reactors will affect the combined performance, as it effects residence times and hence the exponential factors.

Figure 4.3 shows the effect of a non-uniform distribution of flow through two parallel reactors (of equal volume) as compared with a single reactor of volume equal to the sum of the parallel reactors. If this were a continuous permeable reactive wall, then the situation would be analogous to a locally increased flow rate, perhaps because of a region of higher-permeability ground.

The effect of unequal flow distribution is greater the higher the designed reduction in concentration across the reactor. The graph in Fig. 4.3 shows the results for a first-order irreversible reaction and design reductions of 0.001, 0.01 and 0.1 (i.e. $k\tau$ values of 6.9, 4.6 and 2.3). The relative effluent concentration is the average concentration in the discharge from the two parallel reactors divided by that from the single reactor. For example, if 20% of the flow passes through one reactor and 80% through the other, then the average outlet concentrations would be 10.7, 4.5 and 1.9 times higher than for the single reactor (or for an equal division of flow between

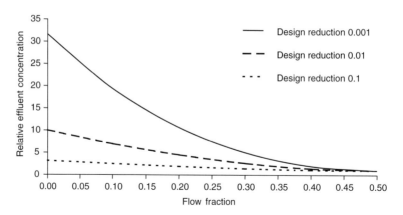

Fig. 4.3. Effect of non-equal flow through parallel reactors of equal size

the two reactors). These figures are for design reductions of 0.001, 0.01 and 0.1, respectively; but with the unequal flow the actual reductions would be 0.094, 0.22 and 0.53 – a significant loss in performance, especially for the higher design reductions.

For a two-reactor model the worst case is no flow through one of the reactors and double flow through the other, thus halving the residence time, so that the concentration reduction drops to the square root of the design value (1000 to 31.6 times, etc.). In the field the effect could be much greater, as there could be the equivalent of more than two parallel reactors. For example, if there were severe short circuiting so that effectively all the flow passed through 25% of a reactor with a design reduction of 0.01 (a reduction of 100 times), then the actual concentration reduction would be only 3.2 times. For second- and higher-order reactions the effect is less severe. For example, for a second-order reaction the concentration reduction would be reduced from 100 to 26 times.

If concentrations vary across an RTZ as well as flows, the situation is more complex than can be considered in this chapter.

Significant results from this modelling exercise are:

- For the special cases of zero-order or first-order irreversible reactions, if the concentration varies across the reactor but not the residence time – for example the groundwater flow is of constant velocity through the reactor, but there are local variations in concentration (e.g. the edge to the centre of a plume) – homogenizing the flow will give no benefit.
- If there is significant variation in flow rate across different areas of an RTZ then the performance will be degraded.
- If there is a significant variation in concentration at different locations it may be more economical to use a funnel-and-gate configuration with multiple gates, rather than funnelling all the flow to a single gate or using a continuous wall of thickness designed for the most contaminated location. The benefit will depend on the cost of renewing the reaction materials compared with the installation cost of further reactors. In general, the reactive material savings may be quite modest but worth investigating.

4.11. Monitoring

The performance of an RTZ may be effectively the sum of the performances of many subsidiary reactors (e.g. individual lengths of a continuous reaction wall). The effluent concentration therefore may vary across the RTZ.

If it is the overall performance of the reactor that is important, local heterogeneity may be acceptable provided variations in flow rate across the reactor are modest. However, this leaves the important question: how should overall performance be monitored in the field?

The effect of non-uniform flow conditions may be exacerbated if the concentration distribution is also non-uniform. Overall, for a small funnel-and-gate reactor the effects may be modest; but for a long continuous wall under adverse conditions they could be significant. It is therefore suggested that designers of RTZs carry out a sensitivity analysis on the effects of varying flow rate and concentration, so as to ensure that the required overall performance is achieved, and that concentration hot spots cannot occur at unacceptable levels.

If an RTZ is used in conjunction with natural attenuation, then the effects of heterogeneity of performance may be important, especially if the attenuation is by biological processes and the concentrations could be locally high enough to be toxic or inhibitory to the degrader bacteria.

The withdrawal of water samples from an RTZ for monitoring purposes must be done with great care. Clearly, 'purging' a well in a wall or adjacent to a wall would cause massive disruption to the local residence time. However, even slow withdrawal of a sample could cause disturbance, and the rate of sample abstraction should be compared with the rate of flow through the RTZ (this may be particularly important for continuous or semi-continuous monitoring systems).

4.12. Conclusions

RTZs can occur naturally, and the accretion of chemical species within an RTZ can result in it becoming a repository, thus requiring specific monitoring and control.

Recycling will seldom improve the performance of an RTZ and generally will decrease it. If a downstream plume is 'recycled' through a reactor, great care must be exercised in setting the flow rate. An increased flow rate will reduce the residence time in the reactor and the likely reduced inlet concentration (unless the concentration downstream in the plume is higher than upstream at the inlet to the reactor) will reduce the mass conversion rate.

The performance of RTZs can be affected by dispersion in the reactor. A first review suggests that this is unlikely to be significant except when high degrees of conversion are required; but there is uncertainty about the dispersion coefficient in reactors, especially at higher flow rates.

Heterogeneous flow can markedly reduce the performance of a reactor, and sensitivity studies should be carried out.

The maximum contaminant degradation (mass reduction) per unit volume of reactor will be achieved at high input concentrations (except for zero-order reactions) and long residence times. However, long residence times increase the potential 'damage' resulting from dispersion and short circuiting. If a design requires long residence times (or a high value of the exponential factor $k\tau$ for first-order kinetics) then these effects should be considered.

If flow or concentration heterogeneities are expected, then an appropriate monitoring strategy must be developed. Further work is required on this, as are field data.

If high design reductions are required within an RTZ, then detailed bench studies may be needed to predict the effects of variations in the inlet concentration and residence time in the reactor. Without such studies, scale up from bench to field scale may be unreliable.

The analyses presented are mostly for simple reaction kinetics – first-order irreversible reactions – as many useful reactions approximate to this. Predictions are more complex for higher-order or reversible reactions, when the design reduction and the effects of recycling, etc., become functions of the input concentration to the RTZ.

Finally, it should be noted that the input concentrations to an RTZ will be very low in comparison with those typical of the feedstocks to chemical process plants. Furthermore, for most chemical plants the design objective will be to maximize production of particular species. For an RTZ the objective is markedly different: to maximize the destruction of species already present at trace levels. Thus, standard chemical engineering procedures may need to be considerably modified.

4.13. References

JEFFERIS, S. A., NORRIS, G. H. and THOMAS, A. O. (1997). Contaminant barriers: from passive containment to reactive treatment zones. *14th International Conference on Soil Mechanics and Geotechnical Engineering, Hamburg*.

LEVENSPIEL, O. (1972). *Chemical Reaction Engineering*. Wiley, New York.

PIERCE, M. L. and MOORE, C. B. (1982). Adsorption of arsenite and arsenate on amorphous iron hydroxide. *Water Research* **16**, 1247–1523.

5. Performance monitoring of a permeable reactive barrier at the Somersworth Landfill Superfund Site

T. Sivavec
GE Corporate Research and Development Center, One Research Circle, Building K-1, Room 5A45, Niskayuna, NY 12309 USA

T. Krug and K. Berry-Spark
GeoSyntec Consultants, 160 Research Lane, Suite 206, Guelph, Ontario N1G 5B2, Canada

R. Focht
EnviroMetal Technologies Inc., 745 Bridge Street West, Suite 7, Waterloo, Ontario N2V 2G6 Canada

5.1. Introduction

As permeable reactive barriers (PRBs) containing zero-valent iron become more widely used to remediate contaminated groundwaters, there remains uncertainty in the prediction of their long-term performance. While a number of accelerated-ageing laboratory and pilot scale tests have not indicated any significant performance issues caused by the build-up of surface precipitates or bio-fouling, there has been relatively little performance data collected in the field at pilot or full scale installations (O'Hannesin and Gillham, 1998; Powell *et al.*, 1998; Vogan *et al.*, 1998, 1999; Mackenzie *et al.*, 1999; Puls *et al.*, 1999).

An increasingly popular and potentially cost-effective construction method for PRBs is the biopolymer slurry technique (Gavaskar *et al.*, 1998; Day *et al.*, 1999; Focht and Vogan, 2001). Installation of an iron-based PRB using biopolymer slurry trenching is similar to constructing a conventional impermeable bentonite slurry wall. As a trench is excavated, biopolymer (e.g. guar gum) is added as liquid shoring to provide stability to the trench walls. Excavation can continue through the bio-polymer without the need for dewatering. Granular iron is then placed into the trench through the slurry using a tremie tube.

While the biopolymer trench construction method had been employed at a limited number of sites prior to the start of the constructability test performed at the Somersworth Landfill Site in New Hampshire in the USA, some uncertainties existed regarding the use of this construction method. Specific objectives of the constructability test at the site, therefore, were to evaluate whether:

(1) the iron–sand mixture could be placed as specified without separation of the iron and sand;
(2) the guar gum could be broken down and/or flushed from the PRB within a reasonable time-frame;
(3) the use and removal of the guar gum would have an adverse effect on the permeability of the sand–iron mixture or aquifer in the vicinity of the PRB;
(4) the use and removal of the guar gum would have an adverse effect on the reactivity of the granular iron;
(5) the use and removal of the guar gum and associated materials would have an adverse effect on the geochemical or microbial conditions in the vicinity of the PRB.

5.2. Site description and characteristics

The Somersworth Sanitary Landfill Superfund Site is a 0.26 ha landfill that was constructed in the early 1930s on the site of a former sand and gravel quarry. The landfill was used to dispose of household trash, business refuse and industrial wastes. Waste was burned at the landfill until 1958. From 1958 to 1981, the waste material was placed in excavated areas, compacted and covered with soil. In 1981, use of the landfill stopped when the city of Somersworth began disposing of its municipal waste at a regional incinerator. In 1981, the city of Somersworth implemented a closure plan for the landfill that involved the covering of a portion of the landfill with clean fill. Volatile organic compounds (VOCs), principally tetrachloroethene (perchloroethylene, PCE), trichloroethene (TCE), 1,2-dichloroethene (1,2-DCE) and vinyl chloride (VC), were found to be present in the groundwater.

The site is characterized by sands and gravels having a hydraulic conductivity of approximately 2×10^{-4} m/s. The hydraulic gradient varies from 0.003 to 0.0012 m/m near the edge of the waste. The top of the water table ranges from less than 0.6 m to about 6.1 m below ground surface. As much as 10% of the waste is located below the water table. The aquifer is 9–12 m thick.

Since the PRB technology had not been implemented at the scale of the Somersworth site or in a landfill setting, an initial pilot scale wall was installed in 1996 using a vibratory-hammer/caisson method. This pilot scale PRB system consisted of a 2.4 m diameter 'gate' of iron between layers of pea gravel. Slurry walls measuring 1.4 m in length funnelled groundwater through the gate. Due to problems with the installation of the funnel-and-gate pilot, the full scale design called for a continuous wall. A second pilot installation that used a biopolymer slurry-supported, open-trench method was therefore undertaken, and is described below.

5.3. PRB installation and development

In late November of 1999, a 7.6 m long trench was excavated using a CAT 330 backhoe with an extended boom and a 0.6 m wide bucket. The depth to bedrock was approximately 10.4 m. Steel beams (0.8 m wide) were placed 6.4 m apart at either end of the trench to control the placement of the sand–iron mixture and to demarcate the ends of the test section. The narrowest width of the trench was measured to be 0.7 m, and the appropriate percentage of iron required was calculated to be 90% by weight. The portions of the test section outside of the steel beam ends were backfilled with soil that had been previously excavated.

Sand and granular iron (source: Connelly-GPM, −8+50 mesh) were mixed using concrete-mixing trucks. The trucks arrived with a premeasured quantity of sand (1570 kg). Granular iron (15 000 kg) was added to the concrete mixing truck at the site, along with city water to wet the iron–sand mixture prior to placement in the trench. Between 1700 and 2270 litres of water was added per truck to wet the mixture.

A plastic-lined slurry containment area was constructed beside the working platform at a lower elevation to hold the excess slurry displaced from the trench as the iron–sand mixture was added to the trench. A shallow diversion trench was dug from one end of the test section to the slurry containment area, allowing the slurry to flow by gravity from the trench to the containment area.

The wetted iron–sand mixture (90%/10% by weight) was placed into the trench using a tremie pipe. The tremie pipe consisted of a 0.30 m diameter metal pipe that was initially lowered to the bottom of the trench. Some care was taken while placing the iron–sand mixture in the trench through the tremie pipe to minimize the potential for segregation and contact between the iron–sand mixture and the biopolymer slurry by placing a paper swab in the tremie pipe before starting to place the iron–sand mixture. Two temporary development wells (15 cm diameter slotted PVC) were placed in the trench prior to the addition of the iron–sand mixture, to allow for later removal of biopolymer and groundwater from the trench.

The trench was then developed to remove and encourage the degradation of the residual biopolymer. Development was initiated by pumping fluid from the two development wells at a rate of approximately 38 l/min. The initial 1135 litres of fluid pumped from the wells was transferred to the slurry containment area. After this, the extracted fluid was discharged to the top of the exposed iron–sand mixture. High-pH enzyme breaker (Rantec LEB-H, 3.8 l) was then added to the fluid being discharged to the top of the exposed iron–sand mixture. Water was extracted from the well and discharged to the top of the iron for approximately 6 h until the viscosity of the fluid was reduced such that 1 US quart (=0.95 l) of fluid passed through a Marsh funnel in 28 s. The two development wells were then pulled out.

Following the development of the trench, the backhoe was used to remove material to expose a clean surface of the iron–sand mixture, and geotextile was placed on this surface. The geotextile was extended vertically along the sides of the trench, and 1.5 m of clay was placed with the geotextile envelope in 0.30 m lifts and compacted by tamping using the backhoe bucket. After the other site work was complete, the biopolymer in the containment area was broken, solidified with the spoils and hauled to the top of the landfill for disposal.

5.4. Cored material testing

Approximately 4 weeks after the trench development was completed, cored samples of the iron–sand mixture in the test section were collected at seven locations (Fig. 5.1) using either a split-spoon or direct-push (Geoprobe) sampling technique. Where possible, samples were collected through the entire vertical profile of the test section. The iron–sand mixtures were visually inspected and photographed immediately after being brought to the surface. The recovery of the iron–sand mixture by the samplers was low (generally 20–75%). The samples collected were considered to be highly disturbed.

A total of five borings were drilled within the test section down to the bedrock surface. The borings confirmed the presence of bedrock at a depth of about 10.4 m below the top of the working platform. The iron–sand mixture was observed in all five borings; however, approximately 0.6 m of sandy material on top of the bedrock surface was observed in CT-5. The samples from boring CT-5 showed a consistent iron–sand mixture throughout the boring down to within 0.5 m of the bedrock surface. Samples from the bottom 0.5 m of the boring indicated the presence of sand with very little or no iron. It is suspected that this sand moved to the bottom of the trench after the last pass with the excavator bucket was made and before the iron–sand mixture was added to the trench. This finding highlights the need to minimize the time between the excavation of the trench and the placement of the iron–sand mixture and to minimize the disturbance of the sides of the trench. The remaining borings suggested that the iron–sand mixture was continuous from the top of the test section to the bedrock surface. The samples from boring CT-3,

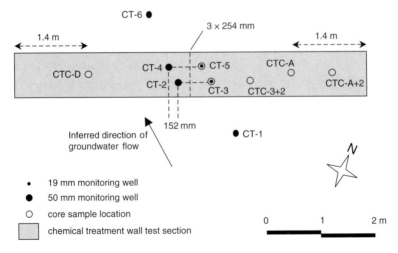

Fig. 5.1. Locations of core samples and monitoring wells in the pilot scale PRB test section at the Somersworth Landfill Superfund Site, New Hampshire, USA

Table 5.1. Summary of chlorinated ethene half-lives from bench scale reactivity tests performed at 10°C

	Unmixed iron		Iron from test section[a]	
	Half-life (h)	r^2	Half-life (h)	r^2
Tetrachloroethene	1.9	0.94	1.6	0.99
Trichloroethene	1.1	0.96	0.6	0.97
cis-1,2-Dichloroethene	2.0	0.96	1.1	0.99
trans-1,2-Dichloroethene	1.0	0.88	0.4	1.00
Vinyl chloride	1.6	0.97	0.9	0.99

[a] Composite of six boring samples

located about 0.25 m away from CT-5, indicated the presence of the iron–sand mixture throughout the entire vertical profile of the test section to the bedrock surface. This finding indicated that the sand above the bedrock surface observed in boring CT-5 was a localized phenomenon.

Nine separate samples collected from the test section and a composite of samples from borings CTC-A and CTC-D were analysed using magnetic separation. The iron content ranged from 72 to 85%, indicating that the composition of the PRB test section was relatively homogeneous. The mean iron content of the nine samples was 81%, which was equal to the iron content of the composite sample. The iron content of the iron–sand mixture was less than expected based on the proportions of iron and sand specified for construction. The reason for this discrepancy between the measured and expected iron contents is believed to be due to bias in the results introduced in the sample collection procedure. It is believed that as a result of its higher density, more iron was lost during sample collection than sand.

Grain size distributions were also determined for unmixed sand and iron, a composite sample of six borings at various depths, and samples from borings CTC-D and CTC-3+2. The grain size distribution for the iron portions of the samples collected from the test section was very similar to that of the unmixed iron. The grain size distribution for the sand portion of samples collected from the test section was, however, consistently finer grained than the unmixed sand. This decrease in grain size distribution in the sand may be the result of abrasion of the sand particles during mixing with the iron in the cement trucks or the result of bias during sample collection. The close agreement of the grain size distributions for iron and sand from the different borings suggested that no significant particle segregation occurred within the test section.

Bench scale reactivity testing was completed using the composite sample from six borings. The degradation rates for the chlorinated ethenes calculated for the results of this testing were higher than those measured in bench scale studies conducted previously using site groundwater. Table 5.1 presents a summary of these half-lives. The degradation rates observed in the composite sample suggest that the degradation rates within the test section were not impaired by the biopolymer construction method. The improvement in the chlorinated ethene degradation rates observed may have resulted from biological processes. The samples submitted for reactivity testing were collected from the test section several weeks after construction and may have been inoculated with the microbiological consortia from the site, which has previously been demonstrated to biodegrade chlorinated ethenes.

A number of samples were also subjected to surface analysis using X-ray photoelectron spectroscopy (XPS) with depth profiling and scanning electron microscopy (SEM). Traces of guar gum residue or its degradation products on the iron surfaces could be detected by XPS (C1s peaks at 286.2 eV). Surface concentrations of carbonate were no higher in the test section iron samples than in the control iron (approximately 2%), as would be expected given the short time the iron was in contact with site groundwater. Similarly, calcium concentrations were very low (0.2–0.3%), and not statistically much higher than that measured in the control iron (0.1%). SEM showed no significant differences between the iron grain boundaries of cored and as-received (virgin) iron (Sivavec, 2001).

5.5. Groundwater wells and monitoring

Six PVC monitoring wells were installed in and around the test section along a transect in the direction of site groundwater flow as shown in Fig. 5.1. Two 50 mm diameter monitoring wells and two 19 mm diameter monitoring wells were installed in the granular iron test section. The 50 mm wells were used for hydraulic testing and in-well monitoring probes, and were installed in borings used for the collection of samples of the iron–sand mixture. The 1.9 cm wells were used for collection of groundwater samples and were located approximately 0.6 m away from the 50 mm wells. In addition, 50 mm diameter monitoring wells were installed in the overburden upgradient and downgradient of the test section. These wells were used for hydraulic testing, in-well monitoring probes and groundwater sampling. The monitoring wells had a screened interval of 4.6 m in length, and the bottoms of the well screens were approximately 1.8 m above the bedrock surface. The wells installed in the test section were constructed without filter packs, while the wells installed in the overburden had filter packs. All wells had bentonite seals and lockable protective casings.

Groundwater samples were collected from four monitoring wells in and around the PRB test section (CT-1, CT-3, CT-5 and CT-6) on three occasions for measurement or analysis of (1) total organic carbon (TOC), total dissolved solids (TDS), pH, redox potential, dissolved oxygen (DO), conductivity, temperature, dissolved iron, viscosity and biomass (by phospholipid fatty acid (PLFA) analysis); and (2) VOCs, to monitor VOC destruction in the sand–iron mixture. The sampling was conducted at approximately 3 week intervals beginning 6 weeks after construction of the test section. Measurements of pH, redox potential, DO, conductivity, temperature and viscosity were conducted in the field using low-flow purge sampling, while other analyses were conducted on samples submitted to an analytical laboratory.

In addition to the monitoring events described above, dedicated, in-well data-logging probes (YSI-600 XLM) were installed in the four 5.1 cm wells set along a transect through the PRB test section. Groundwater elevation, redox potential, pH, temperature, conductivity and DO were monitored at 4 h intervals over a 6 month period. The in-well probes were used to provide greater data density than could be obtained using low-flow purge methods. This data would allow for correlation of groundwater field parameters with changes in barrier performance (e.g. biopolymer breakdown, potential groundwater mounding). In addition, the in-well probes are preferred over low-flow purge methods in that they minimize hydraulic perturbations near the PRB test section.

Field measurement of pH, redox potential and four other groundwater quality parameters collected along the flow path through the test section are shown in Figs 5.2 and 5.3.

Groundwater samples from within the test section, as expected, had a higher pH and lower conductivity than those of the surrounding aquifer. The higher pH in the test section is due to the anaerobic reduction of water by zero-valent iron (equation (1)), which produces hydroxyl ions, resulting in an elevated pH within the test section. The pH change shifts the carbonate equilibrium (equation (2)) and precipitates carbonate minerals (equations (3) and (4)), which decreases the

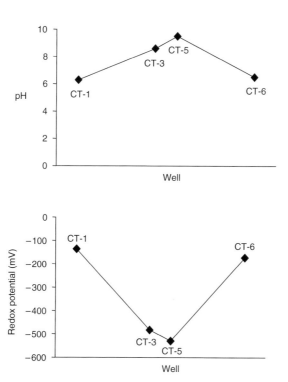

Fig. 5.2. Field parameters measured along the flow path through the PRB test section (low-flow purge/flow-through cell method)

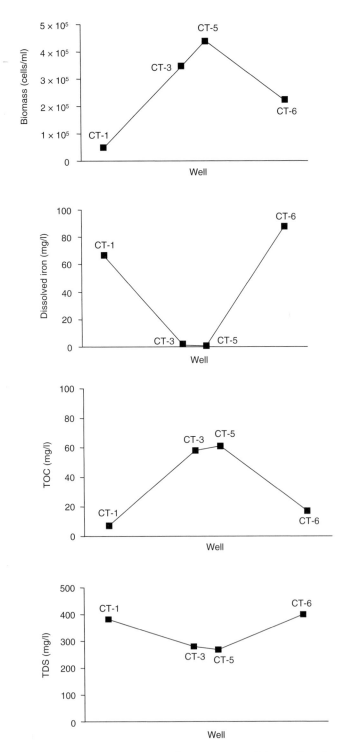

Fig. 5.3. General groundwater quality data measured in wells along the flow path through the PRB test section

conductivity of the groundwater by removing Fe^{2+}, Ca^{2+} and CO_3^{2-} from solution in the iron test section. Downgradient of the test section, groundwater rapidly re-equilibrated with carbonate minerals and resulted in a downgradient conductivity and pH comparable to that measured upgradient.

$$Fe^0 + H_2O \rightarrow Fe^{2+} + H_2(g) + 2OH^- \tag{1}$$

$$HCO_3^- + OH^- \rightleftharpoons CO_3^{2-} + H_2O \qquad (pK_a = 10.3) \tag{2}$$

$$Fe^{2+} + CO_3^{2-} \rightarrow FeCO_3(s) \qquad (K_{FeCO_3} = 3.2 \times 10^{-11}) \tag{3}$$

$$Ca^{2+} + CO_3^{2-} \rightarrow CaCO_3(s) \qquad (K_{CaCO_3} = 2.8 \times 10^{-9}) \tag{4}$$

Groundwater samples from the PRB test section also had lower dissolved iron and TDS concentrations than the samples from the surrounding aquifer, as would be predicted with the precipitation of carbonate minerals in the PRB test section. Groundwater samples from the test section had higher TOC and biomass concentrations than the samples from the upgradient aquifer, most likely indicative of the presence of some residual biopolymer in the test section.

The DO field measurements made using a flow-through cell under low-flow purge sampling conditions were positively biased because the sampling technique introduces oxygen while the groundwater sample is brought to the surface. The DO concentrations measured with the in-well probes were consistently lower than field measurements (Fig. 5.4), and were almost all less than the manufacturer's specified instrument resolution of 0.2 mg/l. The in-well probe data are considered to be more reliable than field measurements, and indicate that the aquifer and the test section were anaerobic.

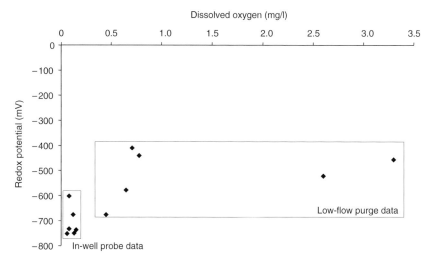

Fig. 5.4. Comparison of redox potential versus DO data sets collected using in-well probes (YSI) and low-flow purge/flow-through cell methods

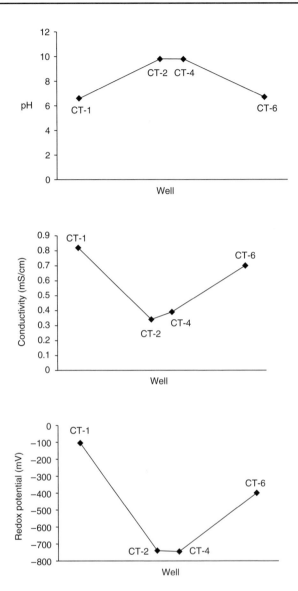

Fig. 5.5. Representative pH, redox potential and conductivity measurements obtained using dedicated, in-well groundwater parameter probes (YSI) in wells along a transect through the pilot PRB

The redox potential of samples collected from the aquifer using a flow-through cell/low-flow purge sampling method was slightly reducing and ranged from –195 to –112 mV over the three monitoring events. The redox potential of samples collected from the PRB test section was more highly reducing than the aquifer and ranged from –676 to –402 mV. The redox potential measurements made using the in-well probes confirmed the reducing conditions within the test section and aquifer;

however, the in-well probes measured even more highly reducing conditions than indicated by field sampling (Fig. 5.5).

This difference appears to be the result of bias in the introduction of oxygen in the purged samples, as similarly low redox potential measurements were measured in the laboratory when the same YSI probes were placed in a mock well set-up containing 90%/10% granular iron–sand by weight. A comparison plot showing the complete data set of redox potential and DO measurements in the PRB test section collected by field sampling and by the in-well YSI probes is given in Fig. 5.4.

No measurable differences in the viscosity of fluid samples from wells installed in the aquifer or the PRB test section were observed. These data indicate that any amount of biopolymer still present in the test section was not significant enough to affect fluid viscosity within levels discernable by this testing method.

Figure 5.6 presents the PCE, TCE, cis-1,2-DCE and VC concentration data in wells along the flow path through the test section for the third monitoring event. The distribution of the chlorinated ethenes suggests that these compounds were degraded within the test section and that the groundwater flowing from the test section (as indicated by monitoring well CT-5) meets the interim clean-up levels (ICL)s. Samples of groundwater from the upgradient well (CT-1) contained elevated concentrations of PCE, TCE, cis-1,2-DCE and VC (up to 200 µg/l, 230 µg/l, 140 µg/l and 33 µg/l, respectively). No detectable concentration of any chlorinated ethene was measured within the test section (CT-3 and CT-5). Samples of groundwater from the down-gradient monitoring well (CT-6) contained some detectable concentrations of chlorinated ethenes, even though no detectable concentrations flowed through the test section. Chlorinated ethenes that became sorbed on to the aquifer solids downgradient of the test section prior to the PRB installation are believed to be desorbing into the groundwater downgradient as groundwater flushes through these contaminated solids. The less chlorinated ethenes (cis-1,2-DCE and VC) are present at higher concentrations because biodegradation processes preferentially degrade the more highly chlorinated ethenes.

5.6. Hydraulic testing

Figure 5.7 summarizes the estimated hydraulic conductivity values calculated from slug tests performed in CT-1, CT-2, CT-4 and CT-6. In general, water level re-sponses to slug addition or withdrawal were very rapid. Oscillatory responses were observed for the wells screened within the PRB test section (CT-2 and CT-4). Log-linear response data were evaluated using the method of Bouwer and Rice (1976), while analysis of oscillatory response data was done using the method of Kipp (1985). The software program ADEPT was used to calculate hydraulic conductivity using these methods (Levy, 1999).

The estimated hydraulic conductivity of the aquifer (CT-1 and CT-6) shows measurement variability but no apparent temporal trends. This variability is attrib-utable to variability inherent to the measurement method. The estimated hydraulic conductivity of the PRB test section, however, decreased between the first and second monitoring events but was stable between the second and third monitoring

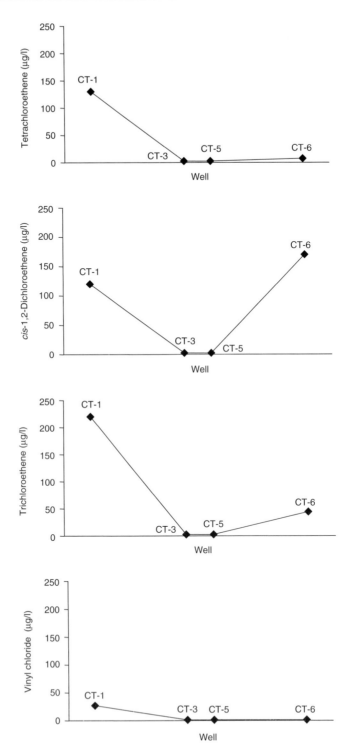

Fig. 5.6. Chlorinated ethenes measured in wells along the flow path through the PRB test section (February 2000 monitoring event)

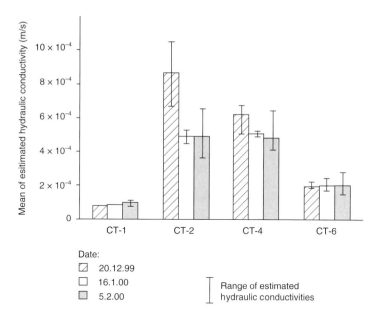

Fig. 5.7. Summary of hydraulic conductivity testing

events. This decrease may be attributable to settling of the iron–sand mixture around the screens of the test section monitoring wells as the wells were developed following installation.

The arithmetic mean of the estimated hydraulic conductivity of the iron–sand mixture in the test section is 5.6×10^{-4} m/s. The estimated hydraulic conductivity of the aquifer ranged from 9.2×10^{-5} m/s at CT-1 to 2.0×10^{-4} m/s at CT-6, and was lower than that of the test section by at least a factor of two.

The slug test data, along with the measurements of the viscosity of the fluid from the wells within the test section, demonstrate that the amount of residual bio-polymer in the test section is not significant enough to have a measurable effect on the hydraulics of the PRB test section.

5.7. Acknowledgements

The authors thank Dr Robert Puls and the National Risk Management Research Laboratory, US Environmental Protection Office, Ada, for use of four YSI-600 XLM sondes in this pilot scale PRB evaluation.

5.8. References

BOUWER, H. and RICE, R. C. (1976). A slug test for determining hydraulic conductivity of unconfined aquifer with completely or partially penetrating wells. *Water Resources Research* **12**(3), 423–428.

DAY, S. R., O'HANNESIN, S. F. and MARSDEN, L. (1999). Geotechnical techniques for the construction of reactive barriers. *Journal of Hazardous Materials* **B67**, 285–297.

GAVASKAR, A. R., GUPTA, N., SASS, B. M. JANOSY, R. J. and O'SULLIVAN, D. (1998). *Permeable Barriers for Groundwater Remediation: Design, Construction and Monitoring*. Battelle, Columbus, Ohio.

FOCHT, R. M. and VOGAN, J. L. (2001). Biopolymer construction techniques for installation of permeable reactive barriers containing granular iron for groundwater remediation. *Preprint Paper of the ACS National Meeting. American Chemical Society, Division of Environmental Chemistry* **41**(1), 1120–1125.

KIPP, K. L. (1985). Type curve analysis of inertial effects in the response of a well to a slug test. *Water Resources Research* **21**(9), 1397–1408.

LEVY, B. S. (1999). *ADEPT Software for Aquifer Data Evaluation*. CHESS, Bethesda, Maryland.

MACKENZIE, P. D., HORNEY, D. P. and SIVAVEC. T. M. (1999). Mineral precipitation and porosity losses in granular iron columns. *Journal of Hazardous Materials* **68**, 1–17.

O'HANNESIN, S. F. and GILLHAM, R. W. (1998). Long-term performance of an *in situ* iron wall for remediation of VOCs. *Groundwater* **36**(1), 164–170.

PULS, R. W., BLOWES, D. W. and GILLHAM, R. W. (1999). Long-term performance monitoring for a permeable reactive barrier at the U.S. Coast Guard Support Center, Elizabeth City, North Carolina. *Journal of Hazardous Materials* **68**, 109–124.

POWELL, R. M., BLOWES, D. W., GILLHAM, R. W., SCHULTZ, D., SIVAVEC, T. M., PULS, R. W., VOGAN, J. L., POWELL, P. D. and LANDIS, R. (1998). *Permeable Reactive Barrier Technologies for Contaminant Remediation*, EPA/600/R-98-/125. US Environmental Protection Agency, Washington DC.

SIVAVEC, T. M. (2001). Long-term performance monitoring of permeable reactive barriers. *Preprint Paper of the ACS National Meeting., American Chemical Society, Division of Environmental Chemistry* **41**(1), 1181–1184.

VOGAN, J. L., BUTLER, B. J., ODZIEMKOWSKI, M. S., FRIDAY, G. and GILLHAM, R. W. (1998). *Designing and Applying Treatment Technologies: Remediation of Chlorinated and Recalcitrant Compounds*. Battelle, Columbus, Ohio.

VOGAN, J. L., FOCHT, R. M., CLARK, D. K. and GRAHAM, D. L. (1999). Performance evaluation of a permeable reactive barrier for remediation of dissolved chlorinated solvents in groundwater. *Journal of Hazardous Materials* **68**, 97–108.

Part III
Sorptive removal and natural processes

6. Metals loading on sorbents and their separation

K. A. Matis, A. I. Zouboulis, G. P. Gallios and N. K. Lazaridis
Department of Chemistry, Aristotle University, GR-54006 Thessaloniki, Greece

6.1. Introduction

From complex abandoned hazardous waste sites to underground petroleum storage tanks and to recent brownfield redevelopment, much assessment and clean-up work has been carried out by the US Environmental Protection Agency (EPA). The expansion of the Internet has increased access to a considerable amount of data, providing information on many types of contamination and the technologies deployed to deal with these problems. Major advances in implementing clean-up programmes have been accomplished. Contaminated soil and groundwater have also been the subject of legislative attention for about 20 years in many countries, for instance the USA (Kovalick, 2000).

Past experiences from the Superfund programme in the USA were used to streamline site investigations and speed up the selection of clean-up actions, leading to an initiative of 'presumptive remedies' which were expected to increase consistency in remedy selection and implementation, and also reduce the cost and time required to clean up similar types of sites (US Environmental Protection Agency, 1996). Instead of establishing one or more presumptive remedies, the final guidance defines a response strategy. The EPA expects that some elements of this strategy will be appropriate for all sites with contaminated groundwater and all elements will be appropriate for many of these sites.

Ion exchange/adsorption (being the first stage of the combined process under investigation) is one of the *ex situ* technologies considered in the feasibility study (in a sample of 25 sites) for the treatment of metals, and is identified, among others, as a presumptive technology. Metals including lead, chromium, arsenic, zinc, cadmium, copper and mercury can cause significant damage to the environment and human health as a result of their mobility and solubility (Evanko and Dzombak, 1997). Sorption of toxic metals is usually performed with fixed beds of common adsorbents, such as goethite (Lehmann *et al.*, 2001).

Nevertheless, ion exchange and adsorption have also been examined for their effectiveness for *in situ* groundwater remediation.

Subsurface treatment walls based on sorption have been tested to intercept and treat migrating contaminants in significant concentrations. Water-permeable treatment walls are generally installed across the flow path of a contaminant plume, allowing the plume to move passively through, while degrading the contaminants (metals, halogenated organic compounds or radionuclides) into harmless by-products. Successful application of the technology requires characterization of the contaminant, groundwater flux and subsurface geology.

For example, in the New Mexico Institute of Mining and Technology, zeolites were developed for permeable barriers, which retain major classes of groundwater contaminants while allowing groundwater to pass through, due to the surface alteration tailored for specific needs (US Environmental Protection Agency, 1995). It was further advocated that this barrier be used selectively for soluble organic compounds, inorganic cations and anions. These methods of *in situ* chemical treatment, as alternatives to conventional pump-and-treat practice, were said to have limited application mainly due to regulatory barriers (Yujun and Allen, 1999).

Thorough understanding of the sorption mechanisms is critical to evaluate the efficiency of *in situ* immobilization techniques using specific sorbents. For example, if sorption occurs by simple electrostatic interaction, the sorbed species may be remobilized by changes in environmental condition. If sorption occurs by inner-sphere complexation or by surface precipitation, the immobilized species tend to be stable and are not easily remobilized.

Several inorganic materials were also shown to act as sorbents for metal ions. Among them, goethite, an abundant iron mineral, was found to exhibit the specific advantage of being able to remove both cations and oxyanions, such as arsenic. The modelling of sorption processes was attempted following the diffuse double-layer model, and good agreement was achieved with the experimental results (Matis *et al.*, 1999).

Sorption is often used as a general term, describing the attachment of charged species from solution to a coexisting solid surface. Defining the method in this way, one can identify many types of sorption phenomena (Matis *et al.*, 1998). The combined process of (bio)sorptive flotation, which is the aim of the present study, falls more readily into the category of pump-and-treat remediation than that of passive technology with permeable reactive barriers.

That pump and treat constitutes one of the most widely used technologies may be due to safety reasons, although it is accompanied by higher costs. The approach has been used at about three-quarters of the Superfund sites; an introductory guide can be downloaded from the Internet (US Environmental Protection Agency, 2001). In a table in that report on the applicability of treatment technologies to contaminated groundwater (No. 3) flotation is described as not suitable for heavy metal contaminants, suitable for polychlorinated biphenyls (PCBs), dioxins, oil and grease/floating materials, and potentially suitable for semivolatile organic compounds and pesticides. An explanation of this statement is that the term 'flotation' was used within the strict sense of its meaning in mineral processing, which is where this selective separation method originated. Bubble aeration (usually the dispersed-air technique) and the addition during conditioning of special chemicals, such as collectors, modifiers and frothers, cause hydrophobic particles, and those rendered hydrophobic, to float.

6.2. Flotation

Another publication (Mulligan *et al.*, 2001) mentions flotation technology (with hydrocyclones and fluidized beds) as a possible soil remediation approach for accomplishing size selection, because this process enables the separation of larger, cleaner particles from smaller, more polluting ones. The KOP Superfund site (USA) was the first full-scale application of soil washing for remediation (US Environmental Protection Agency Superfund, 1995). Soil washing is defined as the technology that uses liquids (usually water, sometimes combined with chemical additives) and a mechanical process to scrub soils, according to the EPA citizen's guide and the respective technology fact sheet. In this case, the soil and sludge washing unit consisted of a series of hydrocyclones, conditioners and froth flotation cells. Performance data showed that the cleaned sand and process oversize met the clean-up levels for 11 metals in this application, including chromium, copper and nickel.

The physical and chemical technologies are well known. An illustrative flowsheet as applied to remediation is shown in Fig. 6.1. If required, liberation is carried out in a series of stages followed by classification, typically as in mineral processing. Perhaps the absence of flotation from laboratory scale washing tests (Neese, 1998) is another reason for the 'low profile' of the process.

Today, flotation has wide applications, as significant know-how has been transferred from the area of mineral processing. Some remarkable examples from the literature include the removal by flotation of algae, viruses and bacteria to produce potable water, and also use in activated sludge thickening. In this field, dissolved-air flotation is used to generate bubbles (Zabel, 1992). A recent flotation conference attempted to bridge the gap between the two flotation techniques (Matis *et al.*, 2001a).

One lesser known but nevertheless highly effective technique is electroflotation or, more correctly, electrolytic flotation. This technique has been used to remove nickel, zinc, lead, copper and cyanide ions from polluted groundwater obtained directly from under a contaminated site, a medium-sized landfill, in order to meet the pretreatment standards of the pollutants imposed by the regulating agency (Poon, 1997). Rock salt solution was used as an electrolyte, generating hypochlorite to oxidize cyanide and hydroxides to form metal precipitates.

Another promising application is to metals recovery in hydrometallurgy: in the zinc pressure leaching plant at Trail, elemental sulphur is separated out from the plumbo-jarosite leach residue in a series of flotation cells (Cominco, 2001). Ion flotation was reviewed by Zouboulis and Matis (1987); here, surface-inactive ions, so-called colligends, are removed from dilute solutions by the addition of a surfactant.

Flotation, considered broadly, also includes precipitate flotation, as an increase in concentration may lead to precipitation of the resulting product (ion–surfactant). In the case of metal ions the usual product is the respective metal hydroxide precipitate obtained by pH adjustment. This process has been compared with metal biosorption (Matis *et al.*, 2001b).

6.3. Biosorption

The removal of metals by biosorption has been extensively studied in recent years (Zouboulis *et al.*, 1997). The process is based on several mechanisms, the most important being physical adsorption (electrostatic forces), ion exchange, surface complexation and surface precipitation. On biomass surfaces several chemical groups may be present, which could attract and subsequently sequester metals from the surrounding environment, such as acetamido groups of chitin, amino and

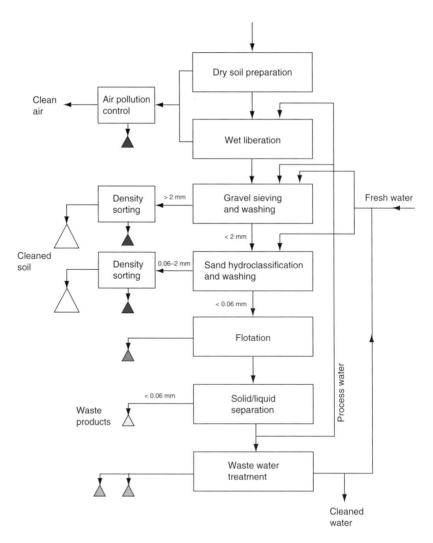

Fig. 6.1. Flowsheet for a soil-washing plant (Neese, 1998)

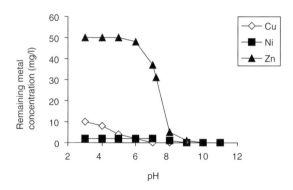

Fig. 6.2. Biosorption from an aqueous multi-metal solution by modified Penicillium chrysogenum *fungal biomass (1.14 kg (dry wt)/m³): effect of pH. The mixture contained as nitrate salts Zn (0.765 mmol/dm³), Cu (0.157 mmol/dm³), Ni (3.408 × 10⁻² mmol/dm³), Ca (2.495 mmol/dm³) and Na (4.348 mmol/dm³). The initial volume to be treated was 500 ml and the contact time 15 min. (Adapted from data presented by Zouboulis et al. (1999))*

phosphate groups of nucleic acids, and sulfydryl (thiol) and carboxyl groups of proteins. The presence of particular functional groups does not necessarily guarantee their accessibility as sorption sites. Microbial metal uptake by non-living cells, which produces metabolism-independent passive binding (adsorption) to cell walls, as well as to other usually fragmented external surfaces, is generally considered to be a rapid process taking place within a few minutes.

Biosorption of toxic metals, commonly found in many industrial wastewaters, from an aqueous mixture was investigated in the laboratory in a continuous stirred-tank reactor (CSTR) configuration, as shown in Fig. 6.2. A theoretical speciation study (with the Mineql+ program) has shown that, in this aqueous mixture of metals, copper is the first metal that precipitates out as hydroxide at a pH of about 5.9, followed by zinc at pH 7.3 and then nickel at pH 7.8. Calcium does not precipitate in the range investigated. It is worth noting that these chemical speciation models based on thermodynamic principles are quite sensitive to the quality of their thermodynamic databases – the default values supplied with the widely available computer programs cannot be relied upon.

Industrial filamentous biomasses of different origin have been tested to see whether they are effective sorbents. The subject of examination was large-scale fermentation waste from the pharmaceutical industry (with the trade name Mycan). Non-living sewage sludge could represent an alternative biosorbent. A sample from the central treatment plant of the Thessaloniki area was successfully tested for the removal of zinc, nickel and cadmium (Solari *et al.*, 1996).

Many aquaphytes are another possibility, for example for Cu^{2+} ions (Schneider *et al.*, 2001). Existing data on the adsorption of metal ions on to such a biomass type were examined in an attempt to evaluate which mechanism is the more likely. It has been reported that adsorption occurs through a specific ion exchange mechanism, and a number of researchers have found experimental evidence supporting such a mechanism. However, there is also evidence of adsorption through simple surface precipitation of metal hydroxide species. These authors suggest that both mechanisms take place, at least in the case studied.

Dead biomass appears to have advantages in comparison to using live micro-organisms: the former may be obtained at low cost, is not subject to metal toxicity, a nutrient supply is unnecessary, etc. Greater binding capacities (for cadmium) were also recorded (Kefala *et al.*, 1999).

When metal removal was calculated from the above experimental runs (given in Figs 6.3–6.6), high values were observed, of the order of 100%, particularly in the alkaline pH region. Dispersed-air flotation (with 3×10^{-4} M dodecylamine as the collector) was used downstream for separation. The biomass recoveries were around 90% at pH 7 and almost 100% at pH 8–9.

6.3.1. Sorptive flotation

Sorption combined with flotation has been investigated by Zouboulis and Matis (1997). Such a combination of processes is necessary to comply with the high regulatory standards now imposed, and the scarcity of water in many places. The process examined (termed biosorptive flotation) involves the abstraction of metal ions on to sorbents, in the fine or ultrafine size range, followed by a flotation stage for the separation of metal-loaded particles. The sorbents used in the first process stage were industrial solid wastes or by-products. In this way, a foam concentrate and treated, clean water as underflow are produced. The recovery of metals is even possible from the concentrate, leading to a clean technology. If required, the sorbents may be recycled after metal desorption.

The choice of cationic surfactant was based on zeta-potential measurements. The influence of a polyelectrolyte presence on the biosorptive flotation of loaded fungi was subsequently examined, using dodecylamine after optimization. The results of

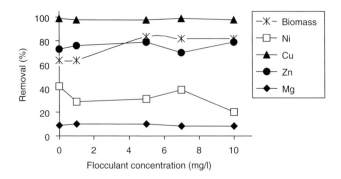

Fig. 6.3. Influence of flocculant concentration on the biosorptive flotation of loaded biomass (6 g/l) and on the removal of toxic metals (1 × 10^{-3} M dodecylamine in 0.6% v/v ethanolic solution, pH 6, 100 mg/l of Mg was also initially present). The flotation retention time was 10 min and the superficial air velocity was 0.27 cm/s

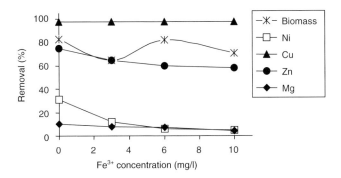

Fig. 6.4. Influence of Fe^{3+} concentration on the biosorptive flotation of fungi and on the removal of toxic metals (Zetag, 5 mg/l; remaining conditions unchanged). The eluate solution was a mixture of sodium sulfate (1 M) and sodium citrate (0.1 M). This part of the work was recently reported at a UEF conference (Matis et al., 2001b)

biomass recovery by flotation were found to be substantially increased (from 64 to 82–84% using 5–10 mg/l of Zetag), while the removal of toxic metals was either unaffected (copper and magnesium) or slightly affected (zinc and nickel) (Fig. 6.3). The residual turbidity (after flotation) was also remarkably improved (it decreased from 126 to 0.4 NTU (nephelometric turbidity unit)), when polyelectrolyte concentrations were increased to 10 mg/l.

Effective flocculation is a prerequisite for dissolved-air flotation, and various flotation techniques have been used. According to the method used for the generation of bubbles, two broad categories of flotation process exist: dispersed-air and dissolved-air flotation; electroflotation is considered to belong to the former. It should be noted that the results of both biomass flotation recovery and metal biosorption removal are presented simultaneously in these figures (although the biosorption results may be a little misleading).

Keeping the previous conditions constant, the influence of coexisting iron ions (at a concentration range of 0–50 mg/l) on biosorptive flotation was examined. The behaviour of fungi in respect of this was found to be different from that of stalks. The presence of ferric iron was not observed to influence substantially the effective separation of biomass by flotation (which remained at about 80%). The removal of Cu and Mg was found to be unaffected, and that of Zn and Ni was decreased (Fig. 6.4). The removal of Fe^{3+} was almost complete (98–99%), i.e. different to the other metals (data not presented in the figure). Residual turbidity was also found to be rather low after flotation (1.2–26.6 NTU) for the iron concentrations studied.

Cost is also an important factor when comparing sorbent materials. Potentially low-cost sorbents, such as natural materials and industrial wastes or by-products, are increasingly in demand owing to their suitability for removal of heavy metals. Certainly, improved sorption capacity may compensate for the cost of additional processing.

A two-stage countercurrent scheme was selected, with one pass of elution, based on extensive studies (funded by an EU environmental research programme).

Flotation has been included following each respective biosorption stages. This scheme is similar to a conventional activated-carbon system for the separation of liquids. The partially loaded biomass was guided from the polishing to the leading biosorption stage, where the wastewater feed was introduced.

It can be concluded that (bio)sorptive flotation appears to be a viable and effective separation process for the aqueous systems (mixture of metal ions) under investigation, no matter what sort of biosorbent is applied. Another advantage of the proposed process is that its base processes are conventional and widely used. The application of the individual techniques may be novel but the underlying methodology is familiar.

6.3.2. Electroflotation

This unconventional flotation technique has been effectively applied to the removal of metal ions from dilute aqueous solutions (effluents). Nebera *et al.* (1980) examined multicomponent systems of ferrocyanides, oxyquinolates and hydroxides of metals. Romanov (1998) has commented on the extraction of heavy metal ions in the form of hydroxides from wastewaters in the metallurgical and metal-working industries. Since the discovery that the activated sludge treatment process can also remove metals from wastewaters, much progress in research has been made on the immobilization of metals by waste biomass, and in related areas. The microbiology of activated sludge in relation to metal binding has also been reviewed (Kasan, 1993). Early work focused on metal ions in biological wastewater treatment processes and mainly emphasized the toxicity of metals to bacteria and the subsequent inhibition of organic matter reduction.

Using aluminium alloy electrodes in place of the usual iron ones improves the results (Fig. 6.5), and biomass recovery is possible from a pH value of approximately

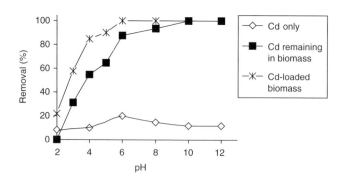

Fig. 6.5. Biosorptive flotation of cadmium on anaerobically digested activated-sludge biomass: effect of solution pH. Electroflotation was used with aluminium alloy horizontal electrodes (130 A/m² current density, 10–20 V (DC current), 0.5 cm interelectrode gap). The initial Cd concentration was 5 mg/l, the biomass concentration was 0.5 g/l, the pH was not modified, and 0.05% v/v ethanol was used as the frother. The electroflotation was carried out in a 1 l beaker over a time-scale of 600 s

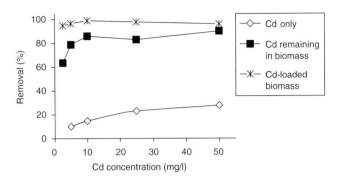

Fig. 6.6. Effect of initial cadmium concentration on 1 g/l of activated sludge biomass (at 65 A/m², 900 s). These figures were recently presented as a conference poster (Zouboulis and Matis, 2000)

7. The results were not affected significantly by a shorter process duration (10 min) or even by a lower current density (65 A/m²). The plot of cadmium removal on biomass and the plot of floated (cadmium-loaded) biomass correlate well with pH variation.

Water chemistry and the aqueous speciation of cadmium are also of significance in this context. Considering the formation of cadmium hydroxide for a cadmium concentration of 1×10^{-4} M (about 11 mg/l), the solution pH value of precipitation is theoretically around 10 (Zouboulis and Matis, 1995). The isoelectric point of $Cd(OH)_2$ is at pH 11.6, and the minimum solubility of cadmium was reported at the same value.

In this work, collectorless (electro)flotation technique was applied to the biomass suspension, and it proved to be an effective solid/liquid separation procedure for the metal-loaded biosorbent particles. Experiments were also conducted to investigate cadmium concentrations (Fig. 6.6) at lower current densities. Very good results were obtained, as the initial cadmium concentration was increased to 50 mg/l.

It was concluded that the cadmium was removed from aqueous solutions at significant rates only by the use of biomass at pH 7, while the metal-loaded biosorbent particles were efficiently recovered by electroflotation in the absence of a collector. A certain stoichiometric relationship was observed to exist between the three main factors influencing the process, i.e. current density, biomass addition and cadmium concentration.

In electroflotation the need for charge neutralizers and coagulating agents may be reduced; in certain applications, due to the surface charge of gas bubbles, they are not even required. The electric field gradient between the electrodes aids the self-flocculation of suspended matter (electrocoagulation). The use of electrolytic cells with consumable anodes (such as aluminium) has attracted scientific interest in the past, since their slow dissolution provides a continuous injection of coagulating metal cations into the treatment system. Various types of electrode design have been used, according to the literature.

The cost of an electroflotation device plus a sand filter compares favourably with a conventional treatment system using cyanide oxidation/alkaline precipitation/polymer-aided clarification (Poon, 1997). A saving of approximately 43% was found when cyanide is present in the groundwater. Based on 1995 values, the total cost of the electroflotation installation (excluding contingencies and engineering) was reported to be US $38 945.

In an article posing a crucial question in its title and trying to explain the disappointing lack of applications of this advanced technology, Eccles (1999) stated that only a small number of pilot plant studies have been carried out to investigate the potential of micro-organisms to remove metals from liquid wastes. We hope that our work will contribute to further improving this technology.

6.4. Conclusion

Although pump-and-treat technology has recently been strongly criticized (US Environmental Protection Agency, 2001), the process of sorptive flotation looks promising as it is able to reduce dissolved contaminant concentrations in groundwater, and ensure that the aquifer complies with clean-up standards or that treated water withdrawn from the aquifer can be put to beneficial use.

6.5. References

COMINCO (2001). *Cominco Trial Operations, Technical Overview*. Cominco, British Columbia.

ECCLES, H. (1999). Treatment of metal-contaminated wastes: why select a biological process? *TIBTECH* **17**, 462–465.

EVANKO, C. R. and DZOMBAK, D. A. (1997). *Remediation of Metals – Contaminated Soils and Groundwater*, TE-97-01. Ground-Water Remediation Technologies Analysis Center, Pittsburgh.

KASAN, H. C. (1993). The role of waste activated sludge and bacteria in metal-ion removal from solution. *Critical Reviews in Environmental Science and Technology* **23**, 79–117.

KEFALA, M. I., ZOUBOULIS, A. I. and MATIS, K. A. (1999). Biosorption of cadmium ions by *Actionomyceces* and separation by flotation. *Environmental Pollution* **104**, 283–293.

KOVALICK, JR, W. W. (2000). Technologies for clean-up of contaminated soil and groundwater in United States: current practice and information resources. International Symposium on Waste Management in Asian Cities, Hong Kong.

LEHMANN, M., ZOUBOULIS, A. I. and MATIS, K. A. (2001). Modelling the sorption of metals from aqueous solutions on goethite fixed-beds. *Environmental Pollution* **113**, 121–128.

MATIS, K. A., ZOUBOULIS, A. I. and LAZARIDIS, N. K. (1998). Removal and recovery of metals from dilute solutions: applications of flotation techniques. In: G. P. Gallios and K. A. Matis (eds), *Mineral Processing and the Environment*, pp. 165–196. Kluwer, Dordrecht.

MATIS, K. A., LEHMANN, M. and ZOUBOULIS, A. I. (1999). Modelling sorption of metals from aqueous solution onto mineral particles: the case of arsenic ions and goethite ore. In: P. Misaelides, F. MacAsek and T. J. Pinnavaia (eds), *Natural Microporous Materials in Environmental Technology*, pp. 463–472. Kluwer, Dordrecht.

MATIS, K. A., ZOUBOULIS, A. I., LAZARIDIS, N. K. and HANCOCK, I. C. (2001a). Sorptive flotation for metal ions recovery. Conference on Froth Flotation/Dissolved Air Flotation: Bridging the Gap, United Engineering Foundation, Tahoe City, California.

MATIS, K. A., ZOUBOULIS, A. I., LAZARIDIS, N. K. and ROUSOU, E. G. (2001b). Heavy metals recovery by biosorption and flotation. 1st European Conference on Bioremediation, TU Crete, Chania.

MULLIGAN, C. N., YONG, R. N. and GIBBS, B. F. (2001). Remediation technologies for metal-contaminated soils and groundwater: an evaluation. *Engineering Geology* **60**, 193–207.

NEBERA, V. P., ZELENTSOV, V. I. and KISELEV, K. A. (1980). Electroflotation of ions from multicomponent systems. In: P. Somasundaran (ed.), *Fine Particles Processing*, pp. 886–894. SME-AIME, New York.

NEESE, T. (1998). Soil remediation: physical and chemical treatment technologies. EU Advanced Study Conference on Soil Remediation: An Overall Approach to a Complex Subject, Vienna (notes available on CD-Rom).

POON, C. P. C. (1997). Electroflotation for groundwater decontamination. *Journal of Hazardous Materials* **55**, 159–170.

ROMANOV, A. M. (1998). Electroflotation in wastewater treatment: results and perspectives. In: G. P. Gallios and K. A. Matis (eds), *Mineral Processing and the Environment*, pp. 335–360. Kluwer, Dordrecht.

SCHNEIDER, I. A. H., RUBIO, J. and SMITH, R. W. (2001). Biosorption of metals onto plant biomass: exchange adsorption or surface precipitation? *International Journal of Mineral Processing* **62**, 111–120.

SOLARI, P., ZOUBOULIS, A. I., MATIS, K. A. and STALIDIS, G. A. (1996). Removal of toxic metals by biosorption onto non-living sewage sludge. *Separation Science and Technology* **31**(8), 1075–1092.

US ENVIRONMENTAL PROTECTION AGENCY (1995). In situ *Remediation Technology Status Report: Treatment Walls, Solid Waste and Emergency Response*, EPA 542-K-94-004. US Environmental Protection Office, Washington DC.

US ENVIRONMENTAL PROTECTION AGENCY SUPERFUND (1995). *Cost and Performance Report: Soil Washing at the King of Prussia Technical Corp. Site*, 68-W3-0001. Office of Solid Waste and Emergency Response, Technology Innovation Office, US Environmental Protection Office, Washington DC.

US ENVIRONMENTAL PROTECTION AGENCY (1996). *Presumptive Response Strategy and* ex situ *Treatment Technologies for Contaminated Ground Water at CERCLA Sites, Final Guidance, Directive 9283.1–12*. US Environmental Protection Agency, Washington DC.

US ENVIRONMENTAL PROTECTION AGENCY (accessed 2001). Introduction to pump-and-treat remediation. Website: http://www.epa.gov/ordntrnt/ORD/WebPubs/pumptreat/pumpdoc.pdf.

YUJUN, Y. and ALLEN, H. E. (1999). In situ *Chemical Treatment. Technology Evaluation Report TE-99–01*. Ground-Water Remediation Technologies Analysis Center, Pittsburgh.

ZABEL, T. F. (1992). Flotation in water treatment. In: P. Mavros and K. A. Matis (eds), *Innovations in Flotation Technology*, pp. 431–454. Kluwer, Dordrecht.

ZOUBOULIS, A. I. and MATIS, K. A. (1987). Ion flotation in environmental technology. *Chemosphere* **16**(2–3), 623–631.

ZOUBOULIS, A. I. and MATIS, K. A. (1995). Removal of cadmium from dilute solutions by flotation. *Water Science and Technology* **31**(3–4), 315–326.

ZOUBOULIS, A. I. and MATIS, K. A. (1997). Removal of metal ions from dilute solutions by sorptive flotation. *Critical Reviews in Environmental Science and Technology* **27**(3), 195–235.

ZOUBOULIS, A. I. and MATIS, K. A. (2000). Cd^{2+} removal by electroflotation onto sewage sludge biomass. *Proceedings of the 4th International Conference on Flotation in Water and Waste Water Treatment*. Finnish Water and Wastewater Works Association, Helsinki.

ZOUBOULIS, A. I., MATIS, K. A. and HANCOCK, I. C. (1997). Biosorption of metals from dilute aqueous solutions. *Separation and Purification Methods* **26**(2), 255–295.

ZOUBOULIS, A. I., ROUSOU, E. G., MATIS, K. A. and HANCOCK, I. C. (1999). Removal of toxic metals from aqueous mixtures: Part 1. Biosorption. *Journal of Chemical Technology and Biotechnology* **74**, 429–436.

7. Sorption mechanisms of heavy metal ions on inorganic solids

M. Fedoroff
Centre National de la Recherche Scientifique, Centre d'Etudes de Chimie Métallurgique (CECM), 15 rue G. Urbain, F-94407 Vitry sur Seine, France

7.1. Introduction

Heavy metals are amongst the most persistent contaminants in natural waters. Metals such as cadmium, mercury and lead are well known for their toxicity. They originate, as a rule, from mining or industrial activities, and from waste repositories lacking proper control. Another type of harmful emission is from radioactive elements released by military or civil nuclear activities and radioactive wastes. An important challenge for the future is the development of methods for the long-term storage of radioactive wastes under safe conditions.

The main goals to solve these problems are:

- the development of methodologies for the decontamination of toxic or radioactive wastes immediately after they have been produced by an industrial process in order to recycle the toxic elements or convert to wastes which can be better stored for long periods of time under safe conditions;
- the development of methodologies for preventing groundwater contamination from old waste repositories under unsatisfactory conditions, by creating reactive barriers;
- the development of methodologies for reprocessing wastes stored under unsatisfactory conditions, which contaminate groundwater at present or may do so in the future, where reactive barriers do not seem to be adequate;
- the reprocessing of contaminated waters;
- the prediction and modelling of the future evolution of polluted groundwater;
- the devising of waste repositories, which can ensure safe conditions of storage over long periods of time through the development of stable storage matrixes, stable containers, efficient reactive barriers around the containers in case of accidental spills or long-term corrosion, and choosing an adequate natural geological and hydrogeological environment (natural barrier) to prevent groundwater contamination by toxic or radioactive elements.

In all these goals, sorption/desorption processes at the water–solid interface play an important role. The use of solid sorbents is a suitable way of decontaminating liquid industrial wastes or reprocessing polluted waters. A solid sorbent is also the basic material for a reactive barrier near an existing or a future waste repository. Similarly, sorption/desorption processes are the main phenomena which control the transport of toxic elements in groundwater in soils or rocks, whether in already contaminated areas or around future waste repositories.

The word 'sorption' is used here for all processes of ion or molecule transfer from a liquid phase to a solid immersed in the same liquid. Sorption may be limited to the actual surface, or may involve a layer of a given thickness. Sometimes, it is possible to observe diffusion into the solid.

Substantial knowledge of sorption processes, both in a qualitative and quantitative sense, is needed for all these purposes. It is necessary to select the most effective sorbents for the retention of certain types of toxic elements, both in columns and in reactive barriers. A quantitative approach is necessary to predict the conditions of use and the efficiency of a column filled with a sorbent in order to decontaminate a polluted solution, or to predict the long-term efficacy and safety of an artificial or natural barrier.

We shall describe, within the scope of this short study, the various quantitative approaches to sorption processes with regard to ion transport in water, discuss how they may influence the selection of a decontamination procedure and give examples of the sorption of heavy elements.

7.2. Distribution coefficients

A first quantitative approach to predicting the migration of an element dissolved in water through a column or through an artificial or natural barrier is the use of the 'distribution coefficient', which is the ratio of the concentration of the element in the solid phase to its concentration in the aqueous phase. This coefficient is generally denoted K_d or R_d. For some authors K_d represents an equilibrium value, while R_d is just a measured value of the above ratio. We shall use K_d as a general term including equilibrium and apparent values, since it is often difficult to experimentally distinguish between them. K_d is used in the calculation of the retardation factor R, which is the ratio of the velocity of water to the velocity of the element through the same volume.

The simplest relation between R and K_d is (Serne and Muller, 1987)

$$R = 1 + \frac{K_d \rho (1 - \varepsilon)}{\varepsilon} \tag{1}$$

where ρ is the volumetric mass of the solid phase (in kg/l), ε is the porosity of the medium and K_d is expressed in litres per kilogram.

Expression (1) is valid for a monodirectional flow in a homogeneous porous medium, with a constant equilibrium K_d. More complicated calculations have to be used where these conditions are not fulfilled. Several models and computer codes have been developed (e.g. Jauzein et al., 1989; Ochs et al., 1998). Our purpose is not to look into that aspect of transport but to discuss the impact of sorption phenomena.

The connection between K_d values and transport explains why many determinations of K_d have been established and databases constituted (Stenhouse and Pöttinger, 1994).

However, experimentally determined K_d values depend very much on the experimental conditions. In many cases the distribution coefficient is not constant and often does not represent an equilibrium value. K_d depends on many factors – the concentration of the element in the solution and in the solid, pH, the influence of complexing agents, the competition of other elements, and the kinetics of the sorption process – and may be subject to long-term variations if the solid phase is transformed with time. Therefore, extrapolation of K_d values to scales larger than those of laboratory experiments, especially to natural systems and to very long periods of time, may be unreliable.

Empirical expressions have been proposed in order to predict the influence of competing elements on the K_d of an element. These expressions use linear or non-linear functions of the concentration of competing elements for the calculation of K_d (Holstetler *et al.*, 1980). However, these expressions are valid only for certain experimental conditions and are not generally based on physicochemical processes.

7.3. Sorption isotherms

The variation of K_d with the concentration of an element can be approached through sorption isotherms. An isotherm represents the variation of the concentration of the element in the solid phase or the variation of K_d as a function of the equilibrium concentration of the element in the aqueous solution, at a constant temperature. An equilibrium of sorption is assumed to be achieved, but apparent isotherms are often used. Experimental isotherms are fitted by several mathematical models which have some theoretical bases but are often used empirically.

The Langmuir model (Langmuir, 1918), derived from the sorption of gas molecules on a homogeneous solid, in which all sites have equal sorption energy, is very often used for sorption of a solute on solids from aqueous solutions. This form then fits the experimental data:

$$C_s = \frac{KSC_1}{1 + KC_1} \tag{2}$$

where C_s is the concentration in the solid, C_1 is the concentration in the liquid, S is the sorption capacity (maximum achievable concentration in the solid) and K is a constant.

Another type of isotherm is the Dubinin–Radushkevich model (Dubinin and Radushkevich, 1947), which does not require homogeneous sorption sites:

$$C_s = S \exp\{-K[RT \ln(1 + 1/C_1)]^2\} \tag{3}$$

where T is the temperature in kelvin and R is the (universal) gas constant ($=$ 8.314510 J/K mol).

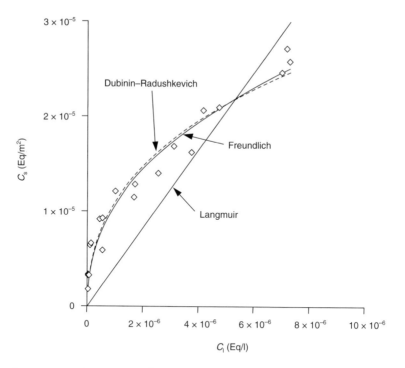

Fig. 7.1. Sorption isotherm of Nd^{3+} on calcite, with fitting by three kinds of isotherm model

A third type is the Freundlich model (Freundlich, 1926):

$$C_s = KC_l^n \qquad (4)$$

where n is another constant.

These models do not all fit the experimental points well when C_l is close to the sorption capacity of the solid.

$K_d = C_s/C_l$ can be calculated from these equations and used in retardation models.

An example, taken from the study of the sorption of neodymium (Nd^{3+}), used as a model for trivalent elements from aqueous solution on calcite, is shown in Fig. 7.1 (Mecherri *et al.*, 1990). The experimental values were fitted by the three models above. The results show that the Langmuir model is inadequate, while both the other models provide a good fit (although the constants obtained are not comparable), indicating that the sorption sites are not homogeneous.

7.4. Sorption models

Several models have been developed in order to quantify sorption data more accurately. These are based on certain hypotheses about the sorption mechanisms and on thermodynamic equilibria. Theoretically, they are able to predict K_d values of an element taking into account competing elements and complexing agents.

7.4.1. Ion exchange model

This model is based on the mass action law, quantifying the exchange of ions between a solid and a solution. No hypotheses are needed on the real mechanisms at the atomic scale.

As an example, we show the case of the exchange of a divalent element M in solution with a monovalent element (protons, H^+) in the solid. The equilibrium is

$$M^{2+} + 2\overline{H^+} \rightleftharpoons \overline{M^{2+}} + 2H^+ \tag{5}$$

where the elements under the bars are in the solid phase and the elements without bars are in the liquid phase. The expression for the mass action law is

$$K = \frac{(\overline{M^{2+}})(H^+)^2}{(M^{2+})(\overline{H^+})^2} \tag{6}$$

where the parentheses represent activities and K is the thermodynamic exchange constant.

In order to determine the constant K and to calculate K_d, we have to derive the activity coefficients. Generally, the activity coefficients in the liquid phase can be calculated, especially for dilute solutions. However, the activity coefficients in the solid phase cannot be calculated *a priori*. Therefore, a 'corrected selectivity coefficient' K_c needs to be defined:

$$K = K_c \frac{f_M}{f_H^2} \tag{7}$$

$$K_c = \frac{|H^+| X_M \gamma_H^2}{|M^{2+}| X_H^2 \gamma_M} \tag{8}$$

where $|H^+|$ and $|M^{2+}|$ are the molalities of the ions in solution, X_H and X_M are the ionic equivalent fractions in the solid, γ_H and γ_M are the activity coefficients in the liquid, and f_H and f_M are the activity coefficients in the solid.

In the general case, K_c does not remain constant when the concentration in the solid is increasing. The thermodynamic constant and the activity constants in the solid can be calculated from the Gaines–Thomas equation (Gaines and Thomas, 1953):

$$\ln K = (Z_H - Z_M) + \int_{X_M=0}^{X_M=1} \ln K_c \, dX_M \tag{9}$$

$$\ln f_M = (X_M - 1) - (1 - X_M) \ln K_c - \int_1^{X_M} \ln K_c \, dX_M \tag{10}$$

$$\Delta G° = -\frac{RT}{Z_M Z_H} \ln K \tag{11}$$

It is possible to interpret the variation of K_c with the concentration in the solid by considering the partition function Q, which is related to the free energy G (Barrer and Falconer, 1956):

$$Q = \sum_j \exp \frac{-E_j}{kT} \tag{12}$$

where E_j is the energy of each level in the sorbent. Or, when many levels have the same energy

$$Q = \sum_j g_i \exp \frac{-E_j}{kT} \tag{13}$$

where E_i is the energy of each family of levels and g_i is the number of equivalent levels (degeneracy factor). Q is related to the free energy by

$$G = -RT \ln Q \tag{14}$$

thus allowing a connection to K_c through equations (9) and (11), and an interpretation of its variation through the energy levels of the sorption sites of the solid. An example of the use of this method is to limit the calculation to the additional energy change in the solid when two adjacent sorption sites are occupied by the ion M (Barrer and Falconer, 1956). It can be shown that the variation of the logarithm of K_c as a function of the ionic fraction X_M of the element in the solid is linear:

$$\ln K_c = A + BX_M \tag{15}$$

where A and B are constants.

However, in several cases K_c is almost constant, simplifying the use of the ion exchange model in transport models. This is the case when the concentration of the sorbed element is always at a trace level or when the interactions between adjacent sorption sites are negligible (coefficients f are almost constant). The constant K can be replaced by an apparent constant K_{ap}, which remains constant within a fairly wide concentration range. Several families of exchange sites with respective apparent constants $K_{ap\,i}$ can also be considered.

As an example, we have studied the sorption of Sr^{2+} on polyantimonic acid, $H_2Sb_2O_6$ (Zouad et al., 1992). The variation of $\ln K_c$ with the Sr concentration in the solid first shows a constant value and then a linear decrease (Fig. 7.2). Using X-ray diffraction, it was shown that sorption proceeds successively through the substitution of two types of site corresponding respectively to each of the linear variations of $\ln K_c$. We can thus predict the variation of K_d with the occupancy of sorption sites.

7.4.2. Surface complexation models

Another family of models, called 'surface complexation models', has been developed for oxyhydroxides. These models are based on the acido-basic properties of the superficial hydroxide groups of these solids. The so-called 'monosite 2-pK model' (Stumm, 1992) assumes that the surface is covered by one type of hydroxyl group with amphoteric properties. These groups lead to the following equilibria:

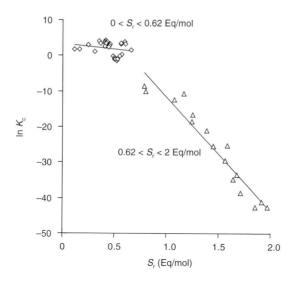

Fig. 7.2. Variation of ln K_c as a function of strontium concentration sorbed in polyantimonic acid, showing the successive occupation of site 1 with almost constant ln K_c, and site 2 with linearly decreasing ln K_c

$$Su\text{–}OH_2^+ \rightleftharpoons Su\text{–}OH + H^+ \tag{16}$$

$$Su\text{–}OH \rightleftharpoons Su\text{–}O^- + H^+ \tag{17}$$

where Su represents the surface. The following two thermodynamic constants correspond to these equilibria:

$$K^+ = \frac{(SuOH)(H^+)}{(SuOH_2^+)} \tag{18}$$

and

$$K^- = \frac{(SuO^-)(H^+)}{(SuOH)} \tag{19}$$

The parentheses represent activities.

The 'surface charge' is defined as $Q = [Su\text{–}OH_2^+] - [Su\text{–}O^-]$.

The point of zero charge (PZC) is the pH value at which the surface charge is equal to zero. The variation of surface charge with pH and the PZC can be determined by titration of a suspension of the solid in a solution by an acid or a base. However, the interpretation of these titration curves and the calculation of the constants K^+ and K^- present several experimental and theoretical difficulties. Owing to lack of space we cannot detail these difficulties here, which have been the subject of many publications. The main theoretical difficulty results from electrostatic interactions in the vicinity of the surface. The activity (H^+) measured in the bulk of the solution is different from the activity in the vicinity of the surface, and the apparent constants K^+ and K^- vary with the surface charge. For the ions present in the electrical field near the surface, it is necessary to consider the electrostatic free energy:

$$\Delta G_{elec} = zF\Psi(x) \tag{20}$$

where z is the charge of the ion, F is the Faraday constant and $\Psi(x)$ is the electric potential, which is dependent on the distance x to the surface. This potential cannot be calculated *a priori*. Several different models, based on the distribution of ions in layers near the surface, have been proposed in order to relate the potential to the surface charge (the constant capacitance model, the Gouy–Chapman model, the Stern model, etc.). Through these models, it is possible to calculate the 'intrinsic' thermodynamic constants.

Ions present in the solution can be sorbed on to the solid by interaction with the surface sites. The sorbed quantity depends on the nature and sign of the ion, together with the sign and charge of the surface and, hence, on pH. Two types of interaction are considered: pure electrostatic interactions lead to 'outer-sphere' surface complexes, whereas formation of covalent bonds with surface groups leads to 'inner-sphere' complexes. Below are some examples of such surface complexes:

$$Su-O^- + M^+ \rightleftharpoons Su-O^-M^+ \qquad \text{(outer-sphere complex)} \tag{21}$$

$$Su-OH + M^{2+} \rightleftharpoons Su-OM^+ + H^+ \quad \text{(inner-sphere complex, monodentate)} \tag{22}$$

$$2Su-OH + M^{2+} \rightleftharpoons (Su-O)_2 + 2H^+ \quad \text{(inner-sphere complex, bidentate)} \tag{23}$$

Thermodynamic sorption constants can be calculated from these models. Finally, K_d values can be predicted from these constants together with their variation with pH and concentration of other species in solution.

However, very often several different ways of modelling can be suggested for the same experimental system, sometimes assuming different types of surface complexes. It is necessary to validate these models by a study of real sorption processes.

The monosite '1-pK model' is a simplified version of the preceding model, in which only one acido-basic equilibrium is considered.

Another surface complexation model is the '1-pK multisite complexation model', or 'Music' model (Hiemstra *et al.*, 1989). It considers that the surface of the solid has several families of oxide and hydroxide sites, each with a specific acidity constant pK. Although each type of site cannot be amphoteric (1-pK model), the occurrence of several families of sites, with acid or basic character, may confer an amphoteric property to the whole surface. Another feature of this model is the fact that the charge of a site may be fractional, this charge depending on the coordination sphere of the surface site. In contrast to the 2-pK model, the acidity constants are calculated here *a priori*, from atomic geometrical and electrostatic considerations and by comparison with the monomers of the same oxides.

In conclusion, this model is more realistic than the monosite 2-pK model, but it requires a good knowledge of the crystal structure and surface orientation of the solid. For these reasons its use is limited in transport calculations.

7.5. Speciation in solution

Reactions and equilibria at the solid–liquid interface depend on the concentration of the various species of elements in the solution. It is necessary to calculate the concentration of free ions or other species entering the equilibrium. Metals may form organic complexes, for example methyl-mercury, in natural media. These calculations may be performed using hydrolysis and complexation constants. Computer programs have been developed to undertake these calculations (Westall *et al.*, 1986).

Colloids, which are nearly always present in natural waters, may modify the transport of elements in groundwater. The fixation of elements on colloids, and transport of these colloids in macroporous media, may accelerate the migration of elements, in comparison to that predicted for water without colloidal particles. In other cases, fixation on colloids may decrease the migration through 'filtration' effects in microporous media.

7.6. Sorption kinetics

All the models discussed above assume that equilibrium is achieved. In fact, kinetic effects play an important role in sorption chemistry. A substantial knowledge of sorption kinetics is very important, especially for industrial or environmental applications. Kinetics may be a critical factor, limiting the sorption process in a column operation, even when equilibrium K_d values are very high.

Kinetics may be modelled by considering the processes controlling the whole sorption mechanism. Such processes include diffusion in the superficial liquid layer, adsorption/desorption, diffusion in the solid, ion exchange, formation of a new superficial solid phase, diffusion in a newly formed superficial phase, etc. For modelling, it is convenient to consider an achievement factor F equal to the ratio of the sorbed quantity at time t to the sorbed quantity at infinite time:

$$F = Q_t/Q_\infty \tag{24}$$

It is possible to develop a numerical model for each of the above processes. Fitting of experimental data by these models can help identify the most probable limiting process, but with the necessity of validating this process by other techniques.

As examples, consider the forms of the equations corresponding to the formation of a new superficial solid phase (equation (25)) and diffusion in a newly formed superficial phase (equation (26)), respectively (Loos-Neskovic and Fedoroff, 1989):

$$t = \frac{r_0}{K_1}[1-(1-F)^{1/3}] \tag{25}$$

$$t = \frac{r_0^2}{DK_2}\left(\frac{1}{6}+\frac{1}{3}(1-F)-\frac{1}{2}(1-F)^{2/3}\right) \tag{26}$$

where t is the time of contact between the solid and the liquid phases, r_0 is the initial radius of solid particles, assumed to be spherical, K_1 and K_2 are constants associated with the experimental conditions, D is the diffusion coefficient in the new phase and F is the achievement factor.

Models have been developed to include the kinetic effect in transport (Jaulmes and Vidal-Madjar, 1991).

Slow kinetics may be a critical factor for column operation. Its impact may be negligible for reactive barriers, in which the liquid flow rate is very slow.

7.7. Sorption mechanisms

As shown above, sorption is generally quantified either by empirical models or by models based on assumed equilibrium mechanisms. We have already pointed out that the experimental data of a single system can be fitted by several models. The question is: do the models represent the real sorption mechanisms? If not, what is the validity of such models? As verified in some examples, the quantitative data from modelling is generally valid if applied within the very strict range of experimental conditions which were used for the establishment of these models. The difficulty is knowing how to extrapolate to other conditions, especially those encountered when enlarging the scale of an application either in space or time, or occurring in uncontrolled natural systems? For radioactive waste repositories, our predictions must be valid for thousands of years.

It is necessary to determine as accurately as possible real sorption mechanisms in order to validate models or develop new ones. To pursue these goals we must use methods that are able to localize the sorbed elements, and determine the nature of their species and types of bonding. The determination of sorption mechanisms is not an easy task: the sorbed elements are very often at trace level concentrations, and systems are heterogeneous, often with several types of sites and species. Solutions can be achieved through a multidisciplinary approach, with several different methods providing complementary results. From the amalgamation of these results, it is generally possible to deduce the sorption mechanisms. Box 7.1 provides an overview of this strategy.

The sorption of several elements on to hydroxyapatites has been studied as an example. These were either toxic elements, or models for radioactive elements. Apatites (stochiometric calcium hydroxyapatite, $Ca_{10}(PO_4)_6(OH)_2$) are the inorganic constituents of bones and teeth. Due to the numerous possible substitutions in the crystalline framework of apatites, these compounds have been proposed for use in

Box 7.1. General strategy for establishing sorption mechanisms

---

- Complete characterization of solid and liquid phases
- Experimental sorption data, with complete balance of sorbed elements and dissolved species, including sorption kinetics and reversibility/irreversibility
- Microscopic methods (e.g. electron microscopy, atomic force microscopy and microprobes)
- Surface spectroscopy (X-ray photoelectron, Auger and ion-scattering techniques)
- Bulk spectroscopy (infrared, Raman, luminescence, X-ray and X-ray absorption fine-structure techniques)
- Equilibrium models and kinetic models

Table 7.1. Main sorption mechanisms on hydroxyapatites

Ion	Sorption mechanism
Cd^{2+}	Exchange with Ca^{2+} in the crystal framework of apatite in two sites (Ca^{II} and Ca^{I}). Equilibrium not achieved within 10 days at 75°C. Sorption controlled by diffusion
	Partial reversibility
Pb^{2+}	Formation of new solid phases plus incorporation in the apatite framework
UO_2^{2+}	Formation of a new amorphous solid phase
Sr^{2+}	Exchange with Ca^{2+}
	Less diffusion than for Cd^{2+}
Eu^{3+}	Superficial sorption (site A) plus incorporation in the apatite framework (site B)
SeO_3^{2-}	Exchange with superficial PO_4^{3-} groups of the apatite framework

reactive barriers. Table 7.1 shows the main sorption mechanisms which can be determined through a multidisciplinary approach (Fedoroff *et al.*, 1999).

The main feature of apatites is that sorption mechanisms depend strongly on the element: sorption takes place through exchange with elements in the crystal framework, with diffusion or formation of new solid phases. We have observed that equilibrium is often not achieved. Thus, as in many cases, the models described above cannot be used. In the absence of good mechanistic models, empirical solutions have to be turned to. For the decontamination of cadmium in water, columns can be used, but their practical capacity is limited by sorption kinetics. However, apatites do seem to be well suited to reactive barriers. Decontamination of lead is not feasible in columns, since the formation of new phases modifies particle size and distribution, which in turn clogs the columns and causes a reduction in the flow rate.

7.8. Influence of other factors

An important factor which is generally not evaluated during sorption experiments is the transformation of the solid during the sorption process, which may require a long period of time, whether in natural systems or in reactive barriers. It is well known from geological processes that mineral species are subject to alteration, dissolution and recrystallization. Although these processes are known, their quantitative impact on the long-term efficiency of artificial or natural sorption barriers is difficult to estimate, since such data cannot be easily collected experimentally. However, an exchange of knowledge between chemists and geologists may enable the impact of such long-term processes on sorption barriers to be deduced, using data on the sorption on mineral species as well as alteration of minerals and formation of new phases (natural analogues).

An example of this effect is the decrease in density of sorption sites for γ–alumina as a function of time, when this solid is kept in water (Fig. 7.3). In fact, investigation

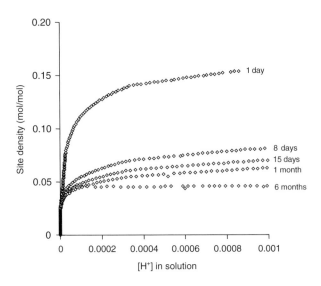

Fig. 7.3. Variation of the density of sorption sites on γ-alumina as a function of (H⁺) activity in water, for several different durations of treatment in water

of the solid by X-ray diffraction, scanning electron microscopy, thermogravimetric analysis and differential thermal analysis shows that γ-alumina is progressively transformed into bayerite (Duc *et al.*, 2002). Crystallites of bayerite can clearly be seen by scanning electron microscopy on the surface of alumina. These bayerite crystallites will modify the sorption capability of the solid.

7.9. Conclusion

This short chapter has not described all aspects of sorption processes, which are fairly complex. Its main goal has been to guide future users of sorption data in the prediction of transport of elements in water and in the selection of improved decontamination systems. It is evident that the scale and location of contaminated media is one of the selection criteria between pump-and-treat and reactive barrier approaches. Another criterion guiding the selection of the sorbent is the nature of the pollutants and the contaminated medium. The next step in selection is knowledge of the sorption mechanism, including sorption kinetics, which can, for example, exclude the use of a fast flow rate in a column. Finally, quantitative models and computer programs can be used to guide the technical development of the decontamination method and its evolution in the future.

7.10. References

BARRER, R. M. and FALCONER, J. D. (1956). Ion exchange in felspathoids as a solid-state reaction. *Proceedings of the Royal Society, Series A* **236**, 227.

DUBININ, M. M. and RADUSHKEVICH, L. V. (1947). Equation of the characteristic curve of activated charcoal. *Proceedings of the Academy of Science, Physical Chemistry USSR* **55**, 331.

DUC, M., LEFÈVRE, G., LEPEUT, P., CAPLAIN, R. and FÉDOROFF, M. (2002). Hydration of γ-alumina in water and its effects on surface reactivity (unpublished).

FEDOROFF, M., JEANJEAN, J., ROUCHAUD, J. C., MAZEROLLES, L., TROCELLIER, P. MAIRELES-TORRES, P. and JONES, D. J.(1999). Sorption kinetics and diffusion of cadmium in calcium hydroxyapatites. *Solid State Sciences* **1**, 71.

FREUNDLICH, H. (1926). *Colloid and Capillary Chemistry*. Methuen, London.

GAINES, G. L. and THOMAS, H. C. (1953). Adsorption studies on clay minerals. A formulation of the thermodynamics of exchange adsorption. *Journal of Chemistry and Physics* **21**, 214.

HIEMSTRA, T., VAN RIEMSDJIK, W. H. and BOLT, G. H. (1989). Multisite proton adsorption modelling at the solid-solution interface of (hydr)oxides: a new approach. *Journal of Colloid Interface Science* **133**, 91.

HOLSTETLER, D. D., SERNE, R. J., BALDWIN, A. J. and PETRIE G. M. (1980). *Status Report on Sorption Information Retrieval Systems*. Pacific Northwest Laboratory, Richland, Washington.

JAULMES, A. and VIDAL-MADJAR, C. (1991). Split-peak phenomenon in nonlinear chromatography. *Analytical Chemistry* **63**, 1165.

JAUZEIN, M., ANDRÉ, C., MARGRITA, R., SARDIN, M. and SCHWEICH, D. (1989). A flexible computer code for modelling transport in porous media: Impact. *Geoderma*, **44**, 95.

LANGMUIR, D. (1918). The adsorption of gases on planar surfaces of glass, mica and platinum. *Journal of the American Chemical Society* **40**, 1361.

LOOS-NESKOVIC, C. and FEDOROFF, M. (1989). Fixation mechanisms of cesium on nickel and zinc ferrocyanides. *Solvent Extraction and Ion Exchange* **7**, 131.

MECHERRI, O. M., BUDIMAN-SASTROWARDOYO, P., ROUCHAUD, J. C. and FEDOROFF, M. (1990). Study of neodymium sorption on orthose and calcite for radionuclide migration modelling in groundwater. *Radiochimica Acta* **50**, 169.

OCHS, M., BOONEKAMP, M., WANNER, H., SATO, H. and YUI, M. (1998). A quantitative model for ion diffusion in compacted bentonite. *Radiochimica Acta* **82**, 437.

SERNE, R. J. and MULLER A. B.(1987). A perspective on adsorption of radionuclides onto geological media. In: D. G. Brookins (ed.), *The Geological Disposal of High Level Radioactive Wastes*. Theophrastus, Athens.

STENHOUSE, M. J. and PÖTTINGER J. (1994). Comparison of sorption databases used in assessments involving crystalline host rock. *Radiochimica Acta* **66/67**, 267.

STUMM, W. (1992). *Chemistry of the Solid–Water Interface*. Wiley-Interscience, New York.

WESTALL, J. C., ZACHARY, J. and MOREL, F. M. M. (1986). *Mineql, a Computer Program for the Calculation of the Chemical Equilibrium Composition of Aqueous Systems*. Department of Chemistry, Oregon State University, Corvallis.

ZOUAD, S., JEANJEAN, J., LOOS-NESKOVIC, C. and FEDOROFF, M. (1992). Structural study and thermodynamics of the fixation of strontium on polyantimonic acid. *Journal of Solid State Chemistry* **98**, 1.

8. Experience with monitored natural attenuation at BTEX-contaminated sites

W. Püttmann and P. Martus
*Institute for Mineralogy, Johann Wolfgang Goethe University Frankfurt,
D-60054 Frankfurt am Main, Germany*

R. Schmitt
HYDR.O. Geologists and Engineers, D-52070 Aachen, Germany

8.1. Introduction

Monitored natural attenuation (MNA) is currently used for the remediation of groundwater at numerous benzene, toluene, ethylbenzene and xylene (BTEX) contaminated sites in the USA following detailed technical guidelines developed for its implementation (Wiedemeier *et al.*, 1995; US Environmental Protection Agency, OSWER, 1999). The application of natural attenuation for the remediation of contaminated soil is not considered in these protocols. Nevertheless, natural attenuation is also relied upon for the remediation of small-scale contaminations of the unsaturated zone by petroleum hydrocarbons, particularly in the USA (US National Research Council, 1997).

A proposal for the application of analytical methods for the monitoring of natural attenuation with respect to hydrocarbons in groundwater has been made by Martus and Püttmann (2000) in the context of analytical standards used in Germany. However, in Germany, the acceptability of MNA as a remedial measure is still under discussion, and the federal states do not yet have a common position. For example, in Rhineland-Palatinate the application of MNA for groundwater remediation has not been recommended (Federal State Rhineland-Palatinate (Germany), 2001), whereas in North Rhine-Westphalia the environmental authority accepts its use, particularly in combination with the reduction of the contaminant source by active measures (Odensass, 2001). So far, only a few well-documented case studies are available dealing with the application of MNA for groundwater remediation in Germany.

The process of natural attenuation of petroleum hydrocarbons in contaminated soil and groundwater has been well known for more than 30 years. In the mid-1960s, the Geological Survey of Baden-Württemberg conducted leakage experiments with gas oil in the area of the Upper Rhine valley (Niederterrasse) (Bartz *et al.*, 1969). In 1986, 19 years after starting the leakage experiment, excavations and drillings were carried out in the formerly contaminated area. Hydrocarbons were no longer detectable, either within the area of the formerly contaminated soil, or in the groundwater downstream. The results of this experiment were presented in detail by Käss and Schwille (1993) at the 10th DECHEMA Symposium entitled 'Valuation and Remediation of Soils Contaminated with Petroleum Hydrocarbons'. The authors attributed the limited 'survival time' of petroleum hydrocarbons in soil and ground-water to sorption and biodegradation.

For the treatment of such contamination, Käss and Schwille (1993) suggest first the removal of the remaining oil phase from the subsurface as far as possible. Thereafter, a decision has to be taken whether *in situ* remediation methods are to be used or the contaminated area can lie fallow. In this connection, existing and potential water use downstream from the contaminated area has to be taken into account. In principle, this strategy is similar to the procedure later called natural attenuation. Müller (1952) had recognized earlier that nitrate is consumed in groundwater contaminated by petroleum hydrocarbons, possibly due to microbial degradation of the hydrocarbons. Nevertheless, in the following decades the rapid microbial degradation of hydrocarbons in soils and groundwater was still thought to be possible only under aerobic conditions (Delaune *et al.*, 1980). But in recent years it has become more and more obvious – at least with respect to groundwater – that the potential of anaerobic processes (particularly sulfate reduction, iron reduction and methanogenesis) for biodegrading petroleum hydrocarbons has been under-estimated (Zeyer *et al.*, 1999).

8.2. Analytical monitoring of natural attenuation

Experience with the analytical monitoring of hydrocarbons in soil and ground-water originates from the scientific discipline known as 'forensic environmental geochemistry'. This has been defined by Kaplan *et al.* (1997) to be a 'scientific methodology developed for identifying petroleum-related and other potentially hazardous environmental contaminants and for determining their sources and time of release'. With respect to petroleum hydrocarbons, the methods of this discipline are best used for identification of the type of petroleum or refinery products causing contamination and for the dating of contamination. Insurance companies are par-ticularly interested in such data as evidence in court cases on liability.

The preferential use of n-alkanes as an energy source for bacteria for the dating of gas oil contamination, by measuring the ratio of n-alkane concentrations relative to the concentrations of isoprenoid hydrocarbons (pristane and phytane), which remain largely unattacked by the bacteria. Measurement of the relative intensities of the compounds heptadecane, octadecane, pristane and phytane can be used for the dating of gas oil contamination when the boundary conditions for biodegradation in the subsurface are known. A successful application of this method has been demonstrated by Christensen and Larson (1993). Similarly, Senn and Johnson

(1987) have used gas chromatography analysis and hydrocarbon ratios to determine the age of petroleum contaminations. Since the dating of petroleum contamination is based on the preferential biodegradation of individual hydrocarbons, the method can also be used for MNA. Similar methods based on the relative intensities of individual compounds in gas chromatography analysis are important for the water-soluble aromatic petroleum hydrocarbons.

8.3. Biodegradation of saturated hydrocarbons under aerobic conditions

The biochemical degradation processes of n-alkanes under aerobic conditions have been investigated in detail and are largely understood. First, the n-alkanes are oxidized at the terminal carbon atom, via alcohols and aldehydes, to fatty acids. Under conditions of sufficient oxygen supply, β-oxidation of the fatty acids to β-keto esters occurs, followed by cleavage of an acetyl group by the enzyme acetyl-CoA. The resulting reaction products are fatty acids with a chain length shortened by two carbon atoms. These compounds can again be oxidized in the β-position, and so on, until complete degradation of the carbon chain occurs. However, under conditions of insufficient oxygen the intermediate alkanols and fatty acids generated can react to form esters with structures similar to wax esters but with shorter carbon chain lengths (Gibson, 1984). For example, such esters, dominated by C-14 to C-16 fatty acid and alkanol components, are sometimes enriched in sewage sludge and freshwater sediments.

The biodegradation of branched alkanes under aerobic conditions is less likely than the degradation of n-alkanes. The persistence of the compounds increases with an increasing degree of branching. As a consequence, in the case of biodegradation of soil and groundwater pollution involving highly branched alkanes, the isoprenoid hydrocarbons in particular are enriched compared with the n-alkanes. Finally, the isoprenoid hydrocarbons become the dominant constituents of the residual aliphatics (Püttmann, 1990). Laboratory experiments (without soil) with crude oils and microorganisms isolated from oil fields have shown that n-alkanes are completely degraded after only 3 days. Moreover, after 8 days the concentrations of the isoprenoid hydrocarbons pristane and phytane are significantly reduced (Teschner and Wehner, 1985).

In contrast, extracts from soil samples contaminated with gas oil reveal the isoprenoids pristane and phytane still to be the dominant constituents, even when the contamination of the soil had occurred more than 30 years previously. It is not yet fully understood why in contaminated soil the isoprenoid hydrocarbons (C-15 to C-20) are so persistent whereas in laboratory experiments without a soil matrix the compounds are readily biodegraded. The mineral constituents of soils could possibly act as a physical trap, particularly for isoprenoids, to prevent them from being biodegraded.

Cyclic hydrocarbons are also major constituents of petrochemical products. Under aerobic conditions, cyclic alkanes are not degraded until the n-alkanes are largely depleted. For this reason, cyclic alkanes are enriched compared with n-alkanes during biodegradation. Laboratory experiments have shown that biodegradation of cyclic alkanes is initiated by hydroxylation of the ring system followed by ring cleavage. Aliphatic dicarboxylic acids are thereby generated, and these are further degraded by β-oxidation, similar to the degradation of n-alkanes (Perry, 1984). The aerobic biodegradation of saturated hydrocarbons is discussed in more detail by Kästner *et al.* (1993).

8.4. Biodegradation of saturated hydrocarbons under anaerobic conditions

Micro-organisms can also use saturated hydrocarbons as a substrate in an anaerobic aquifer. The degradation processes are linked to the reduction of electron acceptors such as NO_3^-, Fe^{III} oxides, Mn^{IV} oxides, SO_4^{2-} and CO_2. Under nitrate-reducing conditions, saturated hydrocarbons are readily biodegraded in groundwater (Holliger and Zehnder, 1996).

Sulfate-reducing bacteria had previously been thought not to be able to feed on saturated hydrocarbons. Jobson *et al.* (1979) first conducted a laboratory experiment with crude oil from the North Cantal deposit and reported that sulfate-reducing bacteria were not able to degrade saturated hydrocarbons. However, when the crude oil was pretreated with aerobic bacteria, the sulfate-reducing bacteria were able to feed on the reaction products. Saturated hydrocarbons were the dominant constituents of the crude oil. As a consequence of this experiment, the authors suggested that although sulfate-reducing bacteria are not able to utilize saturated hydrocarbons they are able to feed on the residue (mostly aromatic hydrocarbons) after removal of the saturated constituents (Jobson *et al.*, 1979). More recently, investigations with pure cultures of sulfate-reducing bacteria isolated from petroleum production plants have shown that the bacteria were able to degrade n-alkanes with a chain length from C-12 to C-20 to carbon dioxide (Aeckersberg *et al.*, 1991).

Additionally, thermophilic sulfate-reducing bacteria isolated from sediments of the Guaymas Basin were able to mineralize preferentially the short-chain n-alkanes in samples of crude oil (Rueter *et al.*, 1994). In the case of a gas oil contamination found as a liquid phase on top of groundwater in a formerly used industrial area, some locations displayed an almost complete degradation of the n-alkanes accompanied by almost complete removal of sulfate from the water phase below the oil. A concomitant degradation of aromatic hydrocarbons was not observed in these samples. The observations were interpreted to be a result of saturated hydrocarbon degradation by sulfate-reducing bacteria (Gossel and Püttmann, 1991). Based on these findings there can now be no doubt that sulfate-reducing bacteria are able to mineralize n-alkanes, at least with chain lengths of up to C-20.

With respect to sediments, saturated hydrocarbons were expected to be largely inert, so long as oxygen, nitrate and sulfate are not available. The use of n-alkane distribution patterns for the reconstruction of biogeochemical processes, and for the determination of organic matter input into sediments, is based on this assumption. However, recently it has been reported that enriched cultures of anaerobic bacteria in sulfate-free media under strictly anoxic conditions are able to degrade hexadecane to methane and carbon dioxide (Zengler *et al.*, 1999). This observation suggests that methanogenesis is also possible with n-alkanes. Such processes cannot be expected to be of relevance for hydrocarbons in non-polluted soils and ground-water, due to the presence of oxygen, nitrate or sulfate. However, at severely contaminated sites these electron acceptors can be completely consumed at the centre of the contaminant source. In those circumstances, methanogenesis at the expense of n-alkanes might be possible. To our knowledge, the biodegradation of saturated aliphatic hydrocarbons under iron- and manganese-reducing conditions has not been reported to date.

8.5. Biodegradation of aromatic hydrocarbons under aerobic conditions

For a long time it had been thought that rapid biodegradation of aromatic hydrocarbons could only be expected under unaerobic conditions, and that under aerobic conditions the processes are rather slow (Kästner *et al.* 1993 (and references cited therein)). The degradation pathways have now been determined through laboratory experiments for many aromatic hydrocarbons (Cerniglia, 1984). For monoaromatic hydrocarbons, complete mineralization can be expected, whereas polycyclic aromatic hydrocarbons tend to be transformed into metabolites which are incorporated into humic substances (Sims *et al.*, 1990).

8.6. Biodegradation of aromatic hydrocarbons under anaerobic conditions

As with aliphatic hydrocarbons, aromatic hydrocarbons can also be utilized by micro-organisms in an anaerobic aquifer. The degradation processes are again linked to the reduction of electron acceptors, such as NO_3^-, Fe^{III}-oxides, Mn^{IV}-oxides, SO_4^{2-} and CO_2 (Chapelle, 1993). The processes have been investigated in more detail for aromatic hydrocarbons than for saturated hydrocarbons, because aromatic hydrocarbons are comparatively more water-soluble and are therefore present in groundwater in much higher concentrations at most contaminated sites. Additionally, aromatic hydrocarbons (particularly benzene) are more critical with respect to toxicological risks than saturated hydrocarbons.

Inconsistent results have been reported for the biodegradation of aromatic hydrocarbons under nitrate-reducing conditions. Based on laboratory experiments, Major *et al.* (1988) and Morgan *et al.* (1993) suggested that the biodegradation of benzene in a petroleum-contaminated aquifer could be supported by the addition of nitrate. In contrast, results from experiments carried out by Hutchins *et al.* (1991) and Lovley (1997) suggest that nitrate alone cannot stimulate the anaerobic degradation of benzene. However, there is no disagreement about the capability of nitrate-reducing bacteria to readily degrade most of the alkylated benzenes with the exception of *o*-xylene and ethylbenzene (Hutchins *et al.*, 1991). Also, contradictory results have been published from field experiments relating to the *in situ* remediation of groundwater contaminated with petroleum, by the addition of nitrate to the aquifer.

From such a case study of *in situ* groundwater remediation, Battermann (1986) reports the almost complete removal of benzene and toluene, associated with only minor degradation of xylenes. In a similar experiment by Lemon *et al.* (1989), toluene was completely eliminated, while benzene, ethylbenzene and the xylenes largely resisted biodegradation. A possible explanation for this discrepancy is given by Hutchins *et al.* (1991): for the biodegradation of benzene, oxygen has to be present in the aquifer in addition to nitrate. This explanation implies that bio-degradation of benzene by nitrate-reducing bacteria is not possible under strictly anaerobic conditions. The published results lead one to conclude that, for *in situ* remediation of BTEX-contaminated groundwater, the addition of nitrate to the aquifer cannot be successful when anaerobic conditions prevail.

The biodegradation of aromatic hydrocarbons by sulfate-reducing bacteria has been studied intensively. Abundant references to this topic have been collected by Wiedemeier *et al.* (1999) and Wilkes et al. (2000). For a long time the biodegradation of aromatic hydrocarbons under anaerobic conditions was thought to be slow compared with aerobic processes. However, Lovley *et al.* (1995) have shown, from laboratory experiments with sediments from San Diego Bay in California, that benzene was degraded within 55 days, largely to CO_2 under strictly anaerobic conditions in the presence of sulfate. In the experiment, sulfate reduction was proven to be the dominant process, by inhibition of the benzene degradation through the addition of molybdate. Furthermore, the benzene degradation declined significantly when the sulfate was used up, and increased again after the addition of more sulfate. Based on these studies, Weiner *et al.* (1998) carried out *in situ* experiments for enhanced benzene degradation in contaminated groundwater by the addition of sulfate to the aquifer. It was concluded that sulfate might be more effective than oxygen for the *in situ* enhancement of such biodegradation.

However, it is not yet clear why, in some contaminated aquifer materials, the addition of sulfate could stimulate the degradation of benzene only after previous inoculation of the sediments with sulfate-reducing bacteria isolated from aquatic sediments (Weiner and Lovley, 1998). The potential for the enhancement of the natural attenuation of aromatic hydrocarbons by the addition of sulfate to a contaminated aquifer should not be overestimated for the following reason. Upon sulfate reduction, H_2S is generated, which is withdrawn from the system by the formation of iron monosulfides, so long as sufficient reduced iron is available. However, H_2S will enrich in the groundwater when insufficient reduced iron is available. In those circumstances the efficiency of sulfate reduction is not dependent on the supply of sulfate, but upon the elimination of H_2S. Additionally, in most urban centres of Germany, abundant sulfate is supplied to the groundwater by leaching from the building rubble used as filling material in the past.

The importance of iron-reducing bacteria for the biodegradation of aromatic hydrocarbons had been underestimated until Lovley *et al.* (1989) reported that micro-organisms are able to oxidize aromatic compounds using Fe^{III} as an electron acceptor. In the case of an oil spill in Bemidji, Minnesota, it was shown that iron reduction combined with methanogenesis was the dominant process for the degrad-ation of BTEX aromatics (Baedecker *et al.*, 1993; Bennet *et al.*, 1993; Eganhouse *et al.*, 1993). At this location nitrate and sulfate were present in only low amounts in the groundwater upstream of the contamination, and could therefore provide only minor contributions to the biodegradation. Fe^{III} species in an aquifer can generally be utilized by micro-organisms as electron acceptors since they are not soluble in water. In laboratory experiments using sediment material from an aquifer contaminated with petroleum hydrocarbons, Lovley *et al.* (1994) demonstrated that organic ligands such as nitrilotriacetic acid (NTA) are capable of formation of water-soluble Fe^{III} complexes, which can then be utilized for the biodegradation of benzene and toluene combined with iron reduction.

Generally, the reduction of manganese occurs parallel to the reduction of iron but is of minor importance due to the much lower availability of manganese compared with iron in most aquifers. The importance of iron reduction for aromatic hydrocarbon degradation at contaminated sites is often underestimated, since the reaction product Fe^{II} is generally removed from the water phase by precipitation as iron monosulfides through the H_2S originating from the concomitant sulfate reduction. Sulfate reduction can be easily quantified, by measuring the decline of the sulfate concentration in the groundwater. By contrast, iron reduction is difficult to quantify, since the educt Fe^{III} is water-soluble only as an intermediate complexed species, and the reaction product is largely removed from the water by precipitation as sulfides. The quantification of aromatic hydrocarbon degradation by iron reduction would require the measurement of the total iron sulfides precipitated in the contaminated zone of the aquifer, which is *de facto* not possible.

The complete degradation of BTEX aromatics under methanogenic conditions has been reported by Wilson *et al.* (1986). Additionally, Grbic-Galic (1990) was able to show that condensed aromatic hydrocarbons, such as naphthalene, indene and acenaphthene, can be degraded by methanogenic bacteria to CO_2 and CH_4 via several intermediate products. In the case of the crude oil-contaminated aquifer at Bemidji, Minnesota, methanogenesis was accompanied by iron reduction (Baedecker *et al.*, 1993). Indeed, the degradation pathways of the aromatic hydrocarbons under methanogenic conditions have been studied intensively. Grbic-Galic and Vogel (1987) showed, by ^{14}C labelling of toluene, that the methyl group is largely degraded to CO_2, whereas the dominant part of the aromatic ring carbon contributes to CH_4 formation. Investigations into the formation of metabolites during biodegradation of aromatic hydrocarbons, under methanogenic conditions, have indicated that phenol is a metabolite of benzene whereas benzoic acid is a metabolite of toluene. Moreover, alkylated aromatic acids have been detected as metabolites from the corresponding higher alkylated benzenes (Grbic-Galic and Vogel, 1987; Cozzarelli *et al.*, 1990).

Box 8.1 shows the chemical processes taking place during biodegradation of hydrocarbons, under aerobic and anaerobic conditions, using benzene as an example. Additionally, the mass ratios of benzene and the respective electron acceptors, for the complete degradation to CO_2, are shown.

The effectiveness of these processes has been documented in many case studies. One such study was conducted at the former gasworks in Düsseldorf-Flingern, Germany, which was shut down in 1968. Initial investigations carried out in 1982

Box 8.1. Redox reactions occurring in an aquifer

	Mass ratio
Aerobic	
$C_6H_6 + 7.5O_2 \rightarrow 6CO_2 + 3H_2O$	1:3.08
Nitrate reduction	
$C_6H_6 + 6NO_3^- + 6H^+ \rightarrow 6CO_2 + 6H_2O + 3N_2$	1:4.77
Iron reduction	
$C_6H_6 + 30Fe(OH)_3 + 60H^+ \rightarrow 6CO_2 + 30Fe^{2+} + 78H_2O$	1:41.1
Sulfate reduction	
$C_6H_6 + 3.75SO_4^{2-} + 7.5H^+ \rightarrow 6CO_2 + 3.75H_2S + 3H_2O$	1:4.6
Methanogenesis	
$C_6H_6 + 4.5H_2O \rightarrow 2.25CO_2 + 3.75CH_4$	1:0.77

revealed severe pollution of soil and groundwater, particularly with BTEX aromatics around the former benzene-washing facility. The aquifer at the study site consists of late Pleistocene fluvial deposits, with medium- to coarse-grained sand and gravel. Since 1994 more than 20 groundwater observation wells, located downstream from the source of contamination, have been available for groundwater monitoring. Chemical analyses showed that the contaminant plume was largely stationary. Sulfate and iron reduction were revealed to be the dominant processes of aromatic hydrocarbon degradation.

After the removal of most of the contaminant source, a combination of on-site and *in situ* microbiological remediation was carried out by pumping contaminated groundwater. The water was treated in bioreactors, and was re-infiltrated upstream of the former source. The hydrochemical composition of the groundwater between infiltration and pumping wells was investigated during the remediation activities. The results showed that the anaerobic oxidation of aromatic hydrocarbons declined, due to the infiltration of oxygen into the aquifer as a result of the remediation activities. The oxygen was used to preferentially oxidize iron monosulfides, which were precipitated in the aquifer during hydrocarbon degradation by combined sulfate and iron reduction. During the active remediation phase the stability of the contaminant plume had been temporarily disturbed, and concentrations of aromatic hydrocarbons increased in the groundwater downstream from the source area. Further discussion of this case study can be found in the literature (Schmitt *et al.*, 1996, 1998).

8.7. References

AECKERSBERG, F., BAK, F. and WIDDEL, F. (1991). Anaerobic oxidation of saturated hydrocarbons to CO_2 by a new type of sulphate-reducing bacterium. *Archives of Microbiology* **156**, 5–14.

FEDERAL STATE RHINELAND-PALATINATE (GERMANY). (2001). Altablagerungen, Altstandorte und Grundwasserschäden. Informationsblatt 18 des Landes Rheinland-Pfalz. Website: www.gpr.de/News/Gesetze/Gesetze.htm.

BAEDECKER, M. J., COZZARELLI, I. M., EGANHOUSE, R. P., SIEGEL, D. I. and BENNET, P. C. (1993). Crude oil in a shallow sand and gravel aquifer – III. Biogeochemical reactions and mass balance modelling in anoxic groundwater. *Applied Geochemistry* **8**, 569–586.

BARTZ, J., KÄSS, W. and PRIER, H. (1969). Öl- und Benzinversickerungsversuche in der Oberrhein-ebene. – 3. Bericht: Versickerung von Heizöl EL und Rohöl über dem Grundwasser. *GWF- (Wasser/Abwasser)* **110**, 592–595.

BATTERMANN, G. (1986). Decontamination of polluted aquifers by biodegradation. In: J. W. Assink and W. J. Van den Brink (eds), *Proceedings of the 1985 International TNO Conference on Contaminated Soil, Dordrecht*, pp. 711–722.

BENNET, P. C., SIEGEL, D. E., BAEDECKER, M. J. and HULT, M. F. (1993). Crude oil in a shallow sand and gravel aquifer – I. Hydrogeology and inorganic geochemistry. *Applied Geochemistry* **8**, 529–549.

CERNIGLIA, C. E. (1984). Microbial transformation of aromatic hydrocarbons. In: A. M. Atlas (ed.) *Petroleum Microbiology*, pp. 99–128. Macmillan, London.

CHAPELLE, F. E. (1993). *Ground Water Microbiology and Geochemistry*. Wiley, New York.

CHRISTENSEN, L. B. and LARSEN, T. H. (1993). Methods for determining the age of diesel oil spills in the soil. *Ground Water Monitoring Review* **13**, 142–149.

COZZARELLI, I. M., EGANHOUSE, R. P. and BAEDECKER, M. J. (1990). Transformation of monoaromatic hydrocarbons to organic acids in anoxic groundwater environments. *Environmental Geology and Water Science* **16**, 135–141.

DELAUNE, R. D., HAMBRICK, G. A. and PATRICK, W. H., Jr (1980). Degradation of hydrocarbons in oxidized and reduced sediments. *Marine Pollution Bulletin* **11**, 103–106.

EGANHOUSE, R. P., BAEDECKER, M. J., COZZARELLI, I. M., AIKEN, G. R., THORN, K. A. and DORSEY, T. (1993). Crude oil in a shallow sand and gravel aquifer – II. Organic geochemistry. *Applied Geochemistry* **8**, 551–567.

GIBSON, D. T. (1984). *Microbial Degradation of Organic Compounds*. Dekker, New York.

GOSSEL, W. and PÜTTMANN, W. (1991). Untersuchungen zum mikrobiologischen Abbau von Kohlenwasserstoffen im Grundwasserkontaktbereich. *GWF-Wasser/Abwasser* **132**, 126–131.

GRBIC-GALIC, D. (1990). Methanogenic transformation of aromatic hydrocarbons and phenols in groundwater aquifers. *Geomicrobiology Journal* **8**, 167–200.

GRBIC-GALIC, D. and VOGEL, T. M. (1987). Transformation of toluene and benzene by mixed methanogenic cultures. *Applied Environmental Microbiology* **53**, 254–260.

HOLLIGER, C. and ZEHNDER, A. J. B. (1996). Anaerobic biodegradation of hydrocarbons. *Current Opinions in Biotechnology* **7**, 326–330.

HUTCHINS, S. R., DOWNS, W. C., WILSON, J. T., SMITH, G. B., KOVACS, D. A., FINE, D. D., DOUGLAS, R. H. and HENDRIX, D. J. (1991). Effect of nitrate addition on biorestoration of fuel-contaminated aquifer: field demonstration. *Ground Water* **4**, 571–580.

JOBSON, A. M., COOK, F. D. and WESTLAKE, D. W. S. (1979). Interaction of aerobic and anaerobic bacteria in petroleum biodegradation. *Chemistry and Geology* **24**, 355–365.

KAPLAN, I. R., GALPERIN, Y., SHAN-TAN LU and RU-PO LEE (1997). Forensic environmental geochemistry: differentiation of fuel-types, their sources and release time. *Organic Geochemistry* **27**, 289–317.

KÄSS, W. and SCHWILLE, F. (1993). Die 'Lebensdauer' von Mineralöl-Kontaminationen in porösen Medien. In: G. Kreysa and J. Wiesner (eds), *Bewertung und Sanierung mineralöl-kontaminierter Böden*, pp. 533–553. DECHEMA, Frankfurt am Main.

KÄSTNER, M., MAHRO, B. and WIENBERG, R. (1993). Biologischer Schadstoffabbau in kontamin-ierten Böden. In: R. Stegmann(ed.), *Hamburger Berichte*, Band 5, pp. 1–180. Economia, Bonn.

LEMON, L. A., BARBARO, J. P. and BARKER, J. F. (1989). Biotransformation of BTEX under anaerobic denitrifying conditions: evaluation of field observations. *Proceedings of FOCUS – Conference on Eastern Regional Ground Water Issues*, pp. 213–227. NWWA, Dublin.

LOVLEY, D. R. (1997). Potential for anaerobic remediation of BTEX in petroleum-contaminated aquifers. *Journal of Industrial Microbiology* **61**, 953–958.

LOVLEY, D. R., BAEDECKER, M. J., LONERGAN, D. J., COZZARELLI, I. M., PHILLIPS, E. J. P. and SIEGEL, D. I. (1989). Oxidation of aromatic contaminants coupled to microbial iron reduction. *Nature* **339**, 297–300.

LOVLEY, D. R., WOODWARD, J. C. and CHAPELLE, F. H. (1994). Stimulated anoxic biodegradation of aromatic hydrocarbons using Fe(III) ligands. *Nature* **370**, 128–131.

LOVLEY, D. R., COATES, J. D., WOODWARD, J. C. and PHILLIPS, E. J. (1995). Benzene oxidation coupled to sulphate reduction. *Applied Environmental Microbiology* **61**, 953–958.

MAJOR, D. W., MAYFIELD, C. I. and BARKER, J. F. (1988). Biotransformation of benzene by denitrification in aquifer sand. *Ground Water* **26**, 8–14.

MORGAN, P., LEWIS, S. T. and WATKINSON, R. J. (1993). Biodegradation of benzene, toluene, ethylbenzene, and xylenes in gas-condensate-contaminated ground-water. *Environmental Pollution* **82**, 181–190.

MARTUS, P. and PÜTTMANN, W. (2000). Anforderungen bei der Anwendung von Natural Attenuation zur Sanierung von Grundwasserschadensfällen. *Altlasten Spektrum* **2**, 87–105.

MÜLLER, J. (1952). Bedeutsame Feststellungen bei Grundwasserverunreinigungen durch Benzin. *GWF-(Wasser/Abwasser)* **93**, 205–209.

ODENSASS, M. (2001). Beurteilung von 'Natural Attenuation' – Prozesse im Grundwasser. Website: http://www.lua.nrw.de/altlast/odna.pdf.

PERRY, J. J. (1984). Microbial transformation of cyclic alkanes. In: R. M. Atlas (ed.), *Petroleum Microbiology*, pp. 61–98. Macmillan, New York.

PÜTTMANN, W. (1990). Kriterien zur Beurteilung von Sanierungsverfahren auf mikrobiologischer Basis. *Handbuch Bodenschutz*, No. 6440, BoS 5, Lfg. V/90, pp. 1–25. Eric Schmidt, Berlin.

RUETER, P., RABUS, R., WILKES, H., AECKERSBERG, F., RAINEY, F. A., JANNASCH, H. W. and WIDDEL, F. (1994). Anaerobic oxidation of hydrocarbons in crude oil by new types of sulphate-reducing bacteria. *Nature* **372**, 455–458.

SCHMITT, R., LANGGUTH, H.-R., PÜTTMANN, W., ROHNS, H. P., ECKERT, P. and SCHUBERT, J. (1996). Biodegradation of aromatic hydrocarbons under anoxic conditions in a shallow sand and gravel aquifer of the Lower Rhine Valley, Germany. *Organic Geochemistry* **25**, 41–50.

SCHMITT, R., LANGGUTH, H.-R. and PÜTTMANN, W. (1998). Abbau aromatischer Kohlenwasserstoffe und Metabolitenbildung im Grundwasserleiter eines ehemaligen Gaswerk- standorts. *Grundwasser* **2**, 78–86

SENN, R. B. and JOHNSON, M. S. (1987). Interpretation of gas chromatographic data in subsurface hydrocarbon investigations. *Ground Water Monitoring Review* **7**, 58–63.

SIMS, J. L., SIMS, R. C. and MATTHEWS, J. E. (1990). Approach to bioremediation of contaminated soil. *Hazardous Waste and Hazardous Materials* **7**(2), 117–149.

TESCHNER, M. and WEHNER, H. (1985). Chromatographic investigations on biodegraded crude oils. *Chromatographia* **20**, 407–416.

US ENVIRONMENTAL PROTECTION AGENCY, OFFICE OF SOLID WASTE AND EMERGENCY RESPONSE (1999). *Use of Monitored Natural Attenuation at Superfund, RCRA Corrective Action and Underground Storage Tank Sites* (*OSWER Directive 9200.4–17P*), 32 S. US Environmental Agency, Washington DC. Website: http://www.epa.gov/ Swerust1/directive/9200_417.html.

US NATIONAL RESEARCH COUNCIL (1997). *Innovations in Groundwater and Soil Cleanup. From Concept to Commercialization*, XVIII, 292 S. National Academy Press, Washington DC.

WEINER, J. and LOVLEY, D. R. (1998). Anaerobic benzene degradation in petroleum-contaminated aquifer sediments after inoculation with benzene-oxidizing enrichment. *Applied Environmental Microbiology* **64**, 775–778.

WEINER, J. W., LAUCK, T. S. and LOVLEY, D. R. (1998). Enhanced anaerobic benzene degradation with the addition of sulphate. *Bioremediation Journal* **2**, 159–173.

WIEDEMEIER, T. H., WILSON, J. T., KAMPBELL, D. H., MILLER, R. N. and HANSEN, J. E. (1995). *Technical Protocol for Implementing Intrinsic Remediation with Long-term Monitoring for Natural Attenuation of Fuel Contamination Dissolved in Groundwater*, –2 Bde. Air Force Center for Environmental Excellence, Technology Transfer Division, Brooks Air Force Base, San Antonio, Texas.

WIEDEMEIER, T. H., RIFAI, H. S., NEWELL, C. J. and WILSON, J. T. (1999). *Natural Attenuation of Fuels and Chlorinated Solvents in the Subsurface*. Wiley, New York.

WILKES, H., BOREHAM, C., HARMS, G., ZENGLER, K. and RABUS, R. (2000). Anaerobic degradation and carbon isotopic fractioning of alkylbenzenes in crude oil by sulphate-reducing bacteria. *Organic Geochemistry* **31**, 101–115.

WILSON, B. H., SMITH, G. B. and REES, J. F. (1986). Biotransformation of selected alkylbenzenes and halogenated aliphatic hydrocarbons in methanogenic aquifer material: a microcosm study. *Environmental Science and Technology* **20**, 997–1002.

ZENGLER, K., RICHNOW, H. H., ROSSELLO-MORA, R., MICHAELIS, W. and WIDDEL, F. (1999). Methane formation from long-chain alkanes by anaerobic microorganisms. *Nature* **401**, 266–269.

ZEYER, J., BOLLINGER, C., BERNASCONI, S. and SCHROTH, M. (1999). Intrinsische Bio-remediation von Mineralölkontaminationen im Grundwasser: Fallstudie Studen. In: S. Heiden, R. Erb, J. Warrelmann and R. Dierstein (eds), *Biotechnologie im Umweltschutz. Initiativen zum Umweltschutz*, Vol. 12, pp. 132–138. Schmidt, Berlin.

9. Heavy metal speciation and phytoextraction

D. Leštan
University of Ljubljana, Biotechnical Faculty, Jamnikarjeva 101, SLO-1000 Ljubljana, Slovenia

9.1. Introduction

An unfortunate by-product of industrialization, urbanization and modern agronomic practices has been the contamination of soil with toxic heavy metals (HMs). Pollution of land in Europe from HMs is widespread, and although the area affected is not accurately determined, it is estimated at several million hectares (Flathman and Lanza, 1998). The primary sources of this pollution are burning of fossil fuels, smelting activities, fertilizers, pesticides and sewage sludge, road traffic (leaded gasoline), chemical and other industrial activities (batteries, paints, galvanizing, etc.). Furthermore HM-contaminated sludge was used in agricultural practice until the late 1970s. Toxic HM contamination of agricultural soils poses a major environmental and health problem, which is still in need of a cost-effective and environmentally safe technological solution. The undertaking of agricultural activities on HM-contaminated soil raises important questions regarding sustainable agricultural practices, soil quality assessment and food safety. Furthermore, because of stricter environmental laws limiting food production on contaminated lands, the availability of arable land for cultivation will decrease, with social and economic consequences for rural communities.

HM bioavailability, toxicity and transport in the environment depends on the chemical form of the metals. Total analysis may give information concerning possible enrichment of soil with HMs, but generally and for most elements it is insufficient for estimating their ecological effect. HMs are associated with a number of soil components and are retained as exchangeable species on clay minerals, bound to iron and manganese oxides, adsorbed by organic matter or bound to carbonates. The nature of these associations is commonly referred to as speciation. The types of reaction that are likely to control the speciation of metals are: (1) adsorption and desorption, (2) precipitation, (3) surface complex formation, (4) ion exchange, (5) penetration into the crystal structure of minerals and (6) biological mobilization and immobilization (Levy *et al.*, 1992). These reactions are presumably determined and constrained by pedological soil characteristics: soil texture, content of organic matter, content and type of clay minerals and aluminium, iron and manganese oxides, plus prevailing physicochemical conditions in the soil: soil saturation, soil aeration, pH and redox potential. It is the chemical form of the metal species in the soil that determines the mobility of metals, bioavailability of metals to soil biota and plants, and the amount of metals that can be absorbed by plants during phytoextraction.

Sequential extraction of HMs and other trace metals in soils is used to identify the speciation of heavy metals retained in the soil. This technique consists of submitting the soil to the successive action of a series of reagents with different chemical properties. Being capable of either creating weak acidic conditions, changing the redox potential of the solution or oxidizing the complexing agents, these reagents extract a fraction of HMs linked to a specific form. Several sequential extraction schemes have been designed for the determination of the forms of metals in soil.

To preserve the availability of HM-contaminated arable land for cultivation, suitable remediation technologies need to be developed. However, remediation of large areas of low to medium level HM-contaminated agricultural land is a particularly difficult problem to solve. The use of conventional technologies developed for small areas of heavily contaminated sites is not feasible economically. For example, the costs of the remediation of 1 ha of HM-contaminated soil with soil washing or soil solidification–stabilization or landfill disposal of contaminated soil are higher than 720 000 euros. Recently, HM phytoextraction has emerged as a promising, cost-effective alternative to conventional engineering-based remediation methods (Salt *et al.*, 1995). Secondary benefits include wide public acceptance due to its aesthetic value, while the cultivation of plants preserves and enhances all biological components of a polluted medium. The goal of successful phytoextraction is to reduce HM levels in contaminated soil to acceptable levels within a reasonable time frame. To achieve this, plants must accumulate high levels of HMs and produce high amounts of biomass. Early phytoextraction research focused on hyperaccumulating plants, which have the ability to concentrate high amounts of HMs in their plant tissues. However, hyperaccumulators often accumulate only a specific element, and are as a rule slow-growing, low-biomass-producing plants, with little known agronomic characteristics. Moreover, there is no effective hyperaccumulating plant for lead, one of the most widespread and toxic metal pollutants in soil.

9.2. Materials and methods

9.2.1. Soil samples and analysis

For sequential extraction, soil samples (0–5 cm) were collected from 30 locations in Celje County, Slovenia. This region is a topographic basin surrounded by Alpine foothills where assorted industries are located: a zinc smelter, brickworks, steel mills, etc. Temperature inversions within the basin are common and spread emissions from industrial sources over the basin.

Soil samples for phytoextraction experiments were collected from two different locations in Mezica valley, at the industrial site of a lead and zinc smelter in Slovenia (soils 1 and 2), and from a location in Domzale exposed to lead contamination from traffic (soil 3). The properties of soils 1 and 2 and 3 are summarized in Table 9.1.

Table 9.1. HM content in soils and selected properties of soils used in the phytoextraction study

	Soil 1 depth (cm)	Soil 2 depth (cm)			Soil 3 depth (cm)			
	0–25	0–15	15–25	25–35	0–15	15–25	25–35	35–45
Texture	Sandy loam	Loam	Loam	Sandy loam	Loam	Loam	Loam	Loam
Sand (%)	55.4	49.2	48.0	52.4	39.9	39.5	34.7	32
Silt (%)	30.9	40.2	41.0	31.8	45.1	43.2	44.3	44.3
Clay (%)	13.7	10.6	11.0	15.8	15.0	17.3	21.0	24.7
Dry bulk density (g/cm^3)	1.15	1.07	1.06	1.39	1.16	1.24	1.39	1.40
pH (CaCl$_2$)	6.8	6.8	6.7	6.8	6.9	7.1	7.1	7.1
Organic matter content (%)	5.2	13.0	8.7	4.3	8.3	6.2	3.3	2.5
Total nitrogen content (%)	0.25	0.54	0.47	0.25	0.37	0.31	0.19	0.18
P (mg P$_2$O$_5$/100 g soil)	37.3	101.5	63.8	12.7	274.2	234.2	93.2	30.7
K (mg K$_2$O/100 g soil)	9.2	29.2	13.2	7.1	61.7	48.9	55.9	52.8
Cation exchange capacity (mmol$_c$/100 g)		39.1	36.3	23.5	39.1	37.2	36.2	23.7
Pb (mg/kg)	1100	1415	933	236	773	298	191	66
Zn (mg/kg)	800	1087	751	300	613	352	147	104
Cd (mg/kg)	5.5	9.9	8.0	5.7	5.5	4.0	3.2	3.1

Soil samples were analysed for organic matter by the Walkley–Black method, total nitrogen content by the Kjeldahl digestion method, cation exchange capacity by the ammonium acetate and Melich methods, and soil texture by pipette.

9.2.2. Sequential extraction

A modified analytical procedure (Tessier *et al.*, 1979) was used to determine the speciation of lead and zinc into six fractions. The water-soluble fraction was obtained by extraction from 1 g of 2 mm sieved, air-dried soil with 10 ml of deionized water for 1 h. The fraction present as soil colloids was extracted from a residual soil sample with 10 ml of 1 M $MgNO_3$ for 2 h. The fraction bound to carbonates was extracted with 10 ml of 1 M NH_4OAc (pH 5) for 5 h. The fraction bound to iron and manganese oxides was extracted with 20 ml of 0.1 M $NH_2OH \cdot xHCl$ (pH 2) for 12 h. The fraction bound to organic matter was obtained after heating the soil sample (residual after previous extractions) suspended in 3 ml of 0.02 M HNO_3 and 5 ml of 30% of H_2O_2 for 3 h at 85°C, followed by extraction with 15 ml of 1 M NH_4OAc for 3 h. The last fraction was obtained after digestion of a soil sample with aqua regia. Four determinations of HM concentration were undertaken for each sequence.

9.2.3. Column phytoextraction experiments – disturbed soil profile

Soil 1 was placed in 18 cm high, 15 cm diameter columns, which were equipped with trapping devices for leachate collection. Three-week-old seedlings of *Brassica rapa* L. var. *pekinesis* (Nagaoka F1) were transplanted into the columns and grown for 1 week to adapt, before a chelate application on day 1 of the experiment. Ethylenediaminetetraacetic acid (EDTA) and trisodium (*S,S*)-ethylene diamine disuccinate (EDDS) (in 100 ml of water) were applied in four part-weekly additions of total 5 and 10 mmo/ kg on days 1, 8, 15 and 22 of the experiment or in a single dose of 3, 5 and 10 mmo/kg on day 28 of the experiment. The plants were watered three times a week, with 400 ml of tap water. Above-ground tissues were harvested by cutting the stem 1 cm above the soil surface. The biomass was determined after drying at 60°C to a constant weight. The leachates were sampled on days 7, 14, 21 and 28 of the experiment.

9.2.4. Column phytoextraction experiments – undisturbed soil profile

Undisturbed soil profiles were placed in 15 cm diameter columns, 35 cm high for soil 2, and 45 cm high for soil 3, equipped with trapping devices. *Brassica rapa* was grown, transplanted and watered as described above. EDTA and EDDS were applied in a single aliquot of 10 mmol/kg on day 28 of the experiment. The leachates were sampled on days 55, 70, 85 and 100 of the experiment.

9.2.5. HM analysis

Soil samples were ground in an agate mill, digested in aqua regia and analysed by atomic absorption spectroscopy (AAS). Plant samples were dried, ground in a

Table 9.2. Selected properties for the soil samples used in the sequential extraction procedure

Soil parameter	Range of 30 samples
pH	4.6–7.5
Organic matter (%)	3.0–17.6
Clay (%)	10.0–31.9
Silt (%)	26.4–73.0
Sand (%)	11.7–61.8
CEC (mmol$_c$/100 g)	24.4–56.8
Total HM content	
Pb (mg/kg)	43.4–1690
Zn (mg/kg)	170.0–6178

titanium mill, digested in 65% HNO_3 in a microwave oven, and analysed by flame AAS (lead and zinc) and by electrothermic AAS (cadmium). HMs in leachates were determined by inductively coupled plasma–atomic emission spectroscopy.

9.2.6. Phospholipid analysis

The toxicity of EDTA and EDDS additions on a microbial population was tested on soil 1 (200 g dry soil/jar, 70% field water capacity) in a 10 week growth chamber experiment (24°C, 85% relative humidity, three replicates). After 3 weeks of stabiliz-ation, 10 mmol/kg of EDTA and EDDS were added. Five grams of soil were sampled on days 1 and 56 after chelate addition and analysed for phospholipid fatty acids (PLFAs) (Frostegard *et al.*, 1991). Fatty acids were designated using the nomenclature described by Frostegard and Baath (1996). The ratio between the *trans*-18:1ω9 and *cis*-18:1ω9 acids was used to monitor stress levels to micro-organisms (Guckert *et al.*, 1986). The structure of microbial groups in soil was presented in relative shares of microbial groups, determined as the mole per cent of PLFAs indicative of particular microbial groups against total PLFAs.

9.3. Results and discussion

9.3.1. Sequential extractions

The range of selected pedological properties of the topsoil samples from 30 locations in Celje County are presented in Table 9.2. As shown, the soil samples differed substantially in pH, organic matter content, texture, cation exchange capacity and HM contamination. Table 9.3 lists the distribution of the various species of lead and zinc. Lead in all soils was mostly bound to carbonate and organic matter, or remained in the residual fraction. Little or none of the total lead in soils was present in a form directly available to plants. Zinc in the soil samples was distributed among all fractions of the sequential extractions; however, the soluble zinc fraction was quite small. In the soil samples tested the variations in the distribution of both lead and zinc among the different soil fractions were considerable and can be explained by differences in soil properties which exert an effect on metal speciation.

The relationships and interdependence of lead and zinc speciation, lead and zinc uptake into the test plant (narrow leaf plantain, *Plantago lancelolata*) and soil pedological properties were first determined by cluster analysis.

For lead the dendrogram for 12 variables, describing lead speciation, uptake by plants and the main pedological properties, indicated four clusters (Fig. 9.1). Pedological properties (pH, organic matter content, cation exchange capacity and clay content) were in cluster 1; total lead content, the fraction of lead bound to organic matter, the residual fraction and the fraction bound to carbonates were in cluster 2; lead uptake by plants was in cluster 3; the water- soluble fraction, the fraction bound to iron and manganese oxides, and the fraction exchangeable from soil colloids were in cluster 4. The cluster analysis revealed that the lead uptake by the test plant narrow leaf plantain was not influenced by lead speciation or soil properties and was not even related to the total lead content in the soil. This indicates that the narrow leaf plantain is not a good indicator plant for lead bioavailability in soil and that substantial variations in the distribution of lead among different soil fractions are probably controlled by soil properties and processes which were not measured or included in our experiment.

The relationships between lead uptake by plants, speciation and soil properties were further examined by stepwise multiple regression analysis using multiple linear regression models. Again, uptake of lead by plants could not be related to any of the variables measured. Of the lead species, a statistically significant relationship ($p < 0.01$) was found between the fraction bound to carbonates and the total content of lead in the soil (90.6% of explained variations). An additional 2.1 and 1.8% of variations in this lead fraction were explained by clay and organic matter content, respectively. The fraction of lead bound to organic matter also depended on the total lead content (96.9%); small variations were explained by the organic matter content (2%). As expected, a significant relationship was found between the residual fraction and the total lead content (95.9%).

Table 9.3. Speciation of lead and zinc in soil samples obtained in the sequential extraction procedure

Speciation of HM	Total HM content in soil (range of 30 samples) (%)	
	Pb	Zn
In solution	0.0–0.08	0.0–2.3
Exchangeable	0.0–1.6	0.0–28.5
Bound to carbonate	2.04–43.5	3.9–35.1
Bound to Fe and Mn oxides	0.0–16.1	1.39–25.4
Bound to organic matter	35.8–71.1	14.8–56.2
Residual fraction	10.4–53.4	14.2–75.3

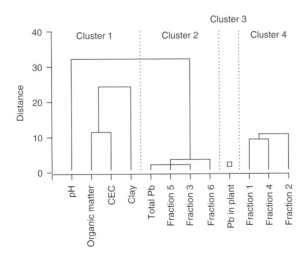

Fig. 9.1. Dendrogram of cluster analysis of variables on soil properties: lead concentration in soil, lead speciation, lead uptake in plants. Fraction 1: soluble. Fraction 2: exchangeable from soil colloids. Fraction 3: bound to carbonates. Fraction 4: bound to iron and manganese oxides. Fraction 5: bound to organic matter. Fraction 6: residual

Cluster analysis of zinc speciation, uptake by plants and the main soil pedological properties revealed an expected pattern that was much clearer than with the lead analysis. The dendrogram of zinc data indicated three clusters (Fig. 9.2). In cluster 1 the fractions of zinc bound to soil organic matter, bound to carbonates, in the residual fraction and bound to iron and manganese oxides were closely correlated with the total zinc content in the soil, and associated with soil pH. Cluster 2 comprised the main soil pedological properties: organic matter and clay content and cation exchange capacity. Cluster 3 related zinc uptake by the test plant (narrow leaf plantain) to two bioavailable fractions of zinc: the soluble fraction and the fraction exchangeable from soil colloids.

Stepwise multiple regression analysis of the zinc data confirmed those relationships indicated by cluster analysis. Statistically significant relationships ($p < 0.01$) were found between zinc uptake by the narrow leaf plantain and the soluble zinc fraction (64.15% of explained variations), soil pH (additional 22% of explained variations) and the fraction of zinc exchangeable from soil colloids (additional 2.2% of explained variations). The soluble zinc fraction and zinc exchangeable from soil colloids were slightly correlated with soil pH (18.3 and 39.8% of explained variations). The fraction of zinc bound to carbonates was found to be dependent on the total content of zinc in the soil (91.2% of explained variations), on organic matter content (additional 3.3% of explained variations) and on clay content (additional 1.6% of explained variations). The fractions of zinc bound to iron and manganese oxides, of zinc bound to organic matter and of residual zinc were found to be dependent only on the total zinc content in the soil (75.9, 93.2 and 93.4% of explained variations, respectively).

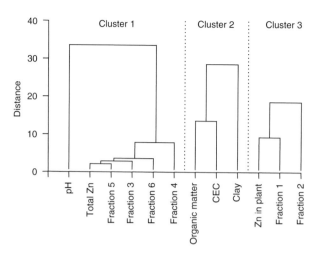

Fig. 9.2. Dendrogram of cluster analysis of variables on soil properties: zinc concentration in soil, zinc speciation, zinc uptake in plants. Fraction 1: soluble. Fraction 2: exchangeable from soil colloids. Fraction 3: bound to carbonates. Fraction 4: bound to iron and manganese oxides. Fraction 5: bound to organic matter. Fraction 6: residual

9.3.2. Chelate-induced phytoextraction of lead, zinc and cadmium

Many hydroponic studies have revealed that the uptake and translocation of HMs in plants are enhanced by increasing HM concentration in the nutrient solution (Huang *et al.*, 1997). The bioavailability of HMs in the soil is, therefore, of paramount importance for successful phytoremediation. Lead has limited solubility (Table 9.2) and bioavailability due to complexation with organic and inorganic soil colloids, sorption on oxides and clays, and precipitation as carbonates, hydroxides and phosphates (Ruby *et al.*, 1999). Therefore, a successful phytoremediation must include mobilization of HMs into the soil solution that is in direct contact with plant roots. In most soils capable of supporting plant growth, the phytoavailable levels of HM and particularly of lead are low and do not allow substantial plant uptake if HM ligands (organic chelates) are not present.

As shown in Fig. 9.3, chelate desorbed lead from the soil matrix (mainly from fraction 3: lead bound to carbonates) into fraction 1 (soluble lead), and therefore increased the share of phytoavailable lead. Chelates also facilitate HM transport into xylem and increase HM translocation from the roots to the shoots of several fast-growing, high-biomass-producing plants (Huang *et al.*, 1997). The literature to date reports a number of chelates that have been used for chelate-induced hyper-accumulation. These include EDTA, cyclohexanediaminetetracetic acid (CDTA), diethylenetriaminepentaacetic acid (DTPA), ethylene glycol-bis-(2-aminoethyl)-tetraacetic acid (EGTA), ethylenediaminedi(*o*-hydroxyphenylacetic)acid (EDDHA), hydroxyethylethylenediaminetriacetic acid (HEDTA) and nitrilotriacetic acid (NTA) (Blaylock *et al.*, 1997; Cooper *et al.*, 1999; Wu *et al.*, 1999). These chelates, however,

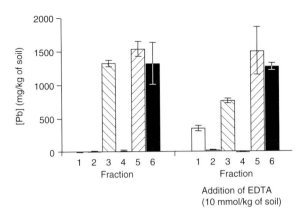

Fig. 9.3. Sequential extraction of lead-contaminated soil with and without addition of chelate (EDTA). Fraction 1: soluble. Fraction 2: exchangeable from soil colloids. Fraction 3: bound to carbonates. Fraction 4: bound to iron and manganese oxides. Fraction 5: bound to organic matter. Fraction 6: residual

are not specific to HMs and interact with other cations present in soil at much higher concentrations. Many synthetic chelates and their complexes with HM are toxic (Sillanpää and Oikari, 1996; Dirilgen, 1998) and poorly photo-, chemo- and bio-degradable in soil environments (Nörtemann, 1999). A combination of widespread use in fertilizers and slow decomposition has led to background concentrations of EDTA in European surface waters in the range of 10–50 mg/l (Kari *et al.*, 1995). Authors investigating chelate-enhanced phytoextraction have also pointed out the risk of possible mobilization of lead and other HMs from soil to groundwater and consequent promotion of off-site migration (Huang and Cunningham, 1996; Cooper *et al.*, 1999; Kulli *et al.*, 1999). Therefore, all the potential risks of the use of chelates for phytoextraction should be thoroughly evaluated before steps are taken towards further development and commercialization of this remediation technology.

In the present study, we compare the efficiency of two chelating agents: EDTA as one of the more efficient and widely tested chelates (Blaylock *et al.*, 1997; Huang *et al.*, 1997; Wu *et al.*, 1999), and a new biodegradable chelate, EDDS. EDDS was reported by Jones and Williams (2001) to be a strong, biodegradable, transition metal and radionuclide chelate. Jaworska *et al.* (1999) assessed the environmental risks of its use in detergent applications. Mineralization of EDDS in sludge-modified soil was rapid and complete in 28 days. The reported calculated half-life was 2.5 days. No recalcitrant metabolites were found in the degradation profile of EDDS. The toxicity to fish and daphnia was low ($EC_{50} > 1000$ mg/l).

9.3.2.1. HM uptake by plants

The analysis of plant material indicated that both chelates when applied to soils 1, 2 and 3 increased the concentrations of lead in the leaves of the test plants. In column experiments with a disturbed soil profile (soil 1), single and weekly additions of all

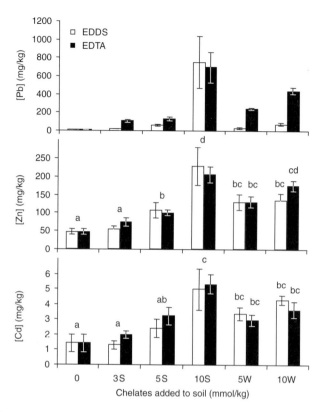

Fig. 9.4. Lead, cadmium and zinc concentrations in leaves of Brassica rapa *grown on contaminated soil (soil 1) in response to different chelate concentrations: 3 mmol/kg single-dose addition of EDDS (EDDS/3S), 5 mmol/kg single-dose addition of EDDS (EDDS/5S), 5 mmol/kg weekly additions of EDDS (EDDS/5W), 10 mmol/kg single-dose addition of EDDS (EDDS/10S), 10 mmol/kg weekly additions of EDDS (EDDS/10W), 5 mmol/kg weekly additions of EDTA (EDTA/5W), 10 mmol/kg weekly additions of EDTA (EDTA/10W), and control soil with no chelate addition. The mean values of five replicates are presented; error bars represent standard deviation. Statistically different treatments, according to the Tukey test, are labelled with different letters*

concentrations of EDTA were significantly more effective than those of EDDS, except for the high single dose, where the effects of EDTA and EDDS were similar (Fig. 9.4). Lead concentrations in test plant shoots of 757.2 and 697.4 mg/kg of dry biomass, an increase of 94.2- and 102.3-fold relative to the control level, were determined for single additions of 10 mmol/kg of EDDS and EDTA. Weekly additions of EDDS were probably less effective due to the rapid EDDS bio-degradability (Jaworska *et al.*, 1999). Consequently, the availability of mobilized species of HMs in EDDS soil treatments was short-lived compared with EDTA treatments.

In experiments with undisturbed soil profiles, single-dose additions of 10 mmol/kg of chelate caused smaller increases of lead uptake in plants than in experiments with soil 1. Lead concentrations in plant shoots were 70.7 ± 9.9 and 18.1 ± 1.8 mg/kg of dry biomass for soil 2, and 8.1 ± 1.8 and 19.4 ± 5.2 mg/kg of dry biomass for soil 3, in treatments with EDTA and EDDS, respectively. The lower lead uptake from soils 2 and 3 compared with soil 1 can presumably be attributed to a high phosphorus content in soils 2 and 3: phosphates form stable complexes with lead and strongly decrease the phytoavailability of lead in soil.

Both EDTA and EDDS were similarly effective in the enhancement of cadmium and zinc uptake by plants in all three soils. In experiments with soil 1 the addition of 10 mmol/kg of EDTA in a single dose increased the concentrations of zinc and cadmium in leaves 4.3- and 3.8-fold, compared with the control treatment. The EDDS addition caused 4.7- and 3.5-times higher concentrations of zinc and cadmium in plant tissues, respectively (see Fig. 9.1). The ability of EDTA to enhance lead uptake in plants over that of zinc, cadmium and other HMs was also reported by Blaylock *et al.* (1997); EDDS shares similar binding preferences. The stability constants for Pb^{2+} are higher than for Cd^{2+} and Zn^{2+} (Martell and Smith, 1974).

In experiments with undisturbed soil profiles the plant shoot zinc concentration in EDTA and EDDS treatments (10 mmol/kg single dose) was 217.8 ± 25.8 and 192.9 ± 8.29 mg/kg for soil 2, and 184.1 ± 7.5 and 169.5 ± 20.6 mg/kg of dry biomass for soil 3. The shoot cadmium concentration of EDTA and EDDS was 4.38 ± 0.53 and 2.30 ± 0.20 mg/kg for soil 2, and 6.20 ± 0.55 and 2.70 ± 9.40 mg/kg of dry biomass for soil 3, respectively.

9.3.2.2. HM leaching

EDTA had a strong influence on HM leaching. In the disturbed soil profile experiment (soil 1) the addition of EDTA (10 mmol/kg weekly doses) caused leaching of 22.7, 7.0 and 39.8% of the initial total lead, zinc and cadmium in 4 weeks. EDDS caused significantly less leaching of lead than EDTA. The same amount of EDDS leached 0.8, 6.2 and 1.5% of the initial total lead, zinc and cadmium (Table 9.4).

Likewise, much lower leaching of lead, zinc and cadmium by EDDS compared with EDTA treatments was determined in undisturbed soil profile experiments (Table 9.4). One hundred days after chelate application, the total amount of lead leached in EDDS treatments was 113-fold lower for soil 2, and 438-fold lower for soil 3. Less prominent, but substantial, were the differences in the amount of leached zinc and cadmium. The total amount of zinc leached in EDDS treatments was 3.7-fold lower for soil 2, and 4-fold lower for soil 3. The total amount of cadmium leached was 80.3-fold lower for soil 2, and 34.8-fold lower for soil 3.

The biodegradability of EDDS (and apparently also of lead, zinc and cadmium EDDS complexes) is a plausible explanation for the lower leaching of HMs from EDDS-treated soils. Several authors have emphasized the risks associated with leaching of HMs from the root zone due to chelate application, and the importance of assessing this risk. However, there are no reports in the current literature on direct measurements of HMs leaching during phytoextraction experiments. Cooper *et al.* (1999) estimated the risk of leaching using the toxicity characteristic leaching procedure (TCLP) at the end of a phytoextraction experiment with three different chelating agents (CDTA, DTPA and HEDTA). All chelates increased the TCLP of lead compared with controls. The TCLP of lead in soils treated with the high rates (20 mmol/kg) of CDTA and DTPA tested greatly exceeded the US Environmental Protection Agency regulatory limits (5 mg/l).

9.3.2.3. Phytotoxicity

In all treatments where chelates were applied, visual symptoms of toxicity were observed as necrotic lesions on cabbage leaves. Single and weekly additions of 10 mmol/kg of EDTA resulted in rapid senescence of the plant shoots and statistically significantly lowered the yield of cabbage biomass (data not provided). EDDS was less phytotoxic than EDTA, and has shown no statistically significant reduction ($p = 0.05$, data not provided) in cabbage growth. This could be explained by either the inherently lower phytotoxicity of EDDS or as a result of the rapid biodegradation of EDDS in soil, as reported by Jaworska *et al.* (1999).

To evaluate the post-treatment toxicity of the remediated soil, a bioassay with red clover was performed on soil 1. The total biomass of the shoots per pot revealed a strong negative effect from 5 mmol/kg and 10 mmol/kg weekly EDTA additions. No

Table 9.4. *Lead, zinc and cadmium leached through the columns with soils 1, 2 and 3 as percentages of initial total HMs in soil. Treatments were: control, 4 weekly (W) additions of EDTA and EDDS of 5 and 10 mmol/kg in total for soil 1; control and single (S) addition of EDTA and EDDS of 10 mmol/kg in total for soils 2 and 3. The mean values and standard deviations of five replicates for soil 1 and four replicates for soils 2 and 3 are presented*

Treatment		HM leached (%)		
		Pb	Zn	Cd
Control	Soil 1	–	0.001 ± 0.000	–
Control	Soil 2	–	–	–
Control	Soil 3	–	–	–
EDTA (5 mmol/kg), W	Soil 1	12.6 ± 4.3	5.2 ± 1.5	32.9 ± 10.6
EDTA (10 mmol/kg), W	Soil 1	22.7 ± 7.2	7.0 ± 1.6	39.8 ± 8.4
EDTA (10 mmol/kg), S	Soil 2	6.10	3.34	5.14
EDTA (10 mmol/kg), S	Soil 3	2.19	2.66	2.65
EDDS (5 mmol/kg), W	Soil 1	0.05 ± 0.074	2.6 ± 1.6	0.2 ± 0.17
EDDS (10 mmol/kg), W	Soil 1	0.83 ± 0.564	6.2 ± 1.0	1.5 ± 0.73
EDDS (10 mmol/kg), S	Soil 2	0.054	0.90	0.064
EDDS (10 mmol/kg), S	Soil 3	0.005	0.66	0.076

–, below detection limits

negative effect from EDDS was found in the bioassay experiment, regardless of its concentration (data not provided).

9.3.2.4. Influence of chelate addition on soil micro-organisms

Phytoextraction is a long-term technology. Therefore, it is imperative to sustain the quality of soil and enable vigorous growth of phytoextracting plants. Soil micro-organisms are critically important for the normal function of soils, and are common indicators of soil quality. The toxicity of EDDS and EDTA to soil bacteria, actino-mycetes and fungi was studied with PLFA analysis of soil 1. We detected 32 different PLFAs in our treatments, and identified 23 of them. In total, these marker PLFAs represented 43–58% of the total PLFAs.

The shifts in the structure of microbial communities were determined using the concentration of marker PLFAs. As shown in Fig. 9.5, the PLFA concentration for fungi was significantly lower than in the control treatment on days 1 and 56 after EDTA addition. In contrast, no statistically significant ($p = 0.05$) effect of EDDS addition on fungal PLFA markers was observed. Neither the PLFA marker concentration for bacteria nor those for actinomycetes changed significantly because of EDTA or EDDS additions. The share of fungal biomass, which is dominant in most soils (Thorn, 1997) seems to be underrated in Fig. 9.5, since the conversion factors from PLFA markers of microbial groups to actual biomass are lacking.

EDTA addition significantly increased the *trans/cis* ratio of PLFAs on days 1 and 56 after chelate application (Fig. 9.5). An increase in the *trans/cis* ratio is associated with starved or stressed micro-organisms in natural environments (Guckert *et al.*, 1986). EDDS addition had a statistically significant effect ($p = 0.05$) only 56 days after its application, compared with the control treatment (Fig. 9.5).

The results are in accord with phytotoxicity tests. EDDS additions were less toxic to soil fungi than EDTA and caused less stress to soil micro-organisms. The PLFA results indicate that soil fungi are more sensitive to EDTA addition than other soil microbial groups. This is in accord with results of Dahlin *et al.* (1997). These authors reported that the effect of HMs on the PLFA pattern was small, except for $18:2\omega6$ PLFA, which decreased in sludge-modified, cadmium, chromium, copper, lead and zinc contaminated soil. This specific PLFA is indicative for fungi (Frostegard *et al.*, 1993). EDDS is a naturally occurring substance, and this is the likely reason for its low toxicity.

9.4. Conclusions

Despite the increasing importance of HMs due to their environmental consequences, there is very little information in the literature on the speciation of HMs in agricultural and other areas in Slovenia.

The results of this study introduce EDDS as a promising new chelate for enhanced, environmentally safe phytoextraction of soils contaminated principally with lead. Measurements of lead in the leaves of *Brassica rapa* revealed that single doses of EDDS were the most effective, and in some soils were equally effective as EDTA. In contrast to EDTA, EDDS addition caused only minor leaching of lead,

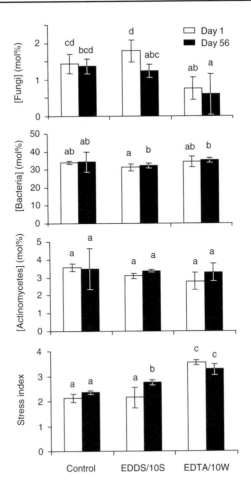

Fig. 9.5. The structure of microbial groups (fungal, bacterial and actinomycetes) determined as mole per cent of PLFAs in soil (soil 1), and stress index (trans-PLFA/cis-PLFA) of microbial populations after 1 and 56 days of chelate addition in a single aliquot of 10 mmol/kg soil (10S). The mean values of three replicates are presented; error bars represent standard deviation. Statistically different treatments, according to the Tukey test ($p \leq 0.05$), are labelled with different letters

and was significantly less toxic to plants and soil micro-organisms. However, even the highest concentrations of HMs in harvestable plant tissues achieved in this study are still far from the concentrations required for efficient phytoextraction procedures.

9.5. Acknowledgement

This work was supported by the Slovenian Ministry for Science and Technology, grant No. J4-0694-0486-98.

9.6. References

BLAYLOCK, M. J., SALT, D. E., DUSHENKOV, S., ZAKHAROVA, O., GUSSMAN, C., KAPULNIK, Y., ENSLEY, B. D. and RASKIN, I. (1997). Enhanced accumulation of Pb in Indian mustard by soil-applied chelating agents. *Environmental Science and Technology* **31**, 860–865.

COOPER, E. M., SIMS, J. T., CUNNINGHAM, S. D., HUANG, J. W. and BERTI, W. R. (1999). Chelate-assisted phytoextraction of lead from contaminated soils. *Journal of Environmental Quality* **28**,1709–1719.

DAHLIN, S., WITTER, MARTENSSON, A. M, TURNER and BAATH, E. (1997). Where's the limit? Changes in the microbiological properties of agricultural soils at low levels of metal contamination. *Soil Biology and Biochemistry* **29**, 1405–1415.

DIRILGEN, N. (1998). Effects of pH and chelator EDTA on Cr toxicity and accumulation in *Lemma minor. Chemosphere* **37**, 771–783.

FROSTEGARD, A. and BAATH, E. (1996). The use of phospholipic fatty acid analysis to estimate bacterial and fungal biomass in soil. *Biology and Fertility of Soils* **22**, 59–65.

FROSTEGARD, A., TUNLID, A. and BAATH, E. (1991). Microbial biomass measured as a total lipid phosphate in soils of different organic content. *Journal of Microbiological Methods* **14**, 151–163.

FROSTEGARD, A., TUNLID, A. and BAATH, E. (1993). Phospholipid fatty acid composition, biomass, and activity of microbial communities from two soil types experimentally exposed to different heavy metals. *Applied and Environmental Microbiology* **59**, 3605–3617.

GUCKERT, J. B., HOOD, M. A. and WHITE, D. C. (1986). Phospholipid ester-linked fatty acid profile changes during nutrient deprivation of *Vibrio cholerae*: increase in the *trans/cis* ratio and proportions of cyclopropyl fatty acids. *Applied and Environmental Microbiology* **52**, 794–801.

HUANG, J. W. and CUNNINGHAM, S. D. (1996). Lead phytoextraction: species variation in lead uptake and translocation. *New Phytologist* **134**, 75–84.

HUANG, J. W., CHEN, J., BERTI, W. R. and CUNNINGHAM, S. D. (1997). Phytoremediation of lead-contaminated soils: role of synthetic chelates in lead phytoextraction. *Environmental Science and Technology* **3**, 800–805.

JAWORSKA, J. S., SCHOWANEK, D. and FEIJTEL, T. C. J. (1999). Environmental risk assessment for trisodium [*S,S*]-ethylene diamine disuccinate, a biodegradable chelator used in detergent application. *Chemosphere* **38**, 3597–3625.

JONES, P. W. and WILLIAMS, D. R. (2001). Chemical speciation used to assess [*S,S′*]- ethylene-diaminedisuccinic acid (EDDS) as a readily-biodegradable replacement for EDTA in radio-chemical decontamination formulations. *Applied Radiation and Isotopes* **54**, 587–593.

KARI, F. G., HILGER, S. and CANONICA, S. (1995). Determination of the reaction quantum yield for the photochemical degradation of Fe(III)-EDTA: Implications for the environmental fate of EDTA in surface waters. *Environmental Science and Technology* **29**, 1008–1017.

KULLI, B., BALMER, M., KREBS, R., LOTHENBACH, B., GEIGER, G. and SCHULIN, R. (1999). The influence of nitrilotriacetate on heavy metal uptake of lettuce and ryegrass. *Journal of Environmental Quality* **28**, 1699–1705.

LEVY, D. B., BARBARICK, K. A., SIEMER, E. G. and SOMMERS, L. E. (1992). Distributioning and partitioning of trace metals in contaminated soils near Leadville, Colorado. *Journal of Environmental Quality* **21**, 185–195.

MARTELL, A. E. and SMITH, R. M. (1974). *Critical Stability Constants*, Vol. 1. *Amino Acids*. Plenum Press, New York.

NÖRTEMANN, B. (1999). Biodegradation of EDTA. *Applied Microbiology and Biotechnology* **51**, 751–759.

RUBY, M. V., SCHOOF, R., BRATTIN, W., GOLDADE, M., POST, G., HARNOIS, M., MOSBY, D. E., CASTEEL, S. W., BERTI, W., CARPENTER, M., EDWARDS, D., CRAGIN, D. and CHAPPELL, W. (1999). Advances in evaluating the oral bioavailability of inorganics in soil for use in human health risk assessment. *Environmental Science and Technology* **32**, 3697–3705.

SALT, D. E., BLAYLOCK, M., KUMAR, P. B. A. N., DUSHENKOV, V., ENSLEY, B. D., CHET, I. and RASKIN, I. (1995). Phytoremediation: a novel strategy for the removal of toxic metals from the environment using plants. *Biotechnology* **13**, 468–474.

SILLANPÄÄ, M. and OIKARI, A. (1996). Assessing the impact of complexation by EDTA and DTPA on heavy metal toxicity using Microtox bioassay. *Chemosphere* **32**, 1485–1497.

TESSIER, A., CAMPBELL, P. G. C. and BISSON, M. (1979). Sequential extraction procedure for the speciation of particulate trace metals. *Analytical Chemistry* **51**, 844–851.

THORN, G. (1997). The fungi in soil. In: J. D. Elsas, J. T. Trevors and E. M. H. Wellington (eds), *Modern Soil Microbiology*, pp. 63–127. Dekker, New York.

WU, J., HSU, F. C. and CUNNINGHAM, S. D. (1999). Chelate-assisted Pb phytoextraction: Pb availability, uptake and translocation. *Environmental Science and Technology* **33**,1898–1904.

9.7. Further reading

EC (1986). Council Directive 86/278/EEC. On the protection of the environment, and in particular of the soil, when sewage sludge is used in agriculture. *EC Official Journal* **L181**.

EPSTEIN, A. L., GUSSMAN, C. D., BLAYLOCK, M. J., YERMIYAHU, U., HUANG, J. W., KAPULNIK, Y. and ORSER, C. S. (1999). EDTA and Pb-EDTA accumulation in *Brassica juncea* grown in Pb-amended soil. *Plant and Soil* **208**, 87–94.

KAYSER, A., WENGER, K., KELLER, A., ATTINGER, W., FELIX, R., GUPTA, S. K. and SCHULIN, R. (2000). Enhancement of phytoextraction of zinc, Cd and Cu from calcareous soil: The use of NTA and sulphur amendments. *Environmental Science and Technology* **34**, 1778–1783.

KLAMER, M., and BAATH, E. (1998). Microbial community dynamics during composting of straw material studied using phospholipid fatty acid analysis. *FEMS Microbiology and Ecology* **27**, 9–20.

NISHIKIORI, T., OKUYAMA, A., NAGANAWA, T., TAKITA, T., HAMIDA, M., TAKEUCHI, T., AOYAGI, T. and UMEZAWA, H. (1984). Production of actinomycetes of (*S,S*)-*N,N'*-ethylenediamine-disuccinic acid, an inhibitor of phospholipase. *Journal of Antibiotics* **37**, 426–427.

PHILLIPS, J. M. and HAYMAN, D. S. (1970). Improved procedures for clearing roots and staining parasitic and vesicular-arbuscular mycorrhizal fungi for rapid assessment of infection. *Transactions of the British Mycological Society* **55**, 58–160.

US ENVIRONMENTAL PROTECTION AGENCY (1995). *Test Methods for Evaluating Solid Waste*. Vol. 1A. *Laboratory Manual of Physical/Chemical Methods*, SW 846, 3rd edn. US Government Printing Office, Washington DC.

US ENVIRONMENTAL PROTECTION AGENCY (1996). *Soil Screening Guidance. Technical Background Document*, USEPA Report 540/R-95/128. US Government Printing Office, Washington DC.

TROUVELOT, A., KONGH, J. L. and GIANINAZZI-PEARSON, V. (1986). Measure du taux de mycorrhization d'un systeme radiculaire. Recherche de methods d'estimation ayant une signification fonctionelle. In: V. Gianinazzi-Pearson and S. Gianinazzi (eds), *Physiological and Genetical Aspects of Mycorrhizae*, pp. 217–221. INRA, Dijon.

VESTAL, J. R. and WHITE, D. C. (1989). Lipid analysis and microbial ecology. *BioScience* **39**, 535–541.

WARD, T. E. (1986). Aerobic and anaerobic biodegradation of nitrilotriacetate in subsurface soils. *Ecotoxicology and Environmental Safety* **11**, 112–125.

WHITE, D. C., FLEMMING, C. A., LEUNG, K. T. and MACNAUGHTON, S. J. (1998). *In situ* microbial ecology for quantitative appraisal, monitoring, and risk assessment of pollution remediation in soils, the subsurface, the rhizosphere and in biofilms. *Journal of Microbiological Methods* **32**, 93–105.

WITSCHEL, M. and EGLI, T. (1998). Purification and characterisation of a lyase from the EDTA-degrading bacterial strain DSM 9103 that catalyses the splitting of [*S,S'*]- ethylene-diaminedisuccinate, a structural isomer of EDTA. *Biodegradation* **8**, 419–428.

XIAN, X. (1989). Effect of chemical forms of cadmium, zinc, and lead in polluted soils on their uptake by cabbage plants. *Plant and Soil* **113**, 257–264.

Part IV
Enhancing the efficiency of remediation processes

10. Development of novel reactive barrier technologies at the **SAFIRA** test site, Bitterfeld

H. Weiss and M. Schirmer
UFZ Centre for Environmental Research Leipzig-Halle, Department of Industrial and Mining Landscapes, Permoserstrasse 15, D-04318 Leipzig, Germany

P. Merkel
University of Tübingen, Chair of Applied Geosciences, Sigwartstrasse 10, D-72076 Tübingen, Germany

10.1. Introduction

Approximately 75% of the drinking water in Germany is extracted from groundwater resources. For this reason the protection of this essential resource is of particular interest. At many industrial sites in Germany, inadequate waste disposal, leakages and war damage have led to pollution of the subsurface with various contaminants, which have in some instances led to groundwater contamination on a regional scale.

For more than a century, open-pit lignite mining and related chemical industries have had a serious impact on soil and groundwater quality in the Bitterfeld region. The groundwater has been contaminated regionally over an area of approximately 25 km² with a total volume of approximately 200 million m³. After decommissioning of the mines and closure of large parts of the local chemical industry the groundwater level has started to rebound as groundwater management ceased, and the pre-industrial regional groundwater flow directions are being re-established.

10.2. The **SAFIRA** project

The goal of SAFIRA (Sanierungs-Forschung in regional kontaminierten Aquiferen – Remediation Research for Regionally Contaminated Aquifers), a joint project of several research partners, is to develop and provide the necessary technologies and methodologies to demonstrate the successful implementation of *in situ* reactive barriers in regionally contaminated aquifers (Weiss *et al.*, 1997). The

Bitterfeld–Wolfen region was chosen as an example of an area with large-scale and complex groundwater contamination.

Due to the extended time frame of field testing and our limited knowledge of the economic factors involved, the SAFIRA project, in contrast to many other research projects, compares several potentially successful groundwater treatment technologies concomitantly. Research groups from UFZ Leipzig, as well as the Universities of Tübingen, Dresden, Kiel, Leipzig and Halle, are involved in a joint research effort sponsored by the Federal Ministry of Education and Research (BMBF).

Within the framework of the remediation research programme of UFZ Leipzig there are also links to several international cooperating partners. Among them is a consortium from the Netherlands, led by the TNO (The Netherlands Organization for Applied Scientific Research).

10.2.1. The structure of SAFIRA

In order to address all aspects of importance in this innovative remediation approach the SAFIRA project is subdivided into several thematic branches (see Fig. 10.1 and the appendix). In addition to aspects of scientific–technical implement-ation, economic evaluation as well as subsequent land use and legal aspects are being addressed. In the scope of the project, special emphasis is put on the possibility that innovative approaches, and some of the results obtained by the individual research partners, can be transferred to other contaminant scenarios both within and outside of Germany.

The overall aim of the research project includes:

*   development and successive implementation of innovative low-energy or passive water treatment technologies for mixed organic groundwater contamin- ations, from prototypes to full-scale technical *in situ* applications;
*   technical–economic optimization of the individual technologies as well as possible combinations, including structural aspects of their implementation for *in situ* reactive barriers;
*   demonstration of the longevity of the new technologies in the field and the determination of their operation costs;
*   assessment of the legal and ecological aspects of the implementation of a reactive barrier system, using the Bitterfeld region as a case study;
*   development of economic application strategies for the most promising technologies.

10.3. The SAFIRA site

10.3.1. Geology and hydrogeology

The SAFIRA test site is underlain by an upper and lower aquifer separated by a lignite seam averaging 13 m in thickness (Fig. 10.2). The upper aquifer extends approximately 20 m below the ground surface (BGS) and is comprised of Quaternary glacio-fluvial sand and gravel with intercalated silt. The lower aquifer is about 28–50 m BGS and is comprised of Tertiary sands.

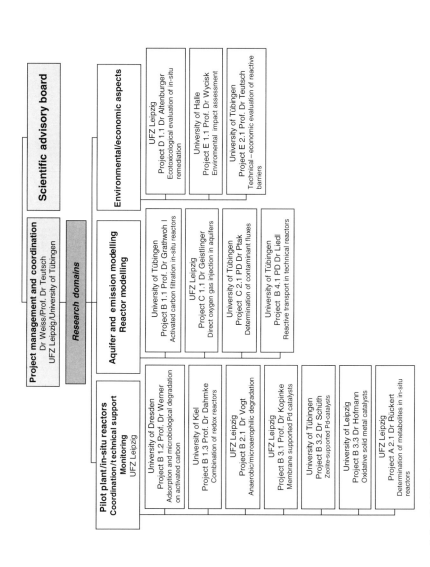

Fig. 10.1. Structure of the SAFIRA project

The water table in the area averages 6 m BGS. The hydraulic conductivity of the Quaternary aquifer ranges from 1×10^{-5} to 1×10^{-2} m/s (average 10^{-3} m/s), and in the Tertiary aquifer from 1×10^{-5} to 3.5×10^{-5} m/s. These strata are underlain by 20 m of Rupelian clay and Paleozoic bedrock.

Both aquifers are confined. At the site both aquifers appear to be hydraulically separated. Differences in the potentiometric surfaces of about 0.2 m lead to a downward vertical hydraulic gradient across the lignite seam.

The groundwater flow direction in the Quaternary aquifer is directed west–east, with hydraulic gradients of about 0.3–0.4‰, resulting in a flow velocity of about 0.1 m/day. The Tertiary aquifer shows a hydraulic gradient of approximately 1‰ in the same direction.

Due to the small hydraulic gradients it is assumed that existing minimal groundwater extraction rates will have a major impact on the local groundwater flow regime.

10.3.2. Groundwater contamination

As a consequence of pesticide and paint production processes, chlorinated hydrocarbons are the dominant groundwater contaminants in the Bitterfeld region. The

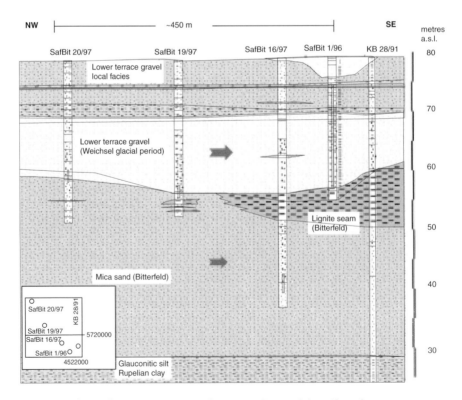

Fig. 10.2. Geological cross-section in the upgradient of the pilot plant

Table 10.1. Pollutants of the groundwater from the Quaternary aquifer at the Bitterfeld test site

Substance	Concentration (µg/l)
Monochlorobenzene	3130–33 000
1,4-Dichlorobenzene	90–1000
1,2-Dichlorobenzene	20–180
Benzene	20–180
Trichloroethene	<10–460
1,2-*cis*-Dichloroethene	10–280
1,2-*trans*-Dichloroethene	10–60
Chloromethylphenols	43.5
Trichlorophenols	9.2
2,4-Dichlorophenol	3.3
Dimethylphenols	1.4
Detected by mass spectroscopy after enrichment	
1,1,2-Trichloroethane	
1,1,2,2-Tetrachloroethane	
Bromobenzene	
1-Chloro-2-methylbenzene	
1,3-Dichlorobenzene	
Vinylchloride	
Tetrachloroethene	
Tetrachlorobenzene	
Toulene	
m/p-Xylene	

main contaminants are chlorobenzene, chloroethane, hexachlorocyclohexane, vinyl-chloride, phenols and toluene. These contaminants migrate from landfills, leaking pipelines or sewage systems in the subsurface.

High concentrations of chlorinated hydrocarbons were encountered in both the Quaternary and Tertiary aquifers, but with distinct differences in concentrations and composition.

In the lower aquifer, aliphatic chlorinated hydrocarbons are the main contaminants (1,2-*cis*-dichloroethane (DCE), *trans*-DCE and trichloroethane (TCE)) whereas the major contaminant in the upper Quaternary aquifer is monochlorobenzene (Table 10.1).

10.4. Pilot facility

The pilot facility in Bitterfeld houses a total of 20 reactors plus an on-site laboratory (Fig. 10.3). The reactor range consists of five shafts with a depth of 23 m, a diameter of 3 m and a shaft-to-shaft distance of 19 m, placed in a line perpendicular to the

Fig. 10.3. (a) The SAFIRA test site at Bitterfeld with the five well shaft housings on the left-hand side and the on-site laboratory building on the right-hand side. (b) The well shafts and horizontal collection wells

local groundwater flow direction. The groundwater for each shaft is collected by two horizontal inlet wells, 10 m in length, which were drilled at an angle of 60°.

Construction at the site was completed in March 1999, and the pilot facility started operation in the autumn of 1999.

Depending on the technologies of interest, the sizes of the reactors range from 1 to 6 m in length and 150 to 1400 mm in diameter. The reactors are designed for a permanent system pressure of 3 bar and are operated in a flow-through mode from bottom to top. Flow rates can be varied, up to 400 l/h.

The on-site laboratory analyses the major contaminants as well as physico-chemical parameters (pH, conductivity, redox potential, temperature and oxygen content), absorbable organic halides (AOX), total organic carbon (TOC) and dissolved organic carbon (DOC) directly at the site, thus minimizing possible alterations of the samples due to transport and storage.

Each experimental reactor has several regularly spaced sampling ports (depending on the size, the expected residence time of the groundwater and the process of interest) to monitor changes in groundwater composition along the passage through the reactor (Weiss *et al.*, 1999, 2000).

10.5. Reactor technologies

The technologies which are being tested in the reactive columns at the pilot plant at the moment include (Merkel *et al.*, 2000):

- Physicochemical technologies:
  - activated carbon filtration;
  - oxidative solid metal catalysis;
  - membrane- and zeolite-supported palladium catalysis.
- Microbial technologies:
  - anaerobic microbial degradation and biodegradation of chlorinated contaminants in an anaerobic/microaerobic system.
- Coupled technologies:
  - adsorption and simultaneous microbial degradation on activated carbon;
  - combination of redox reactors.

10.5.1. Some preliminary results

10.5.1.1. Activated carbon

Activated carbon filtration is performing well and is well suited to the removal of hydrophobic contaminants from groundwater under *in situ* conditions. Over a period of 1.5 years no biofouling or chemofouling of the reactor occurred under the anaerobic conditions prevailing. The physical and chemical parameters of the groundwater (anions/cations, pH, etc.) have remained unchanged (Fig. 10.4). Particle-induced contaminant loads occurred in the initial stage of reactor operation, and can cause contaminant release from particle-bound phases. Breakthrough behaviour for the various contaminants can be calculated using laboratory derived Freundlich isotherms and Freundlich exponents.

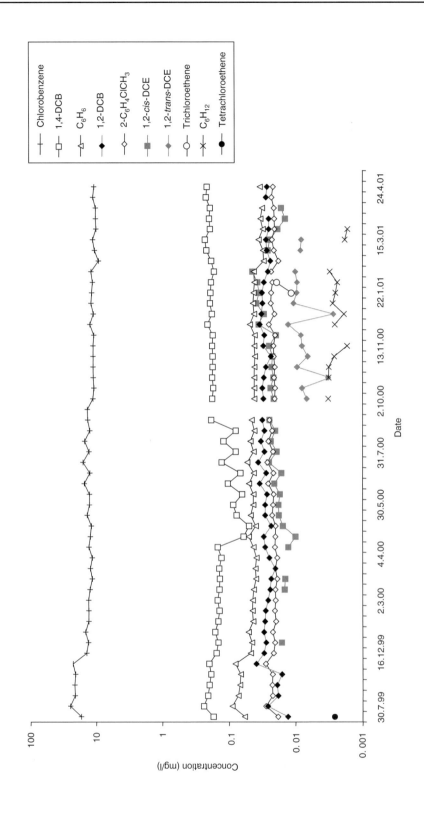

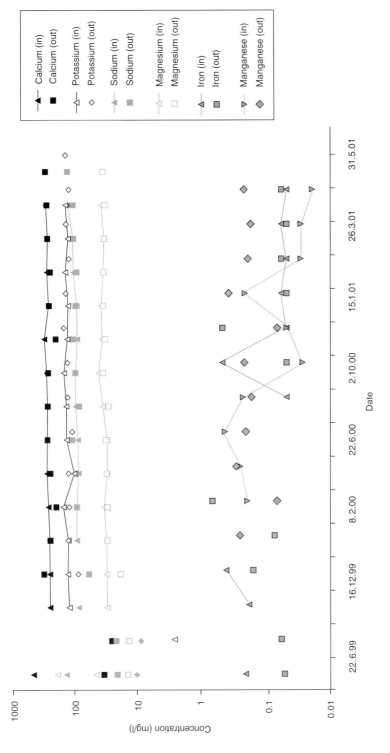

Fig. 10.4. Anion and cation concentrations and contaminant concentrations at the inlet and outlet of the reactor. (Data from S. Kraft and P. Grathwohl, University of Tübingen)

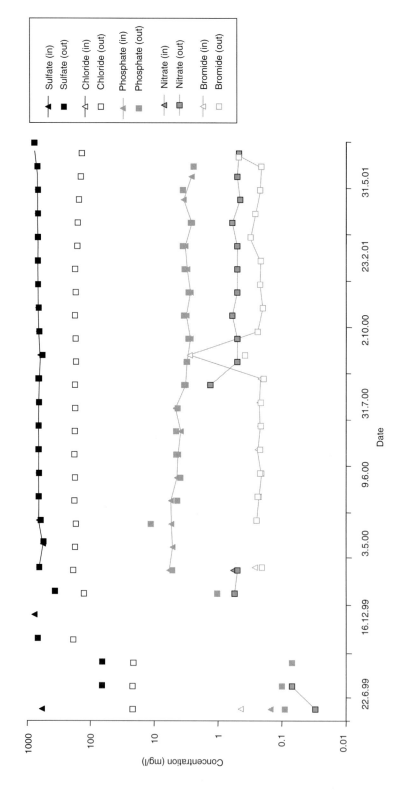

Fig. 10.4 (Contd). Anion and cation concentrations and contaminant concentrations at the inlet and outlet of the reactor. (Data from S. Kraft and P. Grathwohl, University of Tübingen)

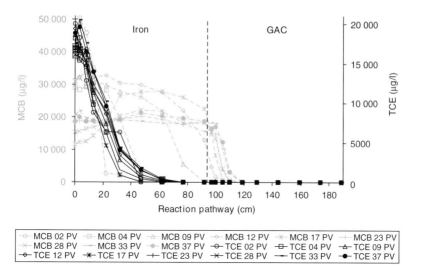

Fig. 10.5. Concentration profiles of TCE and MCB in a sequential Fe^0/GAC system. (Data from R. Köber, D. Schäfer and A. Dahmke, University of Kiel)

10.5.1.2. Microbiology

As opposed to the *a priori* on-site tests, anaerobic degradation of monochlorobenzene could not be initiated in the pilot facility. The focus is now directed towards coupled microaerobic/anaerobic treatment zones.

10.5.1.3. Catalysis

Zeolite- as well as membrane-supported palladium catalysts showed a high efficiency for the degradation of aliphatic as well as aromatic chlorinated hydrocarbons (Schüth *et al.*, 2000). However, palladium-catalysts are deactivated by the production of H$_2$S, due to the microbiological reduction of sulfate. Attempts to suppress microbial activities to increase the longevity by applying periodical H$_2$O$_2$ pulses have so far shown only limited success.

10.5.1.4. Coupled redox reactors

Coupled reactors of iron and granular activated carbon (Fe^0/GAC) are capable of removing mixed contaminations of aliphatic (tetrachloroethylene, TCE) and aromatic chlorinated hydrocarbons (monochlorobenzene, MCB) (Fig. 10.5). A sequential set-up of iron followed by activated carbon has shown to be advantageous compared with a single reactor containing a mixture of iron and carbon. This sequential set-up can increase the longevity of a reactive zone of GAC (Köber *et al.*, 2001b).

Coupled systems of iron- and oxygen-releasing compounds (ORCs) are also capable of removing mixed contaminations of aliphatic (TCE) and aromatic chlorinated hydrocarbons (MCB) (Fig. 10.6). Although the pH of the groundwater is raised significantly (pH 10) in the ORC segment, the MCB is completely degraded microbiologically. Investigations show that the release of oxygen does not depend on the

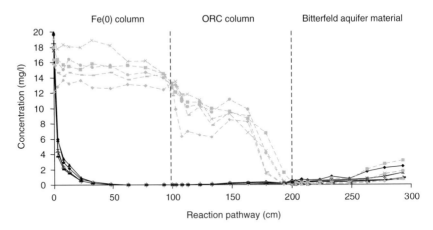

Fig. 10.6. Concentration profiles of TCE and MCB in a sequential Fe⁰ and ORC system. (Data from R. Köber, D. Schäfer and A. Dahmke, University of Kiel)

pH of the injected water. However, the release of oxygen is decreased by an unknown process if Fe^0 and ORC are coupled in sequence. The release of oxygen is not dependent on the pH; therefore, high pH values downgradient of Fe^0 reactive walls will not be rate-limiting to the oxygen release from the ORCs (Köber et al., 2001a).

10.6. Conclusions

During 1 year of large-scale in situ application the different research groups have been adjusting and optimizing their approaches to remediation of the complex groundwater contamination at the Bitterfeld site. Several technologies (activated carbon/coupled redox systems) have proved their general applicability under the anaerobic in situ conditions. The technologies based on catalysis showed their general potential for degrading chlorinated organic compounds but encountered difficulties of catalyst poisoning, due to H_2S production in the process and possible as yet undetermined sulfur species. Over the second year of the project the applied technologies will be tested and optimized based on the experience gained in the initial phase, incorporating new ideas into the existing experimental set-ups where necessary.

New insights into the large variety of contaminant mixtures encountered in the Bitterfeld area (including chlorinated aromatic and aliphatic hydrocarbons as well as herbicides) show that it is most likely that combinations of various technologies will need to be applied in order to address the contaminant problem on a regional scale.

10.7. Appendix

Scientific/technological approaches and location of research groups.

Scientific/technological approach	Location of research group
Project and test site management	UFZ Leipzig University of Tübingen
Determination of metabolites and degradation products of *in situ* technologies	UFZ Leipzig
The influence of groundwater specific parameters on the performance of *in situ* reactors based on the example of *in situ* activated-carbon filtration	University of Tübingen
Elimination of volatile organic compounds by adsorption and simultaneous microbiological degradation on activated carbon	Technical University of Dresden
Development of coupled *in situ* reactors and optimization of the geochemical processes downstream of different *in situ* reactor systems	University of Kiel
Microbiological *in situ* remediation of aquifers contaminated with chloro-organic contaminants at the Bitterfeld study site using autochthonous bacteria	UFZ Leipzig
Dechlorination of chlorohydrocarbons in groundwater by electrochemical and catalytic reactions	UFZ Leipzig
Zeolite-supported catalysts for hydrodehalogenation of contaminants in groundwater	University of Tübingen
Degradation of halogenated organic pollutants by ultrasound-assisted catalytic oxidation	University of Leipzig
Reactive transport in technical reactors	University of Tübingen
Direct oxygen injection into natural aquifer sections and artificial porous media: gas–water dynamics and heterogeneous reactions between mixed fluid phases	UFZ Leipzig
Determination of the contaminant loads in contaminated aquifers for the design of *in situ* reactors	University of Tübingen
Ecotoxicological determination of the performance of *in situ* treatment measures in contaminated aquifers	UFZ Leipzig
Development of model approaches to assess the general effects of groundwater remediation by permeable reactive barriers	University of Halle
Technical and environmental–economic assessment of reactive barriers	University of Tübingen

10.8. References

KÖBER, R., EBERT, M., SCHÄFER, D. and DAHMKE, A. (2001a). Kombination von Fe^0 Reaktionswänden und ORC zur Behandlung komplexer Mischkontaminationen im Grundwasser. *Terra Tech* **3**, 1–6.

KÖBER, R., EBERT, M., SCHÄFER, D. and DAHMKE, A. (2001b). Kombination von Fe^0 und Aktivkohle in Reaktionswänden zur Sanierung komplexer Mischkontaminantionen im Grundwasser. *Altlasten Spektrum* **2**, 91–95.

MERKEL, P., WEISS, H., TEUTSCH, G. and RIJNAARTS, H. (2000). Innovative reactive barrier technologies for regional contaminated groundwater. *Proceedings of the 7th International FZK/TNO Conference on Contaminated Soil, 18–22 September 2000, Leipzig*, pp. 532–540. Thomas Telford, London.

SCHÜTH, C., DISSER, S., SCHÜTH, F. and REINHARD, M. (2000). Tailoring catalysts for hydrodechlorinating chlorinated hydrocarbon contaminants in groundwater. *Applied Catalysis B: Environmental* **28**, 147–152.

WEISS, H., TEUTSCH, G. and DAUS, B. (eds) (1997). Sanierungsforschung in regional kontaminierten Aquiferen. *UFZ Bericht*, No. 27.

WEISS, H., DAUS, B. and TEUTSCH, G. (eds) (1999). SAFIRA, 2. Statusbericht Modellstandort, Mobile Testanlage, Pilotanlage. *UFZ Bericht*, No. 17.

WEISS, H., RIJNAARTS, H., STAPS, S. and MERKEL, P. (eds) (2000). SAFIRA abstracts of the workshop of November 17–18, 1999 at Bitterfeld/Germany. *UFZ Bericht*, No. 23. Website: http://safira.ufz.de/.

11. Electrokinetic techniques and new materials for reactive barriers

K. Czurda, P. Huttenloch, G. Gregolec and K. E. Roehl
Karlsruhe University (TH), Institute for Applied Geology (AGK), Kaiserstrasse 12, D-76128 Karlsruhe, Germany

11.1. Introduction

Passive groundwater remediation using permeable walls is a new and innovative technology. Permeable reactive barriers (PRBs) are subsurface constructions situated across the flow paths of contaminant plumes. The targeted contaminants are removed from the groundwater flow by geochemical processes, such as adsorption, chemical bonding, oxidation/reduction and precipitation (US Environmental Protection Agency, 1998). The technology appears to be a promising approach to groundwater remedi- ation in complex cases where conventional pump-and-treat methods and/or micro- biological techniques have proved unsuccessful. Such cases may include contaminated sites where, for instance, heavy metals are being leached slowly from a contamination source, or where microbial degradation is insufficient (as for polyaromatic hydrocarbons with low bioavailability), or where the contamination is situated in heterogeneous sediments.

Because experience with full-scale installations of PRB systems is limited, little is known about their long-term behaviour and performance effectiveness. Ongoing research by the various groups working on reactive barriers will need to show in the near future whether permeable reactive walls can be established as an accepted, reliable and cost-effective technology for the remediation of contaminated ground- water. Among approaches to develop the PRB technique further are methods to increase their long-term efficacy, and new materials selectively targeting the removal of specific contaminants from the groundwater plume.

11.2. Electrokinetic techniques

Electrokinetic methods have been considered increasingly in recent years for remediation of contaminated clayey soils. Also studied currently is the feasibility of using electrokinetic methods to positively affect the long-term efficiency of reactive barriers. The general aim of the addition of electrokinetic processes to reactive barrier systems is to reduce, by fencing them off at the barrier inflow, the amounts of groundwater constituents that might impair the barrier function by coating or clogging with precipitates.

11.2.1. Electrokinetic soil remediation

The electrokinetic decontamination of polluted sites has become one of the most promising *in situ* treatment technologies, particularly for fine-grained soils where conventional methods such as pump and treat fail. Electrokinetic phenomena in cohesive soil have been used since the beginning of the last century for geotechnical purposes (Casagrande, 1952). In the environmental engineering field, most work on electrokinetics was performed at the end of the 1980s and in the 1990s (e.g. Lageman *et al.*, 1989; Acar *et al.*, 1992; Bruell *et al.*, 1992; Probstein and Hicks, 1993; Acar and Alshawabkeh, 1996; Alshawabkeh and Acar, 1996).

For electrokinetic soil remediation, electrodes are placed in the ground and a direct-current (DC) electric field is applied, which induces movement of the contaminants to the electrode reservoirs (Fig. 11.1). The principal electrokinetic phenomena are electro-osmosis, electromigration and electrophoresis (Probstein, 1994) (Fig. 11.2). The major physical mechanisms of the electric-field-driven transport of contaminants are electromigration and electro-osmosis in fine-grained soils (Probstein and Hicks, 1993), and electromigration and electrophoresis in coarse-grained soils.

11.2.2. Fundamental transport processes

11.2.2.1. Electro-osmosis

Electro-osmosis is defined as the movement of liquid relative to a stationary charged surface. This phenomenon is based on the theory of a diffuse double layer around a charged soil particle (Fig. 11.2). Fine-grained sediments are often dominated by clay minerals. Their usually negatively charged surfaces influence the distribution of ions

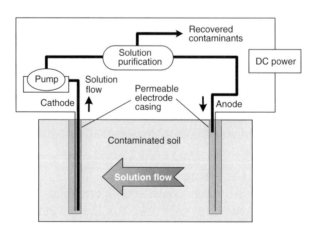

Fig. 11.1. Schematic representation of electrokinetic soil remediation (Haus and Zorn, 1998)

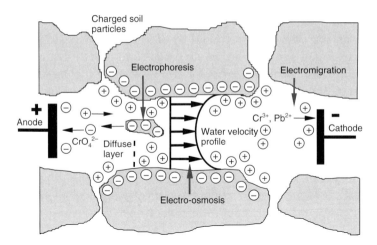

Fig. 11.2. Electrokinetic phenomena induced by an applied electric field

in solution, attracting ions of opposite charge (counter-ions) towards the surface and repelling ions of equal charge (co-ions) away from the surface. The region close to the charged surface is defined as the diffuse double layer and is characterized by an excess of counter-ions over co-ions (Probstein, 1994). By applying an electric field tangentially along the charged surface, this electric field will exert a force on the charge of the diffuse layer.

A movement of the ions will only take place if the force of the applied electric field exceeds the electrostatic force of attraction to the surface. The moving ions in the diffuse double layer are assumed to drag water by viscous interaction through the pores towards the electrodes. Since the double layer contains more cations than anions, there is a net water flow towards the cathode. This process is called electro-osmosis. The electro-osmotically induced water transport is proportional to the electric gradient and to the zeta potential of the charged mineral surface. Due to its occurrence in the diffuse double layer of charged mineral surfaces, electro-osmosis is, on a first approximation, independent of the pore size distribution. In fine-grained, clayey sediments, electro-osmotically induced water flow can become predominant due to their often very low hydraulic conductivity. By applying electric gradients of 100 V/m, electro-osmotic velocities of several centimetres per day can be achieved.

11.2.2.2. Electromigration

When applying an electric field to a soil mass with some moisture content, a force is exerted on the charged ions and molecules dissolved in the pore fluid. This force leads to a mass transfer of these ions and molecules, which is called electro-migration. Anions are moved towards the anode (positive electrode), and cations are moved towards the cathode (negative electrode).

Migration velocity is proportional to the ionic charge, the local electric field, and the ion mobility. The mobility of ions in a free dilute solution – i.e. the velocities of ions in a unit electric field – is in the range of 1×10^{-8} to 1×10^{-7} m^2/Vs (Dean, 1992). However, the effective ionic mobility of ions in soils is usually considerably lower, as the flow paths in soils are much longer and more tortuous than those in aqueous solutions. Practical ranges of effective ionic mobility of ions in fine-grained soils are given by Mitchell (1993).

In fine-grained soils, transport processes by electromigration are typically more efficient than by electro-osmosis, as the driving force is directly affecting the molecules and not the bulk liquid (Jacobs and Probstein, 1996). The transport velocity by electromigration is 5–40 times higher than the electro-osmotic transport velocity (Acar and Alshawabkeh, 1993). In coarse-grained soils the amount of active mineral surfaces is comparatively negligible. As a consequence the effect of electro-osmosis is very low, and electromigration becomes the dominant process.

11.2.2.3. Electrophoresis

Electrophoresis (Fig. 11.2) is defined as the movement of charged particles relative to a stationary fluid under the influence of an electric field (Probstein, 1994). Negatively charged particles will move towards the anode, whereas positively charged particles move towards the cathode. Electrophoresis involves discrete particle transport through water, while electro-osmosis involves water transport through a continuous soil particle network (Mitchell, 1993). As charged particles are of a distinct size, electrophoresis can only take place if the pore sizes are large enough. Hence, electrophoresis is negligible in fine-grained soils. This electro-kinetic phenomenon is used for the separation and analysis of colloids, proteins and nucleic acids (Probstein, 1994).

11.2.3. Electrode reactions

When a DC electric field is applied to a wet soil the system consisting of electrodes, power supply and water-saturated soil behaves as an electrolytic cell. Current flow is from the positive anode to the negative cathode, opposite to electron flow. The power supply acts as an electron pump, pushing electrons from the cathode to the anode. To maintain electrical neutrality, oxidation–reduction reactions occur at the electrodes. Ions or molecules receiving electrons at the cathode are reduced. At the anode, electrons are liberated from ions or molecules that are oxidized. Thus, in an electrokinetic system, not only transport processes but also chemical reactions at the electrodes are induced (Table 11.1).

The principal electrode reaction observed is the electrolysis of water. At the cathode, water is reduced, leading to the production of hydrogen gas and hydroxide ions; whereas at the anode, water is oxidized, and oxygen gas and hydrogen ions are generated.

$$\text{Anode:} \quad H_2O \rightarrow 2H^+ + \tfrac{1}{2}O_2 + 2e^- \qquad E^\circ = 1.23 \text{ V} \qquad (1)$$

$$\text{Cathode:} \quad 2H_2O + 2e^- \rightarrow 2OH^- + H_2 \qquad E^\circ = -0.83 \text{ V} \qquad (2)$$

where E° is the standard electrode potential.

Table 11.1. Electrochemical processes during electrokinetic remediation (Zorn et al., 2000)

	Cathode (−)	Anode (+)
Process	Reduction	Oxidation
Redox processes	$4H_2O + 4e^- \rightarrow 2H_2(g) + 4OH^-$	$2H_2O \rightarrow O_2(g) + 4H^+ + 4e^-$ Possible oxidation of organic substances to CO_2
pH	Alkaline	Acidic
Heavy metals	Precipitation (oxides, hydroxides, carbonates, etc.)	Dissolution

Therefore, a high-pH front is produced at the cathode, whereas at the anode a low-pH front is generated. Both fronts advance towards the opposite charged electrode by electromigration, diffusion and advection (including electro-osmotic flow). When the two fronts meet, the soil between the electrodes is divided into two zones, a low- and a high-pH zone, with a sharp pH jump in between.

The location of the pH jump depends on several factors, and usually lies closer to the cathode. One factor affecting the location of the pH jump is the relative mobility of hydrogen ions and hydroxide ions. The hydrogen ion has an ionic mobility about twice as high as the hydroxide ion. Electro-osmotic flow typically favours transport towards the cathode, and hence it favours the advancement of the acid front. The concentration and mobility of other ions present in the solution will also affect the location of the pH jump, by influencing the distribution of the electric field, and by forming complexes with hydroxide ions or hydrogen ions. Also, the pH buffer capacity and cation exchange capacity of the medium, and interactions of the solution with the soil, may affect the speed of advancement of the acid and base fronts and the location of the pH jump.

The development of the pH gradient can have a significant effect on the magnitude of electro-osmosis, as well as on solubility, ionic state and charge, and level of adsorption of the contaminants (Probstein and Hicks, 1993).

11.2.4. Application

Electroremediation is applicable to both organic and inorganic contaminants, as well as charged and uncharged species. Electrokinetics applied to fine-grained soils is essentially a process of soil flushing, but has several advantages over the usual pressure-driven pumping technology. The transport rate induced by an electric field is not adversely affected by low soil permeability, and the path followed by the contaminants is confined by the electric field to the region between the electrodes. Electroremediation is therefore advantageous in soils of low or variable permeability, and in situations where dispersion of the contaminants must be prevented (Haus *et al.*, 1999).

Compared with conventional remediation techniques, electroremediation is in principle universally applicable with regard to both types of contaminant (organic, inorganic) and process technologies (on site, off site, *in situ*). Due to the confined transport of contaminants between its electrodes, the electrokinetic remediation technique can be applied even where the contamination source is out of reach, e.g. below buildings. Often the contamination source is located at great depth. Here electroremediation provides a well-directed decontamination, whereby the bore-holes previously used for investigation probes can be reused for insertion of the anode and cathode. The problematic situation of alternating layers of clay/silt and coarser-grained soils can also be controlled by this technique. Moreover, in the case of changing geological conditions, the combination of electroremediation with hydraulic and microbiological techniques is possible (e.g. Czurda and Weiss, 1998; Davis-Hoover *et al.*, 1999; Jackman *et al.*, 1999).

Recent research is concerned with the possibility of combining electrokinetics with the *in situ* groundwater remediation technology of PRBs. This research is integrated in the PEREBAR project (Long-term Performance of Permeable Reactive Barriers used for the Remediation of Contaminated Groundwater), a European project initiated within the 5th Framework Programme of the EU. Its task is to study the feasibility of electrokinetics to positively affect the long-term efficiency of PRBs. The principal idea is, by applying an electric field upstream of the barrier, to reduce the amount of groundwater constituents that flow into the barrier and that might impair barrier function by coating the grains of its reactive medium or clogging it with precipitates.

The possibility of coupling electrokinetics with treatment zones within fine-grained soils has been documented in laboratory and field experiments (Ho *et al.*, 1995, 1999a,b). The process has been called 'lasagna', due to its layered configuration of electrodes and treatment zones (Fig. 11.3). Here the contaminants are transported into treatment zones by electrokinetics, with the aim of avoiding

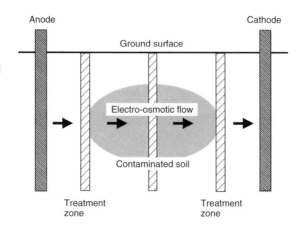

Fig. 11.3. Typical configuration of electrodes and treatment zones used in the lasagna technique (Ho et al., *1995)*

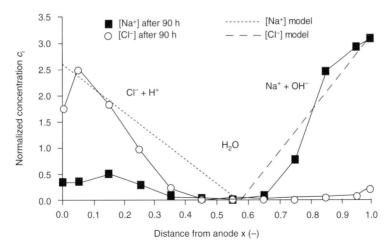

Fig. 11.4. Experimental and modelled separation of Na^+ and Cl^- in sandy soil caused by the application of an electric field

additional treatment steps, for instance the recycling of the water collecting in the electrode reservoirs. Laboratory and field experiments with the lasagna technique have shown promising results, for instance for the removal of *p*-nitrophenol (PNP) from clayey soils using activated carbon in the treatment zone, and trichloroethylene (TCE) using activated carbon and zero-valent iron.

PRBs are used for the remediation of groundwater, and therefore usually placed in aquifers. For utilization of electrokinetic techniques in combination with reactive barriers the following aspects have to be considered:

- *Electrokinetics in the aquifer material.* Though literature studies show that many laboratory and field experiments have proved the applicability of electrokinetics as an *in situ* remediation technique for the decontamination of fine-grained soils – where electro-osmosis plays an important role – little information is available on electrokinetic phenomena within coarse-grained soils (e.g. Kim and Lee, 1999).
- *Combined electric and hydraulic gradients.* No quantitative experience is reported concerning the electromigrative transport of charged species in relation to both electric and hydraulic gradients within aquifer materials or coarse-grained soils.

Thus, as a first step, the fundamental electromigration behaviour of ions under the combined influence of both electric and hydraulic gradients in a coarse-grained sediment had to be investigated. Small-scale experiments were conducted with a simple soil–solution system. Initial results show that ions are transported towards the oppositely charged electrodes under the influence of an electric field and that the relevant electrode reactions are taking place (Fig. 11.4). These results were validated by theoretical simulations based on work by Dzenitis (1997). Further details of the experiments are reported by Gregolec *et al.* (2001).

11.3. Innovative sorbents for PRBs

Sorption barriers are PRBs that utilize retention mechanisms leading to the fixation of the target contaminants to the matrix of the reactive barrier material. The choice of reactive materials and their retention mechanisms are dependent on the type of contamination to be treated by the barrier system. Possible materials for use as reactive components in sorption barriers are activated carbon, natural zeolites, fly-ash zeolites, iron oxides/oxyhydrates, diatomite, phosphate minerals and clay minerals, among others. The potential of zeolites and surface-modified minerals is discussed in the following section.

11.3.1. Zeolites

Zeolites are tectosilicates with a three-dimensional alumosilicate structure containing water molecules and alkali and alkaline earth metals in their structural framework (Gottardi and Galli, 1985).

Zeolites have a high potential as contaminant sorbents due to their high exchange capacity and their selectivity for certain constituents such as NH_4, Pb, Cd and Sr (Pansini, 1996), especially when they are activated by sodium chloride (Ouki et al., 1993). The selectivity of certain zeolite minerals for specific chemical compounds is defined by the pore size and charge properties of the zeolite structure. The unbalanced substitution of Si^{4+} by Al^{3+} in the crystal lattice leads to a net negative charge and, subsequently, to the high cation exchange capacity of most natural zeolites.

Although good experience exists for the use of natural zeolites for drinking water preparation and for removal of heavy metals from industrial waste waters (Blanchard et al., 1984; Zamzow et al., 1990), there is little knowledge of their use in PRBs. The zeolite minerals chabazite and clinoptilolite show a high sorption capacity for lead and cadmium (Ouki et al., 1993). The use of clinoptilolite in permeable barriers to attenuate strontium was studied by Cantrell et al. (1994). Anderson (2000) showed that certain zeolite minerals with high SiO_2/Al_2O_3 ratios have excellent sorption properties for methyl-t-butyl ether (MTBE), chloroform and TCE. The sorption of MTBE and TCE to these zeolites was significantly better than to powdered activated carbon.

The use of engineered zeolites from fly ash as additives to reactive barriers is discussed by Czurda (1999) and Czurda and Haus (2001). By activation of hard-coal pulverized fly ash, using strong bases and elevated temperature, different zeolite phases were obtained. Zeolite formation was dependent upon experimental conditions such as the molarity of the alkali solution, SiO_2/Al_2O_3 ratio of the fly ash, reaction time and temperature. The neoformation of zeolites was proven by X-ray diffraction and scanning electron microscope analyses. Total specific surface areas and cation exchange capacity (CEC) of the treated fly ash increased significantly, with the CEC reaching 160–210 meq/100 g (Czurda, 1999).

In the following, some basic parameters of a sorptive and permeable material are discussed, using a natural zeolite as an example. A clinoptilolite-rich tuff from

Northern Carpathia was selected for this study, to represent a cost-effective natural zeolite. The mineral content is 90% clinoptilolite, plus additional quartz, feldspar and illite. The clinoptilolite was used both as received (clinoptilolite) and in the sodium-exchanged form (Na-clinoptilolite), pretreated according to the method described by Misaelides and Godelitsas (1995). The CEC, determined by the ammonium acetate method (Huang and Roads, 1989), was 145 meq/100 g for the clinoptilolite, and 180 meq/100 g for the sodium-exchanged form.

For sorption barriers, an essential step in planning is the identification of reactive materials as sorbents for the target contaminants, based on the type of the contamination and the characteristics of the site. Relying on laboratory experiments to evaluate the efficiency of the sorption process under the given conditions, an assessment of the sorption capacity can be made, and dimensions for the reactive barrier can be derived.

To be suitable for use as a reactive component in PRBs, zeolites must meet the following important, and sometimes conflicting, conditions:

- high sorption capacity and selectivity for the target contaminants;
- fast reaction kinetics;
- high hydraulic permeability.

11.3.1.1. Quantitative description of sorption processes

The retardation coefficient R summarizes the sorption processes that lead to immobilization of the contaminant. R can be calculated from the relationship between sorbed and aqueous concentrations of the contaminant:

$$R = 1 + \frac{\rho_d}{n} \frac{\partial f(c)}{\partial c} \tag{3}$$

where ρ_d is the bulk density (g/cm³), n is the porosity of the sorptive material and $f(c)$ is a linear or non-linear relationship between the sorbed and aqueous concentrations of the contaminant that can be described by sorption isotherms of the form $S = f(c)$.

The performance of a sorption barrier depends on the sorption capacity of the sorbent. Once that sorption capacity is exhausted, breakthrough of the contaminant will occur. The sorption capacity of a given material for the target contaminant can be obtained from relatively simple laboratory batch experiments evaluated by means of sorption isotherms. Sorption isotherms are widely used to quantitatively describe observed sorption reactions. They are especially useful for the assessment of the sorption capacities of different sorbents and their selectivity for adsorbing certain chemical compounds.

The isotherms presented in Fig. 11.5 show the sorption of copper from deionized water and from a 0.01 M $CaCl_2$ matrix solution on clinoptilolite and Na-clinoptilolite, respectively. The isotherms indicate a non-linear sorption behaviour between the sorbed and aqueous concentrations of the contaminant, which can be described by the Freundlich equation:

$$c_s = K_F c_w^{N_F} \tag{4}$$

where c_s is the amount of the solute sorbed per unit mass of the sorbent, c_w is the equilibrium solute concentration, and K_F and N_F are empirical parameters specific to the sorption material used. The Freundlich parameters are listed in Table 11.2.

Because of its non-linearity, sorption from the copper solution increases relative to decreasing copper concentration. The sodium-exchanged zeolite generally shows a higher affinity for copper than the untreated clinoptilolite. The adsorption isotherms and the resulting Freundlich parameters indicate a significant decrease in sorption for experiments conducted with $CaCl_2$ matrix solutions, due to competition between copper and calcium ions for the sorption sites on the zeolite surface. This example cautions that, for an effective evaluation of zeolite material for its sorption capacity, it is necessary to use real groundwater conditions.

11.3.1.2. Reaction kinetics

Adsorption isotherms are based on an assumption that the sorption reactions reach equilibrium. In sorption barriers, the residence time of the target contaminant in the sorptive matrix must be sufficient to reach the reaction equilibrium. Therefore, the reaction rate of the sorption process employed in the barrier must be sufficiently high compared with the flow velocity of the contaminated groundwater through the barrier.

Most of the pure surface reactions, such as adsorption and ion exchange, are relatively fast. However, the rate-limiting effect is the migration of the target contaminant from the outer surface of the granular particles that build the

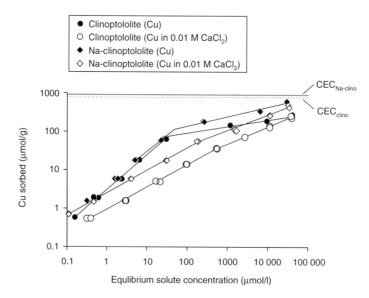

Fig. 11.5. Freundlich adsorption isotherms of copper on clinoptilolite and Na-clinoptilolite in de-ionized water and in 0.01 M $CaCl_2$ solution, respectively (1 g sample in 40 ml copper solutions, contact time 96 h, at 20°C). The CEC is also represented (clinoptilolite, 720 $\mu mol_{eq}/g$; Na-clinoptilolite, 900 $\mu mol_{eq}/g$)

Table 11.2. Freundlich parameters for copper (pure solution) and copper dissolved in 0.01 M CaCl$_2$ matrix solution on clinoptilolite and Na-clinoptilolite

Sample	Low concentration range		High concentration range	
	K_F (μmol^{1-N} g^{-1} l^N)	N_F (–)	K_F (μmol^{1-N} g^{-1} l^N)	N_F (–)
Clinoptilolite				
Cu	3.08	0.90	41.39	0.17
Cu + CaCl$_2$	0.95	0.58	2.10	0.45
Na-clinoptilolite				
Cu	4.25	0.85	43.68	0.25
Cu + CaCl$_2$	2.54	0.58	6.67	0.40

permeable barrier material to fixation sites inside those grains, i.e. within the cages of the zeolite crystal lattice. This migration occurs by diffusion, and is therefore a relatively slow process. The more permeable the wall needs to be, the coarser the reactive material which must be chosen, resulting in a less reactive mineral surface and longer reaction times. For these reasons it is necessary to determine the reaction kinetics of the reactive barrier fillings.

Figure 11.6 shows the adsorption kinetics of copper on clinoptilolite and Na-clinoptilolite, respectively, which is dependent on the size of the clinoptilolite aggregate and the solute concentration. The three different grain sizes studied show different sorption behaviour, with increasing copper uptake as the grain size decreases. The initial sorption rate is higher for small grain sizes than for large grain sizes. The decreasing adsorption rate with increasing particle size is explained by diffusion processes in the particles, which take longer for large particles, therefore increasing the time that is needed to establish equilibrium. Compared with clinoptilolite, Na-clinoptilolite shows a better and a faster sorption behaviour for all grain sizes, due to the preferential exchange of univalent sodium ions for the divalent copper ions.

11.3.1.3. Column studies

An approximation to the environment in reactive barriers is given by column tests, where sorption behaviour can be determined under dynamic flow conditions. The breakthrough of the contaminant is, as a first approximation, defined as the point where half the concentration of the input solution is detected in the column eluate ($c/c_0 = 0.5$). The dimensionless retardation coefficient R_d (see equation (3)) is the number of pore volumes percolated through the column at the breakthrough point.

The breakthrough curves for copper (100 mg/l of copper dissolved in 0.01 M CaCl$_2$ matrix solution) percolated through clinoptilolite and Na-clinoptilolite columns (sand/zeolite mixtures containing 10% zeolite) of different grain sizes and at different flow velocities are shown in Fig. 11.7. The resulting retardation coefficients R_d (also for 10 mg/l copper solutions) are listed in Table 11.3.

As expected, the retardation coefficients resulting from the column tests verify the type of sorption behaviour observed in the batch tests. The following conclusions can be drawn:

- sorption capacity increases with decreasing grain size;
- sorption increases at lower contaminant concentrations (for non-linear sorption) and fewer competing ions (competition effect);
- sorption increases with longer reaction times;
- compared with clinoptilolite, the Na-clinoptilolite shows a higher affinity for contaminants (copper) in combination with faster reaction kinetics.

The studies show that zeolites, especially sodium-activated zeolites, are potential sorbents for reactive barrier fillings due to their high affinity for heavy metals (e.g. copper). For effective evaluation of the zeolite material it is necessary to determine the sorption behaviour under real groundwater conditions.

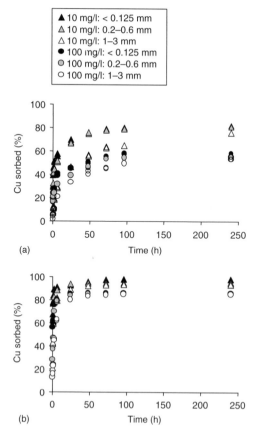

Fig. 11.6. Kinetics with copper (0.01 M CaCl$_2$) for (a) clinoptilolite and (b) Na-clinoptilolite in relation to aggregate size and solute concentration

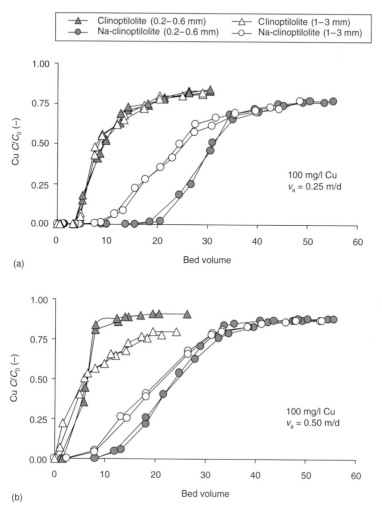

Fig. 11.7. Copper breakthrough curves (100 mg/l of copper in 0.01 M CaCl₂ solution) for (a) clinoptilolite and (b) Na-clinoptilolite, in relation to grain size and apparent flow velocity

11.3.2. Chlorosilane surface-modified natural minerals

The sorptive properties of mineral materials can sometimes be improved significantly by surface modification. The surface modification of minerals by cationic surfactants, to produce, for example, organophilic zeolites and organoclays, is a well-documented process. For instance, the use of hexadecyltrimethyl- ammonium (HDTMA) modified zeolites in reactive barriers is discussed by Bowman (1999). The mechanism of surface modification is based on the exchange of extrastructural cations of the zeolites or other clay minerals by organic cations such as HDTMA up to the external CEC. The affinity of these modified minerals for organic contaminants can be greatly improved by the addition of surfactants.

A different approach is the use of chemical compounds that bind covalently to the mineral surface, leading to high stability of the surface modification.

11.3.2.1. Silanization

Silicon compounds have a high affinity for oxygen. Chlorosilanes, in particular, react readily with free hydroxyl groups to form a thermodynamically stable silicon–oxygen bond. The separation of chlorine from the silane, and hydrogen from the silanol group, forming HCl, enables the covalent attachment of an SiR_3 group (chlorosilane) to a silanol group (mineral surface), forming a Si–O–Si–C moiety. The more free silanol groups that are present on the surface of the material, the more organosilane molecules can be attached. Similar types of organosilane-modified silica have proved to be very effective for chromatographic purposes in reverse-phase high-performance liquid chromatography (HPLC) (Gottwald, 1993). The results of a study on the surface modification of natural zeolites and diatomite with chlorosilanes are discussed below.

Diatomite, also known as kieselguhr, is a siliceous sedimentary rock consisting of the fossilized remains of diatoms. The siliceous skeletons arise from condensing silicic acid that results in a three-dimensional polymerization of silicatetrahedrons and builds an X-ray amorphous opal structure (Barron, 1987). Because of the diatomite structure (free silanol groups) they are suitable for surface modification. The diatomite used in the present study was obtained from United Minerals. It is a calcined product with a particle size range of 0.25–0.85 mm, consisting of 90% X-ray amorphous silica.

The natural zeolite used in the study was a clinoptilolite-rich tuff from Northern Carpathia containing 90% clinoptilolite, plus additional quartz, feldspar and illite. The clinoptilolite was transformed into the H-form (Chen, 1976) before surface modification in order to obtain more free silanol groups. This was achieved by dealumination with 2 M HCl for 1 h (30 ml of acid per gram of zeolite) and subsequent drying at 750°C. The particle size range was 0.2–1.0 mm.

Table 11.3. Retardation coefficients for clinoptilolite and Na-clinoptilolite (sand–zeolite mixtures containing 10% zeolite) resulting from column test with copper (in 0.01 M CaCl_2 matrix solution) in relation to grain size, apparent flow velocity and concentration

Concentration (mg/l):	10		100	
Apparent flow velocity (m/day):	0.25	0.50	0.25	0.50
Clinoptilolite				
0.2–0.6 mm	21.75	16.5	9.0	6.5
1–3 mm	18.50	12.7	7.75	5.75
Na-clinoptilolite				
0.2–0.6 mm	48.0	30.5	36.5	25.0
1–3 mm	46.0	24.75	32.0	22.0

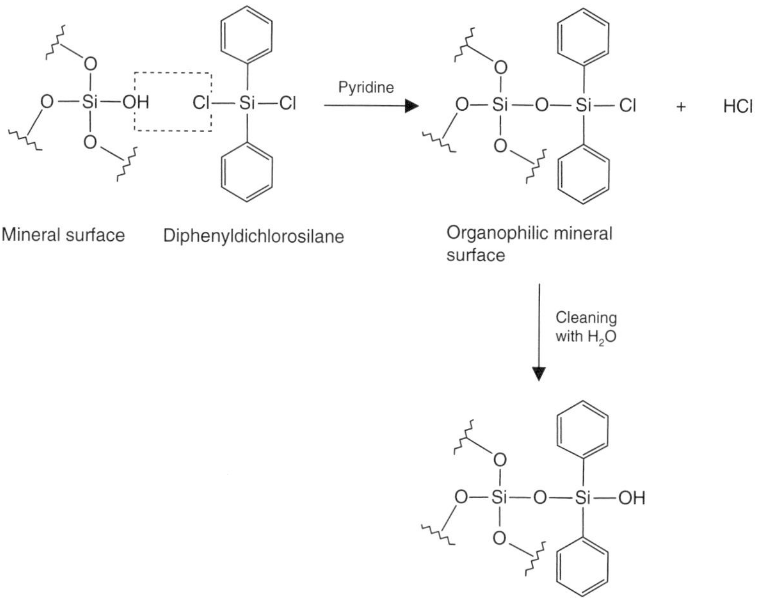

Fig. 11.8. Silanization reaction of mineral surface silanol groups with DPDSCl

The chlorosilane used was a diphenyldichlorosilane (DPDSCl), which in pre-liminary experiments showed the best surface modification results compared with other chlorosilane types. Before surface modification, the diatomite and zeolite materials were dried at 105°C. The materials were treated with DPDSCl in pyridine at 80°C under a nitrogen atmosphere. Pyridine was used as both an organic solvent and a buffer for developing the hydrochloric acid. The whole silanization procedure was repeated to improve surface loading and sorption capacity. The silanization reaction with DPDSCl for diatomite is shown in Fig. 11.8. Details of the procedure are reported by Huttenloch *et al.* (2001).

The success of the surface modification was documented by measurement of the physical and chemical properties of the minerals, which changed dramatically after surface modification (Table 11.4).

The total organic carbon (TOC) content increased from 0.1% for the untreat-ed material up to 2.8% for diatomite and 1.7% for clinoptilolite modified with DPDSCl. The TOC content – resulting from DPDSCl – is 60% higher for diatomite compared with clinoptilolite, due to a higher amount of surface silanol groups. The water adsorption capacity (Enslin/Neff) was reduced from 120% for the natural diatomite to zero, due to the lipophilic nature of the modified diatomite surface; for clinoptilolite, the water adsorption capacity was only reduced from 80% to 30%.

Sorption tests (naphthalene sorption from a 10 mg/l aqueous naphthalene solution) confirmed the successful surface modification. Aromatic organic con-taminants showed a high affinity for the phenyl groups of the surface-modified material. Sorption of naphthalene from water was greatly enhanced by the surface modification, compared with the untreated materials, which showed no measurable sorption of this compound (Table 11.4). The aromatic moieties of the organosilane increase the sorption capacity for organic compounds. The possible types of interaction between the head groups and organic contaminants can be of hydrophobic nature (van der Waals forces), or can be enhanced further by the attraction of aromatic moieties.

11.3.2.2. Sorption of organic compounds

Sorption isotherms (0.3 g of sample material in 20 ml of solution, 24 h contact time, 20°C) were prepared for o-xylene, toluene and naphthalene, for the surface-modified diatomite and clinoptilolite, and the untreated raw materials. Concentration ranges were 1–30 mg/l for o-xylene, 1–100 mg/l for toluene and 1–20 mg/l for naphthalene. Kinetic batch sorption experiments showed that a contact time of 24 h was sufficient to achieve quasi-equilibrium.

Freundlich sorption isotherms (see equation (4)) for o-xylene, toluene and naphthalene are shown in Fig. 11.9. For all isotherms the linear correlation coefficients were better than 0.99. The resulting Freundlich parameters and the solubility of the target organic contaminants (Merian and Zander, 1982) are listed in Table 11.5. Untreated diatomite and clinoptilolite did not show any affinity for the organic compounds.

The sorption behaviour of surface-modified diatomite and clinoptilolite with DPDSCl followed the trend naphthalene > o-xylene > toluene. The K_F values indicate that sorption capacity is correlated with the water solubility of the organic compounds (Table 11.5). This is due to the hydrophobic sorption of aromatic compounds, which obviously increases with decreasing solubility of the solvent (Dzombak and Luthy, 1984).

The non-linearity of the isotherm ($N \neq 1$) of the modified material suggests that the diphenyl groups, which are attached covalently to the mineral surface, act as a hydrophobic surface rather than a hydrophobic phase. Therefore, the hydrophobic

Table 11.4. Water adsorption capacity, TOC content and the extent of naphthalene sorption from a 10 mg/l aqueous naphthalene solution for untreated and surface-modified diatomite and clinoptilolite (surface modification with DPDSCl)

Material	Water adsorption capacity (% by weight)	TOC content (% by weight)	Naphthalene sorption (%)[a]
Untreated diatomite	120	0.1	ND
Surface-modified diatomite	0	2.8	70
Untreated clinoptilolite	80	0.1	ND
Surface-modified clinoptilolite (H-form)	30	1.7	54

[a]Naphthalene removed from a 10 mg/l aqueous naphthalene solution (0.3 g of sample treated for 24 h with 20 ml of solution). ND, not detectable

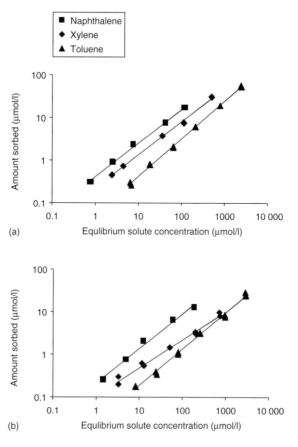

Fig. 11.9. Non-linear adsorption isotherms for naphthalene, xylene and toluene on (a) surface-modified diatomite and (b) clinoptilolite

Table 11.5. Freundlich isotherm parameters for the adsorption of toluene, o-xylene and naphthalene on diatomite and clinoptilolite surface modified with DPDSCl, compared with the water solubility of the organic compounds ($r^2 > 0.99$ for all isotherms). No sorption was detectable for the untreated materials

Compound	Water solubility (mg/l)	Surface-modified diatomite		Surface-modified clinoptilolite	
		K_F ($\mu mol^{1-N}\, g^{-1}\, l^{N}$)	N_F (–)	K_F ($\mu mol^{1-N}\, g^{-1}\, l^{N}$)	N_F (–)
Toluene	515	0.05	0.88	0.03	0.85
o-Xylene	175	0.22	0.78	0.11	0.66
Naphthalene	30	0.41	0.79	0.21	0.81

organic compounds were most likely adsorbed physically on the diphenyl groups of the modified surface, and were not partitioned into the organic phase.

A better coverage of the diatomite surface with DPDSCl (as proved by an increased TOC content) resulted in a better sorption capacity compared with the surface-modified clinoptilolite. In the present study, sorption from an aqueous solution of 1 mg/l each of naphthalene, *o*-xylene and toluene was 84, 73 and 40% for surface-modified diatomite, and 62, 47 and 24% for modified clinoptilolite, respectively.

11.3.2.3. Stability of the surface complex

The surface-modified diatomite and clinoptilolite showed no significant change in their sorption behaviour towards aromatic compounds when exposed to extremes in pH (matrix solutions at pH 3 and 10 adjusted with HCl and NaOH, respectively) and in ionic strength (0.01 M and 1 M $CaCl_2$ matrix solutions). Also, after repeated treatment with pure organic solvent (*o*-xylene) and subsequent regeneration at 60°C for 24 h, no change in sorption behaviour could be observed. It can be concluded that the surface modification of the two materials exhibits great stability even under extreme conditions, which will facilitate their application in PRBs and waste water treatment plants.

11.4. References

ACAR, Y. B. and ALSHAWABKEH, A. N. (1993). Principles of electrokinetic remediation. *Environmental Science and Technology* **A27**(13), 2638–2647.

ACAR, Y. B. and ALSHAWABKEH, A. N. (1996). Electrokinetic remediation. I: pilot-scale tests with lead-spiked kaolinite. *Journal of Geotechnical Engineering* **122**, 173–185.

ACAR, Y. B., LI, H. and GALE, R. J. (1992). Phenol removal from kaolinite by electrokinetics. *Journal of the Geotechnical Division of the ASCE* **118**(11), 1837–1852.

ALSHAWABKEH, A. N. and ACAR, Y. B. (1996). Electrokinetic remediation. II: theoretical model. *Journal of Geotechnical Engineering* **122**(3), 186–196.

ANDERSON, M. A. (2000). Removal of MTBE and other organic contaminants from water by sorption to high silica zeolites. *Environmental Science and Technology* **34**, 725–727.

BARRON, J. A. (1987). Diatomite: Environmental and geological factors affecting its distribution, In: J. R. Hein (ed.), *Siliceous Sedimentary Rock-hosted Ores and Petroleum*, pp. 164–178. Van Nostrand Reinhold, New York.

BLANCHARD, G., MAUNAYE, M. and MARTIN, G. (1984). Removal of heavy metals from waters by means of natural zeolites. *Water Research* **18**(12), 1501–1507.

BOWMAN, R. B. (1999). *Pilot-scale Testing of a Surfactant-modified Zeolite PRB. Ground Water Currents*, EPA-N-99-002. US Environmental Protection Agency, Washington DC.

BRUELL, C. J., SEGALL, B. A. and WALSH, M. T. (1992). Electroosmotic removal of gasoline hydrocarbons and TCE from clay. *Journal of Environmental Engineering* **118**(1), 68–83.

CANTRELL, K. J., MARTIN, P. F. and SZECSODY, J. E. (1994). Clinoptilolite as an *in-situ* permeable barrier to strontium migration in ground water. In: W. Gee and N. R. Wing (eds), *Proceedings of* in situ *Remediation – Scientific Basis for Current and Future Technologies*: 33rd *Hanford Symposium, Columbus*, pp. 839–850.

CASAGRANDE, L. (1952). Electroosmotic stabilization of soils. *Boston Society of Civil Engineering* **49**(1).

CHEN, N. Y (1976). Hydrophobic properties of zeolites. *Journal of Physical Chemistry* **80**(1), 60–64.

CZURDA, K. (1999). Reactive walls with fly ash zeolites as surface active components. In: H. Kodama, A. R. Mermut and J. K. Torrance (eds.), *Proceedings of Clays for our Future: 11th International Clay Conference, Ottawa*, pp. 153–156.

CZURDA, K. and HAUS, R. (2002). Reactive walls with fly ash zeolites as *in situ* groundwater remediation measure. *Applied Clay Science* (in press).

CZURDA, K. and WEISS, T. (1998). Flugasche-Zeolithe als Sorptionsmedium in reaktiven Ton-Barrieren, In: K.-H. Henning and J. Kasbohm (eds), Berichte der Deutschen Ton- und Tonmineralgruppe Vol. 6, pp. 84–87. DTTG, Greifswald.

DAVIS-HOOVER, W. J., ROULIER, M. H., KEMPER, M., VESPER, S. J., AL-ABED, S., BRYNDZIA, L. T., MURDOCH, L. C., CLUXTON, P. and SLACK, W. W. (1999). *Proceedings of Horizontal Lasagna for Bioremediation of TCE, 2nd Symposium, Denmark*, pp. 159–164.

DEAN, J. A. (ed.) (1992). *Lange's Handbook of Chemistry*, 14th edn. McGraw-Hill, New York.

DZENITIS, J. M. (1997). Steady state and limiting current in electroremediation of soil. *Journal of the Electrochemical Society* **144**(3), 1317–1322.

DZOMBAK, D. A. and LUTHY, R. G. (1984). Estimating adsorption of polycyclic aromatic hydrocarbons on soils. *Soil Science* **137**(5), 292–308.

GOTTARDI, G. and GALLI, E. (1985). *Natural Zeolites*. Springer-Verlag, Berlin.

GOTTWALD, W. (1993). *RP-HPLC für Anwender*. Wiley-VCH, Weinheim.

GREGOLEC, G., ZORN, R., KURZBACH, A., ROEHL, K. E. and CZURDA, K. (2001). Coupling of hydraulic and electric gradients in sandy soils. In: K. Czurda, R. Haus, C. Kappeler and R. Zorn (eds), EREM 2001–3rd Symposium and Status Report on Electrokinetic Remediation. *Schriftenreihe Angewandte Geologie Karlsruhe* **63**, 41/1–41/15.

HAUS, R. and ZORN, R. (1998). Elektrokinetische *in-situ*-Sanierung kontaminierter Industrie-standorte. *Schriftenreihe Angewandte Geologie Karlsruhe* **54**, 93–118.

HAUS, R., ZORN, R. and ALDENKORTT, D. (1999). Electroremediation: *in situ* treatment of chromate contaminated soil, In: R. N. Yong and H. R. Thomas (eds), *Geoenvironmental Engineering, Ground Contamination: Pollutant Management and Remediation*, pp. 384–391, Thomas Telford, London.

HO, S. V., SHERIDAN, P. W., ATHMER, C. J., HEITKAMP, M. A., BRACKIN, J. M., WEBER, D. and BRODSKY, P. H. (1995). Integrated *in situ* soil remediation technology, the lasagna process. *Environmental Science and Technology* **A29**, 2528–2534.

HO, S. V., ATHMER, C., SHERIDAN, P. W., HUGHES, B. M., ORTH, R., MCKENZIE, D., BRODSKY, P. H., SHAPIRO, A., THORNTON, R., SALVO, J., SCHULTZ, D., LANDIS, R., GRIFFFITH, R. and SHOEMAKER, S. (1999a). The lasagna technology for *in situ* soil remediation. 1. Small field test. *Environmental Science and Technology* **33**(7), 1086–1091.

HO, S. V., ATHMER, C., SHERIDAN, P. W., HUGHES, B. M., ORTH, R., MCKENZIE, D., BRODSKY, P. H., SHAPIRO, A., SIVAVEC, T. M., SALVO, J., SCHULTZ, D., LANDIS, R., GRIFFFITH, R. and SHOEMAKER, S. (1999b). The lasagna technology for *in situ* soil remediation. 2. Large field test. *Environmental Science and Technology* **33**(7), 1092–1099.

HUANG, C. P. and ROADS, E. A. (1989). Adsorption of Zn(II) onto hydrous aluminosilicates. *Journal of Colloid and Interface Science* **131**(2), 289–306.

HUTTENLOCH, P., ROEHL, K. E. and CZURDA, K. (2001). Sorption of nonpolar aromatic contaminants by chlorosilane surface modified natural minerals. *Environmental Science and Technology* **35**(21), 4260–4264.

JACKMAN, S. A., MAINI, G., SHARMAN, A. K. and KNOWLES, C. J. (1999). The effects of direct electric current on the viability and metabolism of acidophilic bacteria. *Enzyme and Microbial Technology* **A24**, 316–324.

JACOBS, R. A. and PROBSTEIN, R. F. (1996). Two-dimensional modeling of electroremediation. *AIChE Journal* **A42**(6), 1685–1696.

KIM, J. and LEE, K. (1999). Effects of electric field directions on surfactant enhanced electro-kinetic remediation of diesel-contaminated sand column. *Journal of Environmental Science and Health A* **34**(4) 863–877.

LAGEMANN, R., POOL, W. and SEFFINGA, G. A. (1989). Electro-reclamation: theory and practice. *Chemistry and Industry* **18**, 585–590.

MERIAN, E. and ZANDER, M. (1982). Volatile aromatics. In: O. Hutzinger (ed.), *The Handbook of Environmental Chemistry – Anthropogenic Compounds*, Vol. 3, Part B, pp. 117–161. Springer-Verlag, Berlin.

MISAELIDES, P. and GODELITSAS, A. (1995). Removal of heavy metals from aqueous solutions using pretreated natural zeolitic materials: the case of Mercury(II). *Toxicological and Environmental Chemistry* **51**, 21–29.

MITCHELL, J. K. (1993). *Fundamentals of Soil Behaviour*. Wiley, New York.

OUKI, S. K., Cheesman, C. and Perry, R. (1993). Effects of conditioning and treatment of chabazite and clinoptilolite prior to lead and cadmium removal. *Environmental Science and Technology* **27**, 1108–1116.

PANSINI, M. (1996). Natural zeolites as cation exchangers for environmental protection. *Mineralium Deposita* **31**, 563–575.

PROBSTEIN, R. F. (1994). *Physicochemical Hydrodynamics – An Introduction*, 2nd edn. Wiley, New York.

PROBSTEIN, R. F. and HICKS, R. E. (1993). Removal of contaminants from soil by electric fields. *Science* **260**, 498–503.

US ENVIRONMENTAL PROTECTION AGENCY (1998). *Permeable Reactive Barrier Technologies for Contaminant Remediation*, *US EPA Remedial Technology Fact Sheet*, EPA 600/R-98/125. US Environmental Protection Agency, Washington DC.

ZAMZOW, M. J., EICHBAUM, B. R., SANDGREN, K. R. and SHANKS, D. E. (1990). Removal of heavy metals and other cations from wastewater using zeolites. *Separation Science and Technology* **25**, 1555–1569.

ZORN, R., HAUS, R. and CZURDA, K. (2000). Removal of phenol in soils by using DC electric fields. In: R. Herndon and J. John (eds), *Proceedings of Prague 2000*.

Part V
Groundwater remediation following mining activities

12. Kinetics of uranium removal from water

B. J. Merkel
Freiberg University of Mining and Technology, Department of Geology,
Gustav-Zeuner-Strasse 12, D-09596 Freiberg/Saxony, Germany

12.1. Sources of uranium

The environment, and in particular the aquatic environment, receives uranium from different sources, but mainly from the combustion of oil and coal, from phosphate fertilizers, from certain geological layers, from uranium mining and milling activities and, finally, from the nuclear industry. Contamination problems may in particular arise from mining activities, when uranium and related radionuclides endanger groundwater and surface water, which may require clean-up measures. Contamination due to oil and coal combustion or the use of phosphate fertilizers are generally distributed and at a lower level.

According to Merkel and Sperling (1998), uranium concentrations in groundwater vary between 0.1 and 120 µg/l, with a median in the range 0.2–2.2 µg/l. River water has concentrations from 0.03 to 7 µg/l (Langmuir, 1997): the Amazon shows very low concentrations of only 0.03 µg/l; while German rivers are normally in the range of 1–3 µg/l, with the exception of the Zwickauer Mulde (>10 µg/l), due to uranium mining and milling activities in its catchment area. By contrast, river water containing irrigation return flows may exceed 20 µg/l, which is probably mostly due to uranium in phosphate fertilizers (Langmuir, 1997). Phosphate rocks contain between 3 and 400 mg/kg of uranium, some of which is removed during their processing. Some waste from the nuclear industry (uranium enrichment) is also used as a fertilizer, containing 0.1–1 mg/kg of uranium. According to the US Environmental Protection Agency (1999) the uranium concentration in fertilizers varies between 7 and 660 mg/kg and, after 80 years of phosphate fertilizer application in some soils, uranium concentrations have doubled. The concentration of uranium in sea water is significant, in the range of 2.2–3.7 µg/l, which is an indication of the mobility of uranium under oxidizing conditions.

The atmosphere presently receives 50 tonnes/year of uranium from the worldwide combustion of coal (4×10^9 tonnes/year) – assuming 1.3 mg/kg of uranium in the coal, with 99% removed by precipitators. Fly ash and coal ash, however, contain the major part of such uranium (4950 tonnes/year). Both atmospheric and agricultural uranium contaminations are low-level contaminations with respect to the concentrations measured at a single spot. In contrast, uranium mining and milling as well as phosphate mining activities occur only at certain sites but result in high concentrations in the air, soil and water. Groundwater in uranium mines is reported to contain 15–1000 µg/l of uranium, leachate from uranium tailings typically contains 100–20 000 µg/l and process water of *in situ* leaching mines may show concentrations of up to 100 mg/l (Wolkersdorfer, 1996; Langmuir, 1997). Phosphate mining has also to be considered a potential source of uranium contamination, especially if uranium is extracted from the phosphate rock in order to produce uranium as a by-product.

12.2. Toxicity of uranium

Uranium is a naturally occurring radioactive element with several radionuclides and no stable isotope. ^{238}U and ^{235}U are the mother nuclides of two of the three natural radioactive decay chains. Natural uranium contains 99.7% ^{238}U with a half-life of 4.46×10^9 years and only 0.3% ^{235}U with a half-life of 2.45×10^5 years. Nuclear fuel is enriched in ^{235}U to 2–5%, while nuclear weapons contain more than 90% ^{235}U. Both enrichment processes produce huge amounts of depleted uranium, which is mainly used for the production of military projectiles. Because uranium is a predominantly α-emitting radionuclide, there is concern about DNA damage and the promotion of carcinogenesis. However, due to the very long half-life of ^{238}U and the low amount of ^{235}U, which has a much shorter half-life, exposure to natural uranium and depleted uranium at low concentrations is unlikely to be a significant health risk (US Department of Health and Human Services, 1999).

However, in addition to its radiotoxic potential, uranium has considerable chemical toxicity. The kidney has been identified as the most sensitive target of uranium toxicity. This is mediated by accumulation of uranium in the renal tubular epithelium, which induces cellular necrosis and atrophy in the tubular wall. On average, ingested uranium appears to be less toxic than inhaled uranium, since only 0.1–6% of even the more soluble uranium compounds are absorbed in the gastrointestinal tract (US Department of Health and Human Services, 1999). Water-soluble salts appear to be more toxic than insoluble compounds. In general, hexavalent uranium has been found to be more toxic than tetravalent uranium.

Uranium intake from food differs over a wide range (~0.1 µg/l in fruit juice, 0.5–1.3 µg/l in fresh fruit and vegetables, 2–18 µg/l in potatoes, 14 µg/l in beef, 26 µg/l in beef liver, 70 µg/l in beef kidney, 9.5–31 µg/l in shellfish, <0.1–400 µg/l in drinking water (US Department of Health and Human Services, 1999). The US Environmental Protection Agency maximum contamination level (MCL) for uranium was recently changed from 20 to 30 µg/l; however, the World Health Organization (WHO) established a provisional MCL of 2 µg/l, both mainly based on animal experiments. The WHO guideline value for the chemical toxicity of uranium was derived using a total daily intake (TDI) approach and an uncertainty factor of 100 (WHO, 1998).

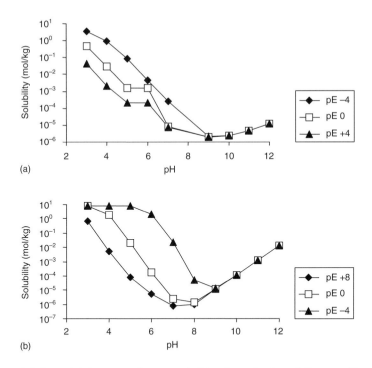

Fig. 12.1. Solubility of (a) uranophane and (b) schoepite for different pE values

12.3. Water chemistry

Uranium occurs in natural aquatic systems in the +4 and +6 oxidation states, as U^{5+} is only metastable. Uranium is a redox-sensitive element, with U^{IV} species predominating in water with very low Eh values. Since the most common uranium minerals are in the U^{IV} oxidation state, and their solubility products are very low, the uranium concentrations under reduced conditions are less than 10^{-9} mol/l. By contrast, U^{VI} species such as UO_2^{2+} predominate in oxidized waters. Only a few U^{VI} minerals (e.g. schoepite and uranophane) are able to limit the solubility of aqueous U^{VI} species even under oxidizing conditions. Uranophane is a uranium mineral which reacts in water according to the following chemical reaction, with a log K_{SP} of 9.42:

$$Ca(H_3O)_2(UO_2)_2(SiO_4)_2{:}3H_2O + 6H^+ \rightleftharpoons Ca^{2+} + 2H_4SiO_4 + 2UO_2^{2+} + 5H_2O \quad (1)$$

Uranophane is of particular interest, since its solubility is positively correlated with the redox potential, and it is more soluble under reducing conditions in water for pH values below 9; most uranium minerals show a higher solubility under oxidizing conditions. Figure 12.1 was calculated by means of PHREEQC and the database Wat4f_U.dat for pure water. If the pE is less than –4 or greater than +4 there is no further change in the solubility behaviour. For pH values greater than 9, the Eh has no effect on the solubility of uranophane.

Figure 12.1 also shows the solubility of schoepite for pE values of –4, 0 and +8. Schoepite, like uranophane, is more soluble under reducing conditions if the pH is less than 9. Under oxidizing conditions there is a solubility minimum between pH 7

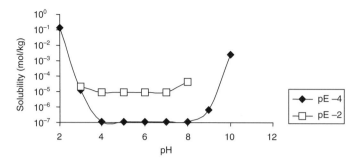

Fig. 12.2. Solubility of amorphous uranium oxide for pE values of –2 and –4

and 8 of about 0.001 mmol/l and 238 μg/l of uranium, respectively, which is still relatively high for natural waters.

By comparison, the solubility of amorphous uranium oxide ($UO_{2(a)}$) is lower under increased reducing conditions but becomes very high with increasing pE. Figure 12.2 shows the calculated $UO_{2(a)}$ solubility for pE values of –4 and –2. For a pE of –4 a solubility of 10^{-7} mol/l was calculated, and for pE values less than –4 even lower concentrations result. Thus, under extremely reduced conditions, $UO_{2(a)}$ is a limiting uranium mineral in the pH range from 4 to 8. If the pH is less than 4 or greater than 8, this is not valid, as can be seen from Fig. 12.2. From this, it can be concluded that at a pH between 6 and 7 and with intermediate pE values, either schoepit, uranophane or amorphous uranium oxide might be the limiting uranium mineral phase controlling the uranium concentrations of groundwater.

12.4. Natural attenuation processes

Unlike organic contaminants, heavy metals are not subject to biological decay. Although uranium is a radioactive element, due to its long half-life, radioactive decay is no solution for the decontamination and clean-up of contaminated sites, particularly since uranium has a number of radioactive and hazardous daughter products (e.g. radium, radon and lead).

Uranium can be removed from water by ion exchange, sorption or precipitation. Since uranium may be present (depending on the ligands available and the pH of the water) as positive, negative or zero-valent species, it may or may not be subject to cation or anion exchange. Sorption, in particular, takes place on freshly precipitated iron hydroxides.

Cation and anion exchange resins can be very effective in uranium removal, and thus this technique is widely used in conventional water treatment. However, both cation and anion exchange resins have to be regenerated from time to time. Under neutral pH conditions, negatively charged uranium species such as $UO_2(CO_3)_2^{2-}$, $UO_2(SO_4)_2^{2-}$, $UO_2(HPO_4)_2^{2-}$ and $UO_2(H_2PO_4)_3^{-}$ predominate. But clay minerals tend to be anion exchangers only at low pH, and thus negatively charged uranium complexes tend to be mobile in groundwater. Only at low pH values does cation exchange on clay minerals occur, since at low pH values the UO_2^{2+} complex is likely to be the dominant species.

Thus, precipitation would appear to be the most probable and sustainable natural attenuation process of uranium removal from water. Since schoepit and uranophane may limit the uranium concentration to only 10^{-6} mol/l (some 100 µg/l of uranium) and not to naturally low levels of several micrograms per litre, the reduction of U^{VI} to U^{IV} is necessary.

Uranium reduction takes place in the pE range when iron(III) has already been reduced to iron(II); thus strong reductants such as CH_4 and Fe^{2+} are required. Initially the decay of organic matter consumes free oxygen. Subsequently, micro-organisms gain energy from reducing nitrate (to N_2), manganese, iron, uranium and sulfate (to HS^-) and, finally, CO_2 (to CH_4). This process depends on temperature, nutrients and an organic source for the micro-organisms.

The kinetically controlled reaction can be described as follows:

$$\frac{dm_i}{dt} = c_{ik} k_k \tag{2}$$

where dm_i/dt is moles per unit time (s), c_{ik} is the concentration of species i (mol/l) and k_k is the reaction rate (mol/(l s)).

The general kinetic reaction rate R_k of minerals is

$$R_k = r_k \frac{A_0}{V} \left(\frac{m_k}{m_{0k}} \right)^n \tag{3}$$

where r_k is the specific rate (mol/(m² s)), A_0 is the initial mineral surface (m²), V is the mass of the solution (kg water), m_{0k} is the initial amount (moles) of the mineral, m_k is the amount of mineral (moles) at time t and n is a surface-related constant. $(m_k/m_{0k})^n$ is a factor that takes into account the change A_0/V during the dissolution process. Assuming uniform dissolution of spheres and cubes, n is 2/3.

Since often not all parameters are available, a simplified equation can be used:

$$r_k = k_k(1 - SR^\sigma) \tag{4}$$

where SR is the saturation ratio (ion activity product/solubility product) and σ is a coefficient related to stoichiometry. Often the exponent σ is 1. An advantage of this simple equation is that it is valid for both over- and undersaturation.

r_k can also be expressed as the saturation index (SI = log SR) (Appelo et al., 1984):

$$r_k = k_k \sigma \, SI \tag{5}$$

Another choice is the Monod equation:

$$r_k = r_{max} \frac{C}{K_m + C} \tag{6}$$

where r_{max} is the maximum reaction rate, K_m is the concentration (when the rate is 50% of the maximum rate), and C is the concentration (mol/l).

The Monod rate is commonly used to simulate the decay of organic matter (Van Cappellen and Wang, 1996), and can be derived from the equation describing first-order reactions. For degradation of organic matter (C), the first-order rate equation is

$$\frac{dS_C}{dt} = -k_1 S_C \tag{7}$$

where S_C is the organic carbon content (mol/l), and k_1 is the first-order decay constant (s^{-1}).

The combined overall Monod expression for degradation of organic carbon in a freshwater aquifer is (Parkhurst and Appelo, 1999)

$$R_C = 6s_C \frac{s_C}{s_{C_0}} \left(\frac{1.57 \times 10^{-9} m_{O_2}}{2.94 \times 10^{-4} + m_{O_2}} + \frac{1.67 \times 10^{-11} m_{NO_3^-}}{1.55 \times 10^{-4} + m_{NO_3^-}} \right) \tag{8}$$

However, the constants are variable and depend on temperature, micro-organisms and nutrients and upon the kind of organic matter.

12.5. Case study

12.5.1. Königstein

The Königstein mine was opened as a conventional underground uranium operation in 1967. It is situated close to the city of Dresden, and about 1–2 km from the River Elbe. The surrounding rocks are Cretaceous sandstones (Fig. 12.3). *In situ* leaching (ISL) was begun in 1984 by applying sulfuric acid to rock, which had been blasted into blocks to increase the permeability of the sandstone. At the end of 1991 mine was abandoned, after uranium production was stopped in the reunified Germany.

After a 10 year period of investigation and testing, flooding of the mine was started at the beginning of 2001. The amount of uranium left in the Königstein mine is roughly estimated at 20 000 tonnes. The acid solution in the leached blocks after flooding shows uranium concentrations of about 60 mg/l. Assuming a groundwater recharge for the mine site of about 42 l/s, the uranium output rate of a down-stream contamination plume can be roughly calculated as 2.5 g/s, or 216 kg/day, or 80 tonnes per year.

There are several options for the rehabilitation of the Königstein mine, which are discussed in detail elsewhere. After flooding of the mine, the uranium-contaminated water could be treated by a pump-and-treat technique for a period, but not indefinitely. At any time some uranium and other contaminants could either migrate into the groundwater (case 1) or be discharged directly to the River Elbe as a result of an intervention measure (case 2). Consequently, the downstream groundwater zone might become a reactive area under case 1, and investigations into the natural attenuation capacities of the aquifer are necessary. Dilution into the River Elbe will not be discussed in this study.

Three groundwater models have been set up by Wismut GmbH over the last decade, and one by Flesch (2000). However, these models were mainly three-

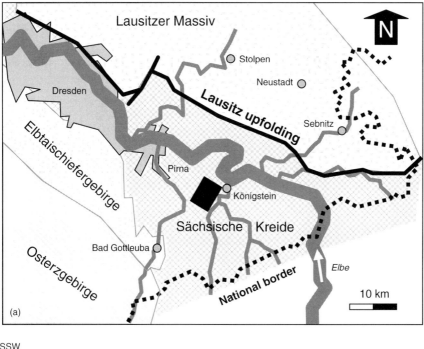

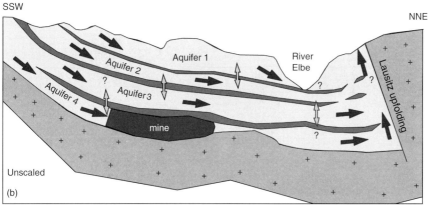

Fig. 12.3. (a) Map of the Königstein mine area south-east of the city of Dresden.
(b) Cross-section showing the mine location and aquifer No. 4

dimensional groundwater flow models, with little or no capacity to model multi-
element, multi-species and reactive transport. A very simple approach is shown
below, which helps clarify the complex chemical processes in the downstream area
of Königstein. The model is based on a one-dimensional streamline, taking into
account dilution and dispersion as well as thermodynamically and kinetically
controlled reactions. It was written in PHREEQC, using a modified database named
Watf4_U (Merkel, 1999), which in addition to a revised uranium database contains
constants for thorium and radium. Thorium and radium data were taken from the

MINTEQ database, and uranium data were predominantly taken from Langmuir (1997) and from our own experience. Only uranium data considered reliable were left in the database.

The one-dimensional model presented here does not simulate the site of the deep mine itself but only the downstream area. Therefore, the upper boundary condition for the geochemical model is the quality of the mine water after flooding has taken place, independent of the question as to whether this water has been treated or not.

Assuming a hydraulic conductivity of some 10^{-5} m/s, an effective porosity of 5–10% and mean hydraulic gradients, an average interstitial velocity of 100 m/year was assumed. The length of the streamline was set to be 1 km, divided into 10 cells, each with a length of 100 m. Dispersion was assumed to be 5 m. The assumed flooding water chemistry is given in Table 12.1. Figure 12.4 displays the distribution of uranium species in this water, with $UO_2SO_4^0$ being the dominant species.

A simple non-linear dilution with uncontaminated groundwater (Table 12.2) was assumed, with no dilution close to the mine and a dilution factor of 0.5 for the last cell. These dilution factors were found by an analytical three-dimensional solution of the situation. In order to have simple boundary conditions, it was assumed that the source strength in the mine is constant over the simulation time of 20 years, which is definitely a worst-case scenario.

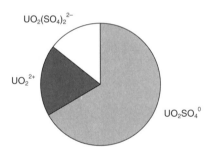

Fig. 12.4. Distribution of uranium species in the mine water

Table 12.1. Assumed quality of mine water

pH	pE	T										
2.3	10.6	12°C										

Anions (mg/l)

HCO_3^-	SO_4^{2-}	Cl^-	NO_3^-									
200	5000	450	100									

Cations (mg/l)

U	As	Pb	Ni	Mn	Fe	Cd	Al	Ca	Mg	Na	K	Si
40	2	0.2	5	20	600	1	200	400	50	500	4	50

Table 12.2. Assumed quality of uncontaminated groundwater

pH	pE	T							
6.6	2.0	12°C							

Anions (mg/l)

HCO_3^-	SO_4^{2-}	Cl^-	NO_3^-						
200	14.3	2.1	0.5						

Cations (mg/l)

U	As	Pb	Mn	Al	Ca	Mg	Na	K	Si
0.005	0.004	0.05	0.07	0.02	36.6	3.5	5.8	1.5	3.6

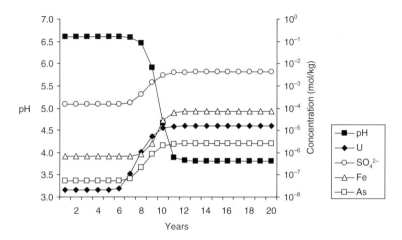

Fig. 12.5. pH and concentrations of total uranium, sulfate, iron and arsenic at a point 1 km downstream of the mine, after dilution of mine water with groundwater. The decrease of uranium compared with an initial concentration of 1.68×10^{-4} mol/l in the mine water is only due to dilution. The same holds for the initial arsenic concentration (2.7×10^{-5} mol/l) and the initial pH (2.3). However, sulfate and iron concentrations are decreased, not only by dilution but also by precipitation of iron hydroxide and Na-jarosite

Figure 12.5 shows the result of a simulation assuming boundary conditions as follows. No pyrite or calcite is assumed to be available to the aquifer. In which case, the only limiting mineral phases are $Fe(OH)_{3(a)}$ in the first 9 years and Na-jarosite $(NaFe_3(SO_4)_2(OH)_6)$ thereafter. Additionally, quartz is assumed to be slightly oversaturated. These minerals were therefore placed in equilibrium within the PHREEQC simulation.

Assuming small amounts of calcite in the aquifer (which is likely), the calcite is dissolved and the pH reaches 6.9. The break-through concentrations for this case are given in Fig. 12.6. Sulfate, compared with its initial concentrations, decreases from 4.1 to 2.2 mmol/l due to formation of Na-jarosite. Iron decreases from 7.2×10^{-5} to 3×10^{-5} mol/l. Due to the increased calcium concentration, uranophane $(Ca(H_3O)_2(UO_2)_2(SiO_4)_2 \cdot 3H_2O)$ becomes supersaturated (SI = –0.25). Assuming that uranophane is precipitated, the uranium concentration decreases from 1.5×10^{-5} to 4×10^{-6} mol/l (all values for the last cell and 20 years).

Certainly, some uranium and arsenic will be co-precipitated or sorbed on to the iron hydroxides, but this was not taken into account within this particular model. The Königstein uranium ore deposit is a typical roll front deposit where oxidized, uranium-rich groundwater encounters a redox interface rich in organic matter and/or sulfide minerals. The Cretaceous sandstones are still rich in organic matter and contain some pyrite as well. Thus, it may be assumed that reduction will occur by the decay of the organic matter. However, this process is kinetically controlled.

To simulate kinetically controlled decay of organic matter, the combined overall Monod expression for the degradation of organic carbon in a freshwater aquifer from Parkhurst and Appelo (1999) was used in the one-dimensional transport model (see the appendix). Depending on the parameters in this equation, biodegradation is either faster or slower. If the biodegradation is fast, iron, uranium and sulfate will be readily reduced, resulting in negative pE values. Using the original parameters of Van Cappellen and Wang (1996), the decay is considerably slower, and no major changes were shown by the model for iron and uranium: the pE was 2.8 for the 1 km downstream point after 20 years. Since Eh values of about 200 mV (pE 3.4) have been measured in the aquifer of the Königstein area, it is unlikely that pE values of less than 3 will occur due to decay of the organic matter embedded in the sandstones. It is thus reasonable to assume that the decay of the organic matter is slow or

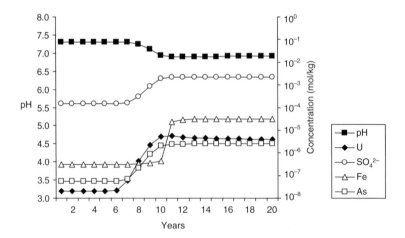

Fig. 12.6. pH and concentrations of total uranium, sulfate, iron and arsenic at a point 1 km downstream of the mine, assuming the presence of calcite in the aquifer. Sulfate, iron and uranium concentrations are lower, due to precipitation of Na-jarosite, amorphous iron hydroxide and uranophane

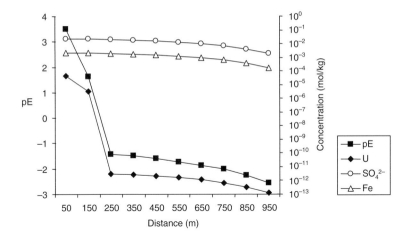

Fig. 12.7. Decrease in uranium concentration resulting from a decrease in pE owing to faster biodegradation of organic matter versus distance from the mine site. Reduction of sulfate and precipitation of iron take more time

inhibited. Furthermore, low pE values, necessary for $UO_{2(a)}$ precipitation, do not occur naturally.

However, if readily biodegradable organic substances were to be added to the groundwater, enhanced natural attenuation could occur. To simulate this, the sulfate reaction rate was changed from 10^{-13} to 10^{-12}, and the model rerun (Fig. 12.7). Uranium is now decreased to concentrations of 10^{-13} mol/l due to precipitation of some uranophane but mainly $UO_{2(a)}$. This also occurs in the model for naturally occurring groundwater, in which case background uranium concentrations of 2–5 μg/l were found. Once again, this indicates that only enhanced natural attenuation is likely to work in this case, since there is insufficient reducing potential available in the aquifer. Pyrite in the aquifer could be a potential reducing factor; however, this would result in lower pE values in the naturally occurring groundwater as well. It should be mentioned that precipitation of amorphous uranium oxide was not modelled kinetically. However, laboratory experiments by Pietzsch (2001) have proven that uraninite is precipitated within a few weeks, and thus amorphous uranium dioxide is likely to be precipitated even faster. Summarizing the results of the PHREEQC simulations, it can be stated that adding readily degradable organic matter to the mine water when flooding the mine could enhance the precipitation of uranium.

12.5.2. Experiments with zero-valent iron (ZVI)

ZVI has been discussed by many researchers as a uranium-removing reagent in permeable reactive walls. However, the mechanism of retardation/fixation of uranium by iron is not understood in detail. Noubactep *et al.* (2001a) studied the removal of uranium using various types of iron and under varying experimental conditions, i.e. iron chips as well as fine-grained iron pretreated with strong hydro-chloric acid. They found that the majority of information in the literature deals with

HCl pretreated ZVI on a laboratory scale. However, even after such pretreatment, iron oxides will readily form, since the oxidation of iron (Fe^0) by groundwater constituents is thermodynamically more favourable than by the uranyl ion (UO_0^{2+}). Furthermore, sorption of UO_2^{2+} to different types of iron oxides is well documented, and the *in situ* formation of iron oxides has also been proposed for scavenging UO_2^{2+} from groundwater.

Based on these considerations, and the fact that uranium reduction by ZVI is an electrochemical process, iron nodules and manganese nodules were combined with well-characterized scrap iron to test the possibility of facilitating the reduction of uranium by accumulating it through sorption in the vicinity of the iron surface (Noubactep *et al.*, 2001b).

Iron nodules accelerate the removal of uranium from groundwater but retard its reduction, since the kinetics are controlled by the diffusion of UO_2^{2+} to the iron surface. Manganese nodules retard the removal of uranium from groundwater but without impact on UO_2^{2+} reduction. This is due to the fact that MnO_2, which is the major constituent of manganese nodules, is able to oxidize Fe^{2+} to Fe^{3+} and thus maintain the corrosion process and prevent the formation of iron oxides at the surface of ZVI (Noubactep *et al.*, 2001a).

In order to verify and evaluate these laboratory findings, columns filled with Fe^0, iron–manganese waste sludge and peat were installed at two abandoned mining sites in the Erzgebirge in East Germany for about 1 year, in order to investigate the removal capacity for uranium, radium and arsenic from water (Schneider *et al.*, 2001). The sorption capacity of peat was exhausted after about 6 months, and no reduction was observed, probably due to the short residence time of the water in the columns. The iron–manganese sludge sorbed about 60% of the radium and 70% of the arsenic, but only 40–70% of the uranium. Some indications were found that part of the uranium was co-precipitated with manganese carbonate. By contrast, Fe^0 removed between 80 and 96% of the uranium, but only about 45% of the radium, and between 70 and 85% of the arsenic. Removal of radium and arsenic is due to co-precipitation with iron hydroxides plus, perhaps, calcite coatings on the iron chips. Investigations by scanning electron microscopy indicated that precipitates of UO_2 and ThO_2 may be present; however, quantitative proof was not possible. Contrasting results were obtained from remobilization experiments with the iron columns. In some cases remobilization with HCl was negligible, in other cases it was significant. More detailed studies on this type of binding are therefore in hand.

12.6. Conclusions

Despite the fact that the German drinking water regulatory framework has no uranium MCL, uranium is a heavy metal of concern in both groundwater and surface water, mainly due to its chemical toxicity – this applies to some naturally occurring groundwaters, but in particular to areas where uranium mining and milling took place (or still does), nuclear energy sites, and heaps where coal and fly ash have been dumped. Uranium in its hexavalent oxidation state is very mobile, and only reduction to tetravalent uranium seems to offer a sustainable means of removing uranium from water. This may be done through strong reductants such as Fe^0 and CH_4, or by biodegradation of organic matter. Sorption is a fast reaction, and thus might be an important first step in the natural attenuation of uranium from water, followed by the slower second step of precipitation of schoepite, uranophane, amorphous uranium oxide or uraninite and/or co-precipitation of iron hydroxides, calcite or siderite. Removal of uranium from water starts with sorption and is then replaced by precipitation and co-precipitation. Modelling uranium removal requires reliable thermodynamic and kinetic data. However, data describing the kinetics of mineral precipitation, in particular, are rarely available.

12.7. Appendix

A simplified example of a PHREEQC input file for one-dimensional transport is shown below, taking into account dilution, dispersion, and thermodynamically and kinetically controlled reactions.

```
SOLUTION 0          Initial solution for flooded mine, example of transport with 10 cells
#Add syntax for solution....

SOLUTION 12-21  stagnant solution, simulates dillution by Groundwater
#Add syntax for solution....

SOLUTION 1-10       mobile water, Groundwater
#Add syntax for solution....

MIX 1                    # dilution defined individually per cell
1.99
12.01
MIX 2
2.98
13.02
MIX 3
3.95
14.05
# Continued until MIX 10....

EQUILIBRIUM_PHASES 1-10
Calcite          0              # calcium may dissolute
Gypsum           0    0         # no dissolution allowed
Pyrite           0    0         # no dissolution allowed
Fe(OH)3(a)       1    0         # allow oversaturation
Al(OH)3(a)       0    0
Jarosite-Na      0    0
Jurbanite        0    0
Uranophane       0    0
Uraninite(c)     0    0

KINETICS 1-10
      Organic_C
                  -formula CH2O
                  -tol          1e-8
                  -m0           1          # mol/L
                  -m1
      -steps 31536000 in 50 steps     # 365 Tage
                  -step_divide 10000000

RATES 1-10
      Organic_C
```

```
-start
10 if (m <= 0) then goto 200
20 mO2 = mol("O2")
30 mNO3 = tot("N(5)")
40 mSO4 = tot("S(6)")
50 rate = 1.57e-9*mO2/(2.94e-4 + mO2) + 1.67e-11*mNO3/(1.55e-4 + mNO3)
60 rate = rate + 1.e-13*mSO4/(1.e-4 + mSO4)
70 moles = rate * m * (m/m0) * time
80 if (moles > m) then moles = m
200 save moles
-end
```

```
TRANSPORT
-cells 10
-shifts 20
-lengths 100
-time_step 3.15e7
-boundary_conditions flux flux
-dispersivities 10*5
-warnings true
-stagnant 1
-punch_cells 1-10      # determine cells for selected output
```

```
SELECTED_OUTPUT
-file 1d_trans.csv
-totals U S(6) Fe As
-saturation_indices Calcite Gypsum Pyrite Al(OH)3(a) Uranophane Uraninite(c)
-equilibrium_phases Fe(OH)3(a) Uranophane Jarosite-Na Jurbanite Calcite
 Gypsum Uraninite(c)
end
```

12.8. References

APPELO, C. A. J., BEEKMAN, H. E. and OOSTERBAAN, A. W. A. (1984). *Hydrochemistry of Springs from Dolomite Reefs in the Southern Alps of Northern Italy. Scientific Publication 150.* International Association of Hydrology.

FLESCH, K. (2000). Hydrogeologisches Modell der Sächsischen Kreide. *Wiss. Mitt. Inst. für Geologie der TU BAF* **14**, 113 S.

LANGMUIR, D. (1997). *Aqueous Environmental Geochemistry.* Prentice-Hall, Englewood Cliffs, New Jersey.

MERKEL, B. (1999) Watf4_U, a modified WATEQ4F database (uranium, radium, thorium). Website: http://www.geo.tu-freiberg.de/~merkel/was_kann_phreeqc.htm.

MERKEL, B. and SPERLING, B. (1998). Hydrogeochemische Stoffsysteme II. *DVWK Schriftenreihe* **117**.

NOUBACTEB, C., MEINRATH, G., DIETRICH, P. and MERKEL, B. (2001a). Mitigation of uranium in groundwater: prospects and limitations. *Proceedings of Migration 2001.*

NOUBACTEP, C., MEINRATH, G., VOLKE, P., PETER, H. J., DIETRICH, P. and MERKEL, B. (2001b). Understanding the mechanism of the uranium mitigation by zero valent iron in effluents. *Wissenschaftliche Mitteilungen des Instituts für Geologie der TU Bergakademie Freiberg* **18**, 36–44.

PARKHURST, D. L. and APPELO, C. A. J. (1999). *User's Guide to Phreeqc (Version 2) – A Computer Program for Speciation, Batch-reaction, One-dimensional Transport and Inverse Geochemical Calculations.* Water-Resources Investigations Report, 99-4529. US Geological Survey, Denver, Colorado.

PIETZSCH, K. (2001). Eliminierung von U(VI) aus wässrigem Medium durch Bioreduktion zu schwer löslichem (IV) (Uraninit). PhD thesis, University of Leipzig.

SCHNEIDER, P., NEITZEL, P. L., OSENBRÜCK, K., NOUBACTEB, C., MERKEL, B. and HURST, S. (2001). *In-situ* treatment of radioactive mine water using reactive materials – results of laboratory and field experiments in uranium ore mines in Germany. *Acta Hydrochimica Hydrobiologica* **29**, 2–3, 129–138.

US DEPARTMENT OF HEALTH AND HUMAN SERVICES (1999). Toxicological profile for Uranium, Atlanta. Website: http://www.atsdr.cdc.gov/toxprofiles/tp150.html.

US ENVIRONMENTAL PROTECTION AGENCY (1999). *Background Report on Fertilizer Use, Contaminants and Regulations*, EPA 747, R98 003. US Environmental Protection Agency, Washington DC.

VAN CAPPELLEN, P. and WANG, Y. (1996). Cycling of iron and manganese in surface sediments. *American Journal of Science* **296**, 197–243.

WHO (1998). *Guidelines for Drinking-water Quality*, 2nd edn, addendum to Vol. 1. *Recommendations*. World Health Organization, Geneva.

WOLKERSDORFER, C. (1996). Hydrogeochemische Verhältnisse im Flutungswasser eines Uranbergwerkes. – Die Lagerstätten Niederschlema/Alberode. *Clausthaler Geowissenschaftliche Dissertationen* **50**.

13. Flooding strategies for decommissioning of uranium mines – a systems approach

R. Gatzweiler[*], A. Jakubick, G. Kiessig, M. Paul and J. Schreyer
Wismut GmbH, Jagdschänkenstrasse 29, D-09117 Chemnitz, Germany

13.1. Introduction

In conventional mining, mines are kept dry by pumping, and the mine waters are discharged into nearby surface waters. Depending on the quality of these waters and the conditions of the receiving streams they often need treatment. The dewatering of the mine results in a cone of depression in the groundwater table. When the extraction process is finished, pumps are normally shut off, and groundwater is allowed to re-enter the mine voids. This process is conventionally called flooding (Gatzweiler, 2000).

The mineral production activities and ventilation and dewatering of the mine during its life tend to increase the solubility of metals and radionuclides in residual ore and in the wall rock. This generally leads to an increase in contaminant concentrations in the mine waters during the flooding process.

Formerly, the normal procedure for closing a mine was to recover all valuable instruments and materials and then to stop pumping. This has changed with the increase in environmental awareness within the mining industry, and the introduction of regulatory measures for improved protection of groundwater and surface water. The flooding process now requires optimized conceptual planning, comprehensive and continuous monitoring for control of the flooding, and the calibration of models to predict the quantities and quality of water which might need pump-and-treat measures, plus options for steering the process, and provision of fall-back options.

An optimized flooding concept also includes the evaluation of potentially applicable water treatment processes and possible alternatives, such as passive water treatment systems, *in situ* (e.g. reactive barriers) or *ex situ* (e.g. wetlands).

---

[*]Retired: formerly at Wismut GmbH.

Due to mining, the original hydraulic conditions of the mine field are disturbed and cannot normally be totally restored. The contaminated flood waters therefore potentially intrude into adjacent aquifers, or outflow to the surface occurs either through mine workings or via hydraulically conductive seepage zones.

The predominant remediation actions to be taken against the risk of contaminating adjacent groundwater aquifers or surface waters during or after flooding are pumping to avoid the flood level rising beyond certain critical levels and treatment of these waters before discharge. In many cases this is a precautionary measure against the risk of contaminating groundwater and surface water, or impact on the stability of the surface (e.g. in the case of construction sites), and is not a true remediation measure. Such pumping action often prevents the achievement of equilibrium flooded conditions, and as a consequence the total duration for treatment of the flood water is extended.

In cases where stable rebound conditions cannot be expected to develop, for instance due to convection currents within the mine, an initial phase of pump and treat can be the optimal strategy, under a presumption that the total dissolved contaminant load is not increased through the pumping action. This initial phase is called the washing or sweeping phase. The aim of this phase is to 'scoop off' an initial contaminant load and to avoid long-term dilution of this contaminant source with less contaminated infiltration water.

The most difficult scenario for optimization of the flooding process is when contaminated infiltration waters originating from high levels within a mine, which cannot be flooded and which hydraulically cannot be separated from the flooded mine, continuously discharge their contaminant load into the flooded portion of the mine. This is often the case with old metal mines, where mining historically started in the rich oxidized zone of the ore body above the groundwater level, and which consequently produces an acid mine drainage problem.

13.2. Mine remediation and flooding

The overall aim of mine remediation is to exclude hazards and risks to life and health of the population and the environment, and thus allow further utilization of the mine area with minimal restrictions (Gatzweiler and Meyer, 2000). In practice, this means to restore, so far as can be justified economically and ecologically, the mine area and to stabilize it geomechanically, hydrologically and geochemically. In almost all cases these aims can only be achieved through flooding.

A well-planned flooding concept further aims to:

- protect economically used and potentially usable groundwater aquifers;
- protect surface waters;
- minimize water volumes used in flooding, and contaminant loads of waters flowing out of the mine.

Particularly with respect to the last aim, the closure and flooding of mines should be initiated and completed as quickly as possible, and the flood level should be maximized as early as possible. This saves on the costs of keeping the mine open, and also stops oxidation and restricts the release of contaminants into the environment through air and water pathways.

Notwithstanding the different conditions prevailing within individual mines, the preparation for flooding generally includes:

- preparation of a flooding prognosis, including modelling the hydrodynamic and hydrochemical conditions;
- evaluation of requirements to manage the flooding process, and to develop fall-back options;
- definition of the technological sequence for abandoning individual mine areas and levels;
- clearing the mine of all toxic and water-soluble materials;
- installation of a monitoring system;
- treatment of overflow or seepage waters and selection of appropriate and acceptable discharge points.

The flooding process can be grouped into:

- preventive measures, e.g. filling of near-surface mine voids or shafts to stabilize the ground or the construction of barriers to avoid the mixing of differently contaminated mine waters from different mine fields;
- establishment of a monitoring system to measure flooding level, contaminant concentrations, seismic activity, etc. These are used as input data for the prognoses of development and the final stages of the flooding level, the volumes of overflow or seepage waters and their chemistry;
- timely establishment of technical installations to influence (manage) the flooding process, such as water treatment, mine access or mine ventilation systems.

In summary, flooding is an integral part of mine remediation. It is a multi-attribute process and thus offers possibilities for optimization. Furthermore, an optimized flooding concept requires a comprehensive definition of the hydraulic, hydrogeological, geochemical and geomechanical conditions of the ore deposit/mine before flooding starts, plus systematic observation, measurement and interpretation while flooding is in progress.

Since the actual chemical and physicochemical processes prevailing within the rising flood waters are poorly understood, it is advisable to carry out flooding tests within representative sections of a mine. The resulting data, though arising from various complex interactions, can provide a basis for improved modelling and more reliable forecasts, in particular regarding the development of contaminant concentrations.

With the rising flood level, oxygen is being depleted in the flood waters, which is reflected in decreasing Eh values. This leads some contaminants (e.g. uranium) to precipitate. Others, however (e.g. arsenic and radium), can be more mobile under increased reducing conditions, and therefore desorption processes must be taken into consideration.

As long-term water treatment costs represent a heavy financial burden to the mine owner and/or the public, the general aim is to keep the period of water treatment required as short as possible. One strategy to achieve this is to retain as much contamination as possible within the mine, i.e. to reduce or stop the mobilization of contaminants, or to facilitate natural or artificial precipitation or sorption in a geochemically stable form, for example by *in situ* lime precipitation or sorption within a reactive barrier.

13.3. Pump and treat versus collect and treat

Based on general experience, it is important to establish equilibrium conditions as early as possible in the flooding process, which means arriving as early as possible at the final flooded level. This facilitates more reliable forecasts regarding the quantity and quality of waters to be treated over the long term, and also improves the chances for the development of a layering of the flood waters. This would keep parts of the contamination at lower levels within the mine, and in the optimal case less contaminated infiltration water would occupy the upper levels, and consequently the discharge. This would shorten the duration of treatment necessary, under the assumption that total restoration of the mine water aquifer is not required.

A pump-and-treat intervention during the flooding process normally increases convection currents. This consequently increases mixing and therefore potentially increases the duration of treatment. Careful consideration is thus needed regarding the placement of pumps within a mine. Furthermore, changes in the pump positions during flooding are generally recommended.

An early stage washing or sweeping action by pump and treat is normally used in the restoration of mining aquifers *in situ*. It can also be advantageous if used in conventional mine flooding, when hydraulically isolated plumes of more contaminated flood waters occur.

For Wismut mines (except the Königstein mine, see below) the optimal flooding strategy is to let the flood level rise to its natural maximum level and collect the seepage at hydraulic low points. Though the required collection systems are costly to build, there are advantages compared with pumping the flood waters out of the mine while keeping the flood level below its final level. In addition to the overall economic advantages, the seepage collection option promises better use of natural attenuation processes. Furthermore, it eases the change from conventional water treatment systems to passive systems such as the use of wetlands. This has an important influence on the overall economics of mine remediation, since the efficiency of conventional treatment of flood waters at a late stage of flooding, when the contaminant levels are comparatively low, is much reduced, and passive systems can then be used at a much higher efficiency (Kiessig and Kunze, 1996; Gatzweiler *et al.*, 2000).

13.4. Flooding strategies at Wismut mines

Uranium mining by the former Soviet–German mining company Wismut comprised a total production of about 230 000 tonnes of uranium. Nearly 80% of this production originated from five underground mines, which were closed between 1989 and 1991.

The most important mines from a production point of view were those at Ronneburg and at Schlema. The mines at Königstein, Gittersee and Pöhla contributed between 10 000 and 15 000 tonnes of uranium. The mines at Ronneburg and Königstein had the most impact on the local hydrosphere. The mine area at Ronneburg (which included a large open pit) is characterized by 22 million m^3 of open mine voids, a cone of groundwater depression of more than 70 km^2 and a very serious acid mine drainage problem. At Königstein, acid underground block leaching was used, resulting in a large contaminant potential within an aquifer which is overlain by other aquifers representing important groundwater resources.

Since 1991, Wismut has undertaken remediation at the former mining and milling sites. Mine remediation is far advanced. Flooding is now in progress at all mines (Gatzweiler *et al.*, 2000; Hagen *et al.*, 2000).

At Pöhla the final flood level was reached at the end of 1995. Since then, the flood waters have left the mine at an average flow rate of 17 m^3/h through the former main haulage adit, at a level of 585 m a.s.l., and are treated for arsenic, radium, iron and manganese. Treatment for uranium was abandoned in 1999, since, due to natural attenuation processes, uranium concentrations had dropped below the discharge limit. Since 1999 a passive water treatment system has been successfully piloted and will replace the present water treatment system in 2003. A further favourable condition at Pöhla is that infiltration waters can be collected separately at the main adit level and do not need treatment.

At Schlema, flooding started in 1991. The flood level is maintained by pumping and treating between 120 and 90 m below the final level where overflow would occur. About 90% of the total of 34 million m^3 of open mine voids are flooded. The progress of flooding is mainly determined by the progress of underground remediation work. The volume of water infiltrating the mine is highly dependent on the meteorological conditions, and varies between 600 and 1200 m^3/h. While the total maximum treatment capacity is 1000 m^3/h, periods of higher infiltration are hydraulically buffered by 1 million to 2 million m^3 of open mine space below the overflow level.

Hydrochemically, the flood waters at Schlema are pH neutral and presently show an intermediate to slightly reduced redox potential. They carry a substantial load of uranium, radium, arsenic, iron and manganese. Since the mine extends to a total depth of about 2000 m, with rock temperatures of more than 50°C (during mining), the flooded mine is characterized by strong convection currents causing significant mixing. This is probably enhanced by the technical necessity to draw the water by pumping from below the 540 m level. The forecast for the need for water treatment is up to 25 years. Conceptual studies and laboratory tests have been carried out to define possible *in situ* treatment methods. But due to the very complicated mine structure, there is little hope of applying *in situ* methods in order to shorten the duration of treatment.

The former uranium coal mine Dresden-Gittersee represents a small part of a large coal field which was abandoned after the Second World War. Flooding started in 1995. The flood level is maintained by pumping from two boreholes in order to

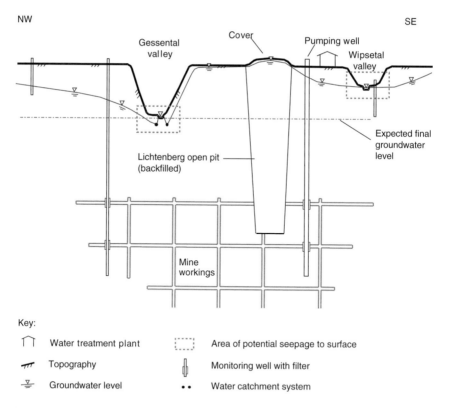

Fig. 13.1. Water management system at the Ronneburg site

keep the level below a toxic waste deposit. The flood waters need treatment for iron, and carry a high sulfate load. Significant layering of iron-rich mine waters and cleaner infiltration waters occurs, which is constantly disturbed through the pumping action. Flooding will be continued, once the toxic waste has been extracted, until the natural groundwater level is reached. An extensive monitoring system is being put in place. The two boreholes will be used as fall-back options in case the flood level needs to be lowered again. After completion of flooding, the discharge of mine waters is expected to occur via various old dewatering adits connecting the mine with the Elbe and Weisseritz rivers and via diffuse seepage into the Weisseritz.

The Ronneburg Mine is the largest Wismut mine site. It consists of six formerly separate mines, which were interconnected in the course of mining which started in 1951 and ceased in 1990. The mine workings, with a total length of about 3000 km, extend in depth between 30 and 940 m. The open mine space has a total volume of about 27 million m³. In addition to the underground mine the Lichtenberg open pit, with a total volume of 160 million m³ and a maximum depth of 240 m, is located in the centre of the mine field. Flooding started in 1998, after construction of a large number of concrete plugs in the main drifts, in order to separate the southern and northern parts of the mine site, as well as isolating different mines in the southern part. This hydraulic separation firstly allows the maximum flood level for the whole mine field to be reached at about the same time, and, secondly, avoids the mixing of differently contaminated flood waters. The final flood level is expected to be reached between 2003 and 2005 (Paul *et al.*, 1998). Figure 13.1 shows the water treatment system at the Ronneburg site.

Since there are no open dewatering adits within the whole of the Ronneburg mine area, it is anticipated that the flood level will rise until natural discharge occurs at topographic low points into creeks, unless pumping or collection measures are taken during the final stages of flooding. This is in fact necessary for the southern part of the mine site, which accounts for about 85% of the mine voids and about 80% of the total inflow into the mine, since, due to natural acid mine drainage generation, the water is strongly mineralized and contaminated by radionuclides, iron, base metals (nickel, cobalt, copper) and arsenic. A water treatment plant was completed in 2002.

In situ measures, such as lime addition where acidic and highly contaminated waters occur or the construction of reactive barriers within mine drifts, were evaluated but shown not to be cost-effective. This is mainly due to the complex geometry of the mine.

To control the rise of the flood water within the southern part of the mine site, and as a fall-back option demanded by the regulatory agency, a 600 mm pumping well with a capacity equivalent to the inflow into the mine has been installed with a hydraulic connection to two main levels of the mine. It will be used to feed water to the treatment plant during the implementation phase. Pumping and treating is expected to last 3–6 months. Thereafter, flooding will continue to the final level.

The aim is to reach a high water inundation level in order to:

- lower the gradient into the depression cone and minimize the catchment area of the mine (which directly influences the amount and contaminant load of mine water to be treated);
- minimize the thickness of the unsaturated zone which is subject to acid mine drainage generation;
- minimize operational costs for water management, including water treatment and sludge disposal as a result of the two previous points.

Water catchment systems have thus to be installed close to the surface, where contaminated flood waters or groundwaters are expected to seep into surface drains. Likely seepage locations lie within the valleys of the Gessenbach, Wipse and Sprotte creeks. As there is substantial uncertainty as to exactly where future seepage will occur, phased construction is planned for these catchment systems. Also, the final flood/groundwater level has to be determined by a stepwise approach. Higher inundation levels increase the possibility of diffuse seepage. A 'safety level' of 240 m a.s.l. has been defined, which corresponds to the topographically lowest point in the Gessenbach valley with potential contact with the flooded mine site.

Another important aspect of the flooding strategy is control of contaminant migration from the flooded part of the backfilled open pit, which is expected to account for the majority of the contaminant load after about 10 years of water treatment.

The final steady-state flow field managed by shallow catchment systems as described above, without intervention by pumping from the flooded mine, will allow density stratification to develop within the mine water column, due to a high salinity gradient.

Flooding of the northern part of the Ronneburg mine area will have a much smaller impact on groundwater and surface water. At present, it is believed that water treatment is not necessary. Nevertheless, contingency plans are being drawn up should unacceptable situations occur which deviate substantially from the predicted scenarios after the maximum flood level is reached. As a fall-back option, a well has been established within the main shaft of the Beerwalde mine, and conceptual plans are in hand for water treatment should it be necessary.

At Königstein a sandstone-type uranium deposit was mined initially by conventional methods, and later by underground block leaching using dilute sulfuric acid. The ore deposit is located in the fourth aquifer, the deepest of four aquifers at the margin of a Cretaceous basin. The third aquifer represents an important groundwater resource for the Dresden region. It is separated by a 10–30 m thick aquitard, which is transected by one major and several minor faults, drillholes and shafts (Schreyer and Zimmerman, 1998).

Due to the reaction with the strongly oxidizing sulfuric acid, the geochemical status of the deposit has been substantially changed. The pollutant potential of the mine, principally from sulfates, base metals, iron and radionuclides, is mainly located in the pore space of the sandstone and in the open mine voids. It includes:

- approximately 2 million m^3 of acid pore solutions with high concentrations of contaminants;
- water-soluble secondary minerals;
- contaminants sorbed at the surface of primary and secondary minerals.

Experimental flooding of a small part of the mine started in 1993, in order to gain information on hydrodynamic, geochemical, rock mechanical and radiation protection aspects, which need to be considered in flooding such a mine. Studies of pollutant output resulted in a two-phase model, i.e. an initial short phase of fast

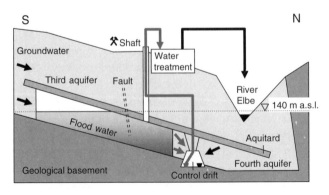

Fig. 13.2. Flooding of the Königstein mine, step 1

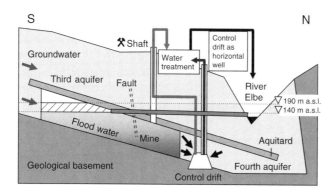

Fig. 13.3. Flooding of the Königstein mine, step 2: regulated flooding at 190 m a.s.l., open control drift

contaminant release followed by a longer phase of much slower contaminant release. The latter phase is controlled by differences in density between that of pore solutions and the flooding medium in the open mine voids, and/or that during diffusion of pore water from the sandstone into the flooded mine openings.

A concept of controlled flooding was developed, with its main objective being the protection of the third aquifer. A major element of this concept is a control drift surrounding the deposit at its hydraulically open sides, which allows the collection of flood waters draining into the drift.

Flooding will be carried out in two major steps. Initially (Fig. 13.2), while the flood level is continuously rising, the flood waters will be collected within the open control drift, pumped to the surface and treated by a conventional high-density sludge lime precipitation process. Once pollutant concentrations within the flooding waters have reached sufficiently low levels, the control drift will be abandoned and flooded, and consequently natural hydrological conditions will be allowed to develop within the mine area (step 2, Fig. 13.3). The flooded control drift can then be used as a horizontal well, from which contaminated flood waters can be pumped to the surface and treated.

Investigations are presently being carried out with a view to using the control drift downstream of the deposit as a reactive barrier. This is based on a prediction that after the initial phase of flooding, during which the easily released contaminants are 'swept out', a long-term phase of slow and minor contaminant release follows, with contaminant concentrations which no longer represent a major risk to the environment and specifically to the third groundwater aquifer. A reactive barrier could then be a more economic and efficient method than conventional pump and treat. Extensive bench scale tests of various materials have shown that a mixture of zero-valent iron and lignite can immobilize the critical contaminants of the Königstein flood water effectively (Klinger *et al.*, 2001). These materials can be used in two ways: either (1) to fill the control drift to form an *in situ* reactive barrier or (2) to construct a solid bed reactor at the surface, which would necessitate continued pumping from the flooded control drift.

A further option to support the flooding process would be the infiltration of strongly mineralized blocks of sandstone by a barium sulfate solution and controlled crystallization of barite. It has been demonstrated in large-scale underground tests that a major part of the contaminants could be immobilized by this method.

Flooding of the Königstein mine started, after 10 years of preparatory work, in January 2001. Flooding is permitted up to the 140 m a.s.l. level, which is expected to be reached during the year 2005. How to proceed with the final flooding beyond this level has not yet been decided. The flooding concept includes four options for the final flooding (Fig. 13.3):

- Continuation of controlled flooding up to the 190 m level (or the highest achievable level) by further use of the open control drift for drainage collection and continued water treatment, until an acceptable water quality is attained. Since the drainage volume would increase, an extension of the capacity of the water treatment plant would be necessary.
- Decommissioning of the underground mine and use of the control drift as a horizontal well, with continuing pump and treat. Treatment could be continued either conventionally or by using reactive materials within a solid bed reactor.
- Decommissioning of the underground mine and construction of a reactive barrier within the control drift. A monitoring system to control the reactive barrier, part of which might be used for pumping as a fall-back option, would need to be established. A treatment plant would need to be kept on stand-by initially.
- Construction of a water release adit at the 140 m a.s.l. level, connecting the mine with the River Elbe, allowing discharge of untreated flood water/groundwater.

13.5. Summary and conclusions

'Pump, treat and discharge' is the normal procedure for mine water management during the production phase of mines. Decommissioning of mines by flooding produces contaminated groundwater which potentially pollutes adjacent groundwater resources and surface waters by seepage or outflow through mine workings. Pumping and treating is also in this phase a basic remediation action, though not the optimal one, since pumping often prevents the development of natural equilibrium conditions. Such equilibrium conditions increase the chances for natural attenuation processes to succeed, but these processes are still poorly understood.

With the steady increase of regulated water quality objectives, water treatment in the decommissioning phase of mines can be very expensive. In the case of the five former Wismut uranium mines, the cost for treating flood waters is estimated to be in the range of 500 million euros incurred over a period of probably more than 25 years. Therefore, flooding a mine needs very careful planning in order to minimize the long-term costs. Various options including basic support measures can be used to optimize the flooding process. Their applicability should be evaluated on the basis of existing conditions at the individual mines, by cost–benefit analyses and careful risk management. This chapter describes the key elements of the flooding strategies for five former Wismut uranium mines. In the case of the Königstein mine, the construction of an *in situ* reactive barrier appears to be a valid option as a support measure to reduce the overall costs of remediation.

13.6. References

GATZWEILER, R. (2000). Remediation of former uranium mining and milling facilities in Germany – The WISMUT Experience. *Restoration of Environments with Radioactive Residue*; *Proceedings of an International Symposium, Arlington, USA*, pp. 477–501. IAEA, Vienna.

GATZWEILER, R. and MEYER, J. (2000). Umweltverträgliche Stilllegung und Verwahrung von Uranerzbergwerken – Fallbeispiel WISMUT. *Tagungsband zu Internationale Konferenz – WISMUT 2000, Schlema*. WISMUT GmbH, Chemnitz.

GATZWEILER, R., JAKUBICK, A. T. and KIESSIG, G. (2000). Remediation options and the significance of water treatment at former uranium production sites in Eastern Germany. *URANIUM 2000 – Proceedings of an International Symposium on the Metallurgy of Uranium, Saskatoon*.

HAGEN, M., GATZWEILER, R. and JAKUBICK, A. T. (2000). Status and outlook for the WISMUT remediation project in the states of Thuringia and Saxony, Germany. *Proceedings of NEA/IAEA Workshop RADLEG-142-O, Moscow*. IAEA, Vienna.

KIESSIG, G. and KUNZE, C. (1996). Wasserbehandlung und Rückstandsentsorgung. *Geowissenschaften* **14**(11), 481–486.

KLINGER, C., JENK, U. and SCHREYER, J. (2001). Efficacy of zero-valent iron and brown coal as reactive materials for reduction of pollutants in acid uranium mine water. *Proceedings of the 8th International Conference on Environmental Management – ICEM'01, Bruges*.

PAUL, M., SÄNGER, H.-J., SNAGOWSKY, S., MÄRTEN, H. and ECKART, M. (1998). Prediction of the flooding process at the Ronneburg site – results of an integrated approach. In: B. Merkel and C. Helling (eds), *Uranium Mining and Hydrology II, Proceedings of the International Conference and Workshop, Freiberg, Germany*, Sven von Loga, Köln.

SCHREYER, J. and ZIMMERMAN, U. (1998). The Königstein flooding concept – status and outlook. In: B. Merkel and C. Helling (eds), *Uranium Mining and Hydrology II, Proceedings of the International Conference and Workshop, Freiberg, Germany*, Sven von Loga, Köln.

14. Investigation into calcium oxide-based reactive barriers to attenuate uranium migration

M. Cővári, J. Csicsák and G. Főlding
Mecsekérc Rt., Esztergár L. u. 19, H-7633 Pécs, Hungary

14.1. Introduction

Past uranium mining and processing by Mecsek Ore Mining Company in Hungary has led to the accumulation of huge amounts of various solid wastes: tailings from conventional milling, wastes from heap leaching and waste rocks. In total, the wastes contain approximately 2800 tonnes of uranium, of which 500 tonnes derive from heap leaching wastes, with an average uranium content of approximately 70 g/tonne. The wastes also contain a small amount of pyrite (0.1–0.2%).

Remediation aims to relocate all heap leaching wastes (a total mass of 7.2 million tonnes) to one of the waste rock piles.

14.2. Leaching of uranium and other heavy metals from the wastes

Relocation of uranium bearing wastes is always accompanied by dissolution of uranium and other heavy metals which are present in the wastes. The main process of this dissolution is associated with the oxidation of pyrite.

Pyrite is oxidized directly and indirectly by ferric ions (Hutchison, 1992; Day, 1994):

$$2FeS_2 + 7O_2 + 2H_2O \rightarrow 2Fe^{2+} + 4SO_4^{2-} + 4H^+ \tag{1}$$

$$4Fe^{2+} + 10H_2O + O_2 \rightarrow 4Fe(OH)_3 + 8H^+ \tag{2}$$

$$2Fe^{2+} + O_2 + 2H^+ \rightarrow 2Fe^{3+} + H_2O \tag{3}$$

$$FeS_2 + 14Fe^{3+} + 8H_2O \rightleftharpoons 15Fe^{2+} + 2SO_4^{2-} + 16H^+ \tag{4}$$

The sulfuric acid formed reacts with the minerals in the waste (dolomite, silicates, etc.):

$$MgCa(CO_3)_2 + 4H^+ \rightleftharpoons Ca^{2+} + Mg^{2+} + 2CO_2 + 2H_2O \tag{5}$$

$$2KAlSi_3O_8 + 2H^+ + H_2O \rightleftharpoons Al_2Si_2O_5(OH)_4 + 4SiO_2 + 2K^+ \tag{6}$$

$$CaAl_2Si_2O_8 + 2H^+ + H_2O \rightleftharpoons Al_2Si_2O_5(OH)_4 + Ca^{2+} \tag{7}$$

CO_2 formed in the dissolution process then reacts with additional carbonates, increasing the concentration of bicarbonate ions in solution:

$$CaCO_3 + H_2CO_3 \rightleftharpoons Ca^{2+} + 2HCO_3^- \tag{8}$$

Sulfuric acid formed in reactions (1)–(4) can react with heavy metal minerals in the wastes. Sulfuric acid has a remarkable effect on uranium oxides, which readily react with the acids, in some cases even at pH 4:

$$UO_3 + H_2SO_4 \rightleftharpoons UO_2SO_4 + H_2O \tag{9}$$

$$UO_3 + 3H_2CO_3 \rightleftharpoons UO_2(CO_3)_3^{4-} + H_2O + 4H^+ \tag{10}$$

Uranium is very often bound to pyrite, hence it is evident that the oxidation of pyrite usually leads to the dissolution of uranium as well.

The presence of some toxic and heavy metals in the leachate is also connected with the oxidation of their sulfides:

$$60CuFeS_2 + 25O_2 + 90H_2O \rightarrow$$
$$60CuSO_4 + 20H\{Fe(SO_4)_2 \cdot 2Fe(OH)_2\} + 20H_2SO_4 \tag{11}$$

$$4FeAsS + 14O_2 + 4H_2O \rightarrow 4Fe^{3+} + 4AsO_4^{3-} + 4SO_4^{2-} + 8H^+ \tag{12}$$

Uranium-containing wastes are subject to the above processes, especially when relocated, i.e. when extensive contact takes place with oxygen. The water seeping through the pile will then contain an elevated concentration of uranium and, depending on the composition of the waste, an elevated concentration of other heavy metals too. Taking into account the composition of the wastes, in particular their pyrite and dolomite content, it becomes clear that the pH of the leachate should be near neutral.

Remediation plans for the former heap leaching piles included the relocation of 7.2 million tonnes of waste to the waste rock pile. Preliminary investigations led to the conclusion that the uranium concentration in the leachate could reach 20–30 mg/l, which was unacceptable from the viewpoint of groundwater contamination. The background concentration of uranium is 0.004 mg/l, i.e. the leachate is highly contaminated with uranium. Because huge amounts of wastes had to be relocated in a rather short period of time, it was evident that the uranium content of the leachate had to be controlled to avoid contaminating groundwater and surface water.

For this reason it was decided in 1996 that protective measures were required, with the aim of decreasing the uranium concentration in the leachate. The chosen method was to construct a lime-based reactive barrier. The results of the investigations are discussed below.

14.3. Results of laboratory experiments

Leachate composition depends on many factors. The following leachate composition was determined after the heap leaching wastes were relocated:

HCO$_3^-$	600 g/l
SO$_4^{2-}$	600 mg/l
Cl$^-$	70 mg/l
Ca^{2+}	180 mg/l
Mg^{2+}	70 mg/l
Na$^+$	200 mg/l
U	**26 mg/l**

In some cases the leachate contains further components originating from the alkaline leaching process used for treating low-grade ores on heap piles.

14.3.1. Main steps of the process

Reactive barriers have been widely used in recent years for the *in situ* treatment of contaminated groundwater (Meggyes, 2000; Roehl, 2001). If active calcium oxide is employed as a reactive material, the following main processes should be considered:

- dissolution of solid calcium oxide;
- decomposition of uranium complexes present in the leachate;
- precipitation of uranium and other di- or higher-valent cations present in the leachate;
- supplementary reactions.

The principal flow chart and chemical reactions of the processes are shown in Fig. 14.1. The main advantage of lime-based reactive barriers is in their low cost compared with other materials. Their main disadvantage is their relatively short lifetime. Therefore, this type of reactive barrier can only be used for short-lived groundwater contamination. This is the case when uranium-containing wastes are relocated: a rapid increase in uranium concentration can be observed immediately after relocation, and the task is to attenuate the increased uranium concentration during a period of 2–3 years.

The main reaction in this treatment by calcium oxide-based reactive barriers is the decomposition of the carbonate complex of uranium and its precipitation in the form of low-solubility calcium diuranate. Some parts of the uranium can be present in the form of uranyl ions; which also precipitates as calcium diuranate.

In Fig. 14.1, equation (3) mainly applies to the precipitation of magnesium present in the leachate. Equations (4)–(8) describe simple water-softening processes resulting in precipitation of calcium carbonate and silicate.

Uranium remains in the barrier until it contains free calcium oxide, i.e. so long as the pH in the barrier is high enough for the precipitation of uranium from carbonate-containing water.

14.3.2. Open-air experiments

The behaviour of calcium oxide-based reactive barriers was investigated in the laboratory and in large-scale tests. Laboratory tests were carried out in the open.

This approach was chosen to simulate natural conditions as closely as possible. Two 1 m³ tanks (in fact, lysimeters) were filled with representative heap leaching wastes: one of the tanks was used for reference purposes, and the test itself was performed in the other one. The reactive barrier was made of commercial lime mixed with solid wastes in the proportion of 1:20 and then placed in lysimeter N2 on top of the drainage gravel. The experiment is illustrated schematically in Fig. 14.2.

Each tank contained 400 kg of heap leaching wastes with 70 g/tonne uranium content.

The tanks were placed on an open area, where water from rainfall and snowfall seeped through the wastes (and, in tank N2, also through the lime-based reactive barrier). Leachate samples were removed from time to time, and measured and analysed for different components.

The experiment lasted for 3 years (from October 1997 to October 2000).

14.3.2.1. Results

The objective of the test was to investigate the performance of lime-based reactive barriers for attenuation of the uranium concentration in the leachate from wastes resulting from the heap leaching process. Some other questions were also addressed. During the test, leachate volumes, uranium concentration, pH, sodium, bicarbonate and carbonate, hydroxide, calcium, magnesium, sulfate and radium concentrations, and other properties were measured.

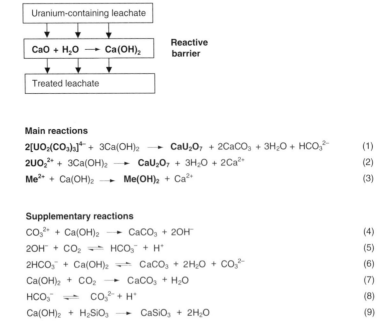

Main reactions

$$2[UO_2(CO_3)_3]^{4-} + 3Ca(OH)_2 \longrightarrow CaU_2O_7 + 2CaCO_3 + 3H_2O + HCO_3^{2-} \qquad (1)$$

$$2UO_2^{2+} + 3Ca(OH)_2 \longrightarrow CaU_2O_7 + 3H_2O + 2Ca^{2+} \qquad (2)$$

$$Me^{2+} + Ca(OH)_2 \longrightarrow Me(OH)_2 + Ca^{2+} \qquad (3)$$

Supplementary reactions

$$CO_3^{2+} + Ca(OH)_2 \longrightarrow CaCO_3 + 2OH^- \qquad (4)$$

$$2OH^- + CO_2 \rightleftharpoons HCO_3^- + H^+ \qquad (5)$$

$$2HCO_3^- + Ca(OH)_2 \rightleftharpoons CaCO_3 + 2H_2O + CO_3^{2-} \qquad (6)$$

$$Ca(OH)_2 + CO_2 \longrightarrow CaCO_3 + H_2O \qquad (7)$$

$$HCO_3^- \rightleftharpoons CO_3^{2-} + H^+ \qquad (8)$$

$$Ca(OH)_2 + H_2SiO_3 \longrightarrow CaSiO_3 + 2H_2O \qquad (9)$$

Fig. 14.1. Chemical reactions in calcium oxide-based reactive barriers

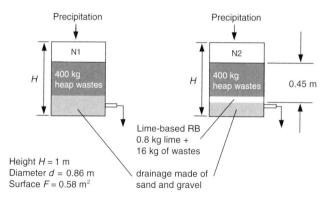

Fig. 14.2. Principal scheme of the experiment

Volume of the leachate

In this measurement it was important to show that the two experimental lysimeters were in identical conditions.

During the test period, approximately 1 m³ of leachate was collected from each lysimeter (972 and 989 litres, respectively). As the surface of the lysimeters was 0.58 m², the annual leachate rate was, at 0.55 m³/m², practically the same for both lysimeters.

The average annual precipitation was approximately 650 mm.

The relatively high volume of leachate was due to low evaporation and other losses because of the absence of run-off and vegetation.

pH of the leachate

Leachates from the two lysimeters had different pHs. The data are presented in Fig. 14.3. In lysimeter N1 the pH of the collected water is lower than that from lysimeter N2. This can be explained by the lime content of the barrier, which results in an elevated pH in the leachate. The average pH for the reference test is 8.33, while for the test with the reactive barrier the average pH is 10.57. The elevated pH was observed even at the end of the test period, i.e. after almost 3 years of leaching. However, the two pH values come closer and closer to each other after 3 years, suggesting that the reactivity of the barrier is approaching its end.

In some respects the elevated pH of the leachate is a disadvantageous feature of the process.

It is worth mentioning that if the leachate is in extensive contact with air, the pH drops to approximately 8.3. This is demonstrated in Fig. 14.4, showing the relation between pH and mixing time. A decrease in pH is connected with the sorption of CO_2 by water.

Specific electrical conductivity of leachate

Electrical conductivity reflects the total salinity of a leachate. Data measured with and without the barrier are close to each other, as shown in Fig. 14.5.

These data are in full agreement with the determination of the dry content (dried at 105°C), which was almost the same in both cases.

The elevated conductivity of the leachate in the initial period is due to the processing solution remaining in the wastes in the form of pore water. These solutes were washed out within 2–3 months, after which the electrical conductivity dropped to 1 mS/cm and below. The data also suggest that the wastes are subject to dissolution and that the leachate from such piles will contain approximately 0.5 g/l of solutes. Figure 14.6. shows the dry content.

Average values of the dry material content are 1.08 g/l for lysimeter N1 and 0.90 g/l for lysimeter N2.

Uranium concentration

Data obtained for the uranium content in the leachate are presented in Fig. 14.7.

It can be seen that the uranium concentration reaches very high values when there is no reactive barrier (as high as 50–60 mg/l at the beginning). After a few months this value starts decreasing steadily, but it is still around 10 mg/l at 1 year. It only

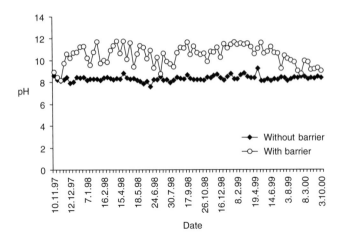

Fig. 14.3. pH of the leachate

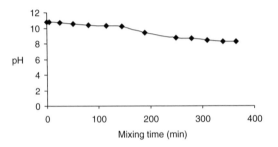

Fig. 14.4. Decrease of pH due to contact with air

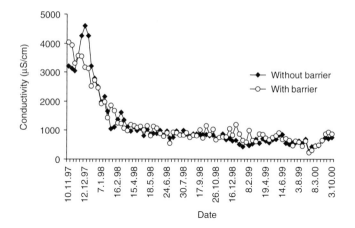

Fig. 14.5. Specific conductivity of the leachate

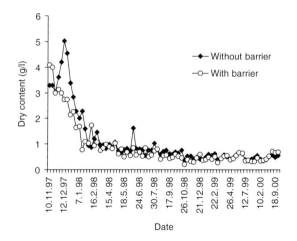

Fig. 14.6. Dry content of the leachate

drops below 5 mg/l in $1\frac{1}{2}$ years, and then remains at that level during the rest of the experiment. The experiment demonstrates that uranium dissolution is especially high in years 1 and 2. At the end of the third year the uranium concentration is 2 mg/l.

Using a lime-based barrier, the uranium concentration is much lower, and remains under 1 mg/l during practically the whole period of the experiment.

The average uranium concentration in the leachate (throughout the whole experiment) was 16 mg/l without the reactive barrier, while it reached only 0.61 mg/l with the barrier. Comparing these values it becomes evident that the calcium oxide barrier is a very effective tool for uranium attenuation.

During the experiment, 8.6 g of uranium was dissolved, of which 8.1 g was retarded by the reactive barrier.

Summing up the results of the 3 year laboratory experiment it can be concluded that active calcium oxide-based reactive barriers can effectively decrease the uranium concentration in the leachate from heap leaching and waste rock piles.

Sulfate–uranium correlation

Figure 14.8 shows the uranium and sulfate concentrations. It can be seen that with the exception of the initial period (~1 month) a good correlation exists between the two components. The somewhat higher concentration of the sulfate at the beginning can be explained by the presence of dissolved sulfates in the waste from chemical leaching processes. After a 1 month washing period the sulfate mainly originates from pyrite oxidation, which has also led to the dissolution of uranium minerals.

(The uranium concentration is given in Fig. 14.8 in units obtained by dividing the real values by 15.) Data is presented for a period somewhat longer than 1 year.

It can be seen that the rate of uranium and sulfate dissolution is likely to be more or less the same.

Effect of barrier arrangement

From practical point of view it is important to know whether the calcium oxide layer has to be located under the wastes or can be placed on the top of the pile.

A separate test has been carried out in two columns to answer this question: in one column the reactive barrier (i.e. the calcium oxide-containing layer) was placed at the bottom of the column, while in the other it was placed at the top. The columns were placed under open-air conditions similar to those of the lysimeters. Both columns were filled with the same heap leaching wastes. The leachate was collected and analysed for different components. The dimensions of the columns used are shown in Fig. 14.9.

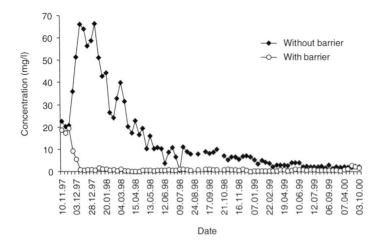

Fig. 14.7. Uranium concentration in the leachate

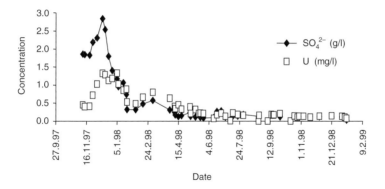

Fig. 14.8. Correlation between uranium and sulfate concentrations (uranium concentration obtained by dividing the actual value by 15)

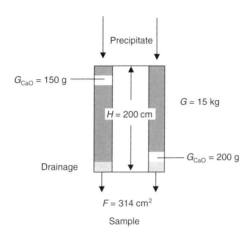

Fig. 14.9. Principal scheme of the experiment with different barrier arrangements

It can be seen that the uranium concentration is much lower if the calcium oxide barrier is placed at the bottom of the column rather than at the top. This means that uranium precipitation occurs predominantly in the barrier itself. Calcium oxide-containing water has much less effect, if any, though this effect cannot be entirely excluded because pH > 9 holds even in this case.

The experiment lasted 8 months. The data obtained are presented in Table 14.1.

Large-scale field test
Based on the very promising results of the open-air laboratory experiments, a field test was carried out on an area of 25 × 7.5 m². The area was divided into two parts: one was for reference purposes, the other for the test itself. The first sub-area was further divided into two parts, and the second sub-area into four parts. In this way, six test fields were built altogether, as shown in Fig. 14.10, each with an area of 7.5 × 4 m².

Table 14.1. Experimental data from the barrier location column experiment

Date	Volume of leachate (ml) Top[a]	Volume of leachate (ml) Bottom[a]	pH Top	pH Bottom	Specific conductivity (µS/cm) Top	Specific conductivity (µS/cm) Bottom	U (mg/l) Top	U (mg/l) Bottom	Dry content (g/l) Top	Dry content (g/l) Bottom	Na (mg/l) Top	Na (mg/l) Bottom
7.1.99	15	100	7.35	7.58	1 448	1 369	0.58	0.43				
5.2.99	160	–	7.83	–	1 195	–	0.44	–				
22.2.99	210	170	8.89	11.72	1 180	3 455	1.69	0.17	0.858	1.994	180	615
23.2.99	900	1 040	9.42	12.25	1 192		2.62	0.04	0.924	1.99	250	520
10.3.99	700	980	9.01	11.26	1 625	3 876	3.29	0.22	1.236	2.516		430
31.3.99	635	280	9.67	10.51	2 212	4 598	3.47	0.16	1.526	2.60	465	875
19.4.99	190	740	9.38	12.15	2 898	4 322	5.53	0.15	2.442	2.574	728	525
26.4.99	730	1 390	9.51	12.26	2 183	3 727	5.78	0.00	1.87	1.562	465	500
30.4.99	1 100	320	9.25	10.36	2 131	2 131	3.7	0.33	1.642	1.404	540	675
20.5.99	300	390	9.20	9.89	3 044	2 485	7.51	0.4	2.606	2.008	825	
10.6.99	310	330	9.47	9.83	3 200	4 250	6.98	0.19				
14.6.99	595	550	9.41	11.74	2 500	2 700	6.15	0.47				
21.6.99	2 080	1 650	9.03	10.57	2 088	1 852	6.27	0.33	1.55	1.262	555	500
23.6.99	840	920	9.46	11.32	2 022	2 190	6.27	0.25	1.48	1.708	540	450
12.7.99	4 000	3 940	9.39	9.86	2 199	1 553	6.21	0.27	1.77	1.408	650	500
16.7.99	340	400	9.72	10.51	2 122	1 632	6.33	0.58	1.57	1.012	588	450
3.8.99	990	970	9.97	9.84	2 336	1 709	6.77	0.13	1.86	1.328	625	500
9.8.99	600	780	9.47	10.24	2 075	1 574	5.87	0.69	1.27	1.27	550	425
6.9.99	900	860	9.63	10.25	2 352	2 034			1.82	1.354		
Average/ sum total	15 595	15 810	9.22	10.61	2 105	2 673	4.29	0.34	1.656	1.83	535	535

[a] Barrier position

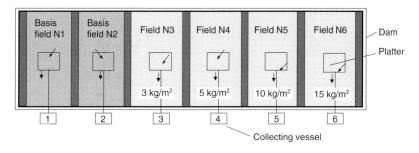

Fig. 14.10. Layout of the field test

The test fields N1 and N2 were not equipped with reactive barriers, and were used as a basis for evaluation, while fields N3, N4, N5, N6 had reactive barriers incorporating horizontal barriers using 3, 5, 10 and 15 kg/m² of lime, respectively.

A platter with a 1 m² surface was first placed on each field for sample collection, from which water could be collected in sample collectors through a pipe. Dams, 50 cm high, separated the fields from each other. Heap leached wastes were then distributed over the whole area more or less homogeneously to a thickness of approximately 3 m, and applying approximately 4–5 tonnes/m² of lime. The lime was finally mixed with the wastes *in situ* by hand, using a rake, at a ratio of approximately 1:20.

Water seeped through the pile, and a part of it was collected in the platters. The experiment lasted for only 7 months, because the area was needed for relocation of heap leaching wastes.

14.3.2.2. Field test results

The main aim of the experiment was to show that a lime-based reactive barrier is capable of working under realistic field conditions. The uranium content in the leachate was thus of primary interest, but other components were also measured.

Uranium concentration

An evaluation of the uranium concentration in the collected samples showed that the leachate from fields N1 and N2 (without barrier) contained much higher uranium concentrations than that from the fields with reactive barriers (N3, N4, N5 and N6). The reactive barrier based on lime proved to be effective for uranium retardation from the leachate. The results are presented in Fig. 14.11. It can be seen that the uranium concentration remained below 1 mg/l only if the lime content was higher than 3 kg/m². This is likely to be due to processing solutions remaining in the wastes (sodium carbonate, bicarbonate, etc.). But even in this case the uranium concentration is substantially lower than for fields N1 and N2.

There was no marked difference between the fields with different lime contents, so for industrial use about 1.5 kg/tonne has been suggested (i.e. 5 kg/m² for a 3 m lift and about 8 kg/m² for a 5 m lift).

pH of the leachate

The pH of the leachate remained practically unchanged during the experiment. The results are presented in Fig. 14.12. It can be seen that the pH of the collected leachate, even from the section with the reactive barrier, was approximately 8.3, i.e. characteristic for a natural equilibrium state. This indicates that the dissolved calcium oxide was neutralized by atmospheric CO_2, as was assumed earlier.

14.3.3. Building reactive barriers in practice

Based on the promising results obtained from the laboratory and field experiments, a lime-based reactive barrier was used in practice for reducing the uranium concentration in leachates during the relocation of heap leaching wastes to the waste rock pile N3. The barrier was built continuously: the wastes being placed in approximately 5 m lifts, with 5–7 kg/m² of lime used under each lift.

The lime was spread on the surface and mixed with the portion of the wastes to be relocated. Agricultural techniques were used for the mixing. The reactive barrier

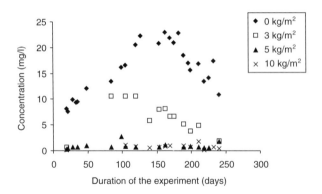

Fig. 14.11. Field test. Uranium concentration in the seepage

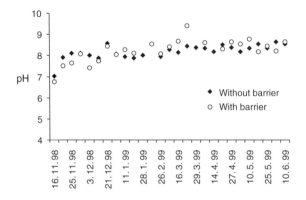

Fig. 14.12. pH of the leachate

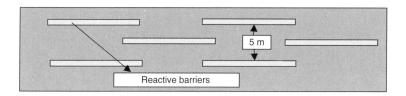

Fig. 14.13. Section of the reactive barriers constructed on waste rock pile N3

was built by overlapping the layers for stability reasons. The principal sequence of the barrier layers is shown in Fig. 14.13.

This protective method has proven to be effective: there have not been any problems with the uranium concentration in the leachate from the relocation of more than 4 million tonnes of heap leaching wastes. It has been controlled to a level of 2 mg/l. (Some elevated values have been caused by leachate from other sources.)

14.4. Conclusions

(1) The uranium concentration in the leachate from uranium-bearing heap leaching wastes can be effectively reduced using lime-based reactive barriers.
(2) The specific amount of lime to be used depends on the composition of the leachate: in the case study described, 1.5 kg/tonne of lime was proposed. Lime should be mixed with waste material in a ratio of at least 1:10 and then uniformly spread on the surface used for relocation.
(3) A temporary increase in the pH of the leachate can be observed, but this tends to decrease to the natural value due to the effect of CO_2 in the air.
(4) The method of staggered horizontal reactive barrier construction developed here was used successfully by the Mecsek Ore Mining Company in Hungary, when uranium containing wastes were relocated from one pile to another.

14.5 References

DAY, S. J. (1994). Evaluation of acid generating rock and acid consuming rock mixing to prevent acid rock drainage. *International Land Reclamation and Mine Drainage Conference and Third International Conference on the Abatement of Acidic Drainage. Conference Proceedings, Pittsburg, Pennsylvania.* Vol. 2, p. 77. Bureau of Mines Special Publications SP 06A-06D-94. US Department of the Interior, Washington DC.

HUTCHISON, I. P. G. and ELLISON, R. D. (1992). *Mine Waste Management.* Lewis, Boca Raton, Florida.

MEGGYES, T., TÜNNERMEIER, T. and SIMON, F. G. (2001). Einführung in die Technologie der reaktiven Wände. In: G. Burghardt, T. Egloffstein and K. Czurda (eds), *ALTLASTEN 2001 Neue Verfahren zur Sicherung und Sanierung, Karslruhe.* Conference Proceedings Vol. 4, pp. 1–25. ICP Eigenverlag Bauen und Umwelt.

ROEHL, K. E. and CZURDA, K. (2001). Reactive Wande – Langzeitverhalten und Standzeiten. In: G. Burghardt, T. Egloffstein and K. Czurda (eds), *ALTLASTEN 2001 Neue Verfahren zur Sicherung und Sanierung, Karslruhe.* Conference Proceedings Vol. 4, pp. 27–36. ICP Eigenverlag Bauen und Umwelt.

Part VI
Groundwater flow modelling

15. Observed and modelled hydraulic aquifer response to slurry wall installation at the former gasworks site, Portadown, Northern Ireland, UK

R. Doherty and R. M. Kalin
Environmental Engineering Research Centre (EERC), The Queen's University Belfast, Belfast BT9 5AG, UK

U. S. Ofterdinger, Y. Yang and K. Dickson
School of Civil Engineering, The Queen's University Belfast, Belfast BT9 5AG, UK

15.1. Introduction

Several studies have shown that permeable reactive barriers (PRBs) provide effective treatment for a variety of groundwater contaminants (Gillham and O'Hannesin, 1994; Benner, *et al.*, 1997). Geochemical investigations as well as detailed numerical modelling studies have focused on the assessment of barrier performance and design (Thomas *et al.*, 1995; Tratnyek *et al.*, 1997; Benner *et al.*, 1999, 2000). However, a key factor for the failure of full-scale implementation of PRB technology is a lack of understanding about how underground installations of the site-specific PRB affect the local and regional groundwater flow systems. As a consequence, special emphasis was given during the implementation of a PRB system at a contaminated former gasworks site to aquifer response and to underground structures in an effort to ensure optimal performance of the reactive barrier.

A sequential biological PRB is currently being used to remediate contaminated groundwater at a former gasworks site in Portadown, Northern Ireland, UK. The majority of soil contamination at this site is close to the surface and is associated with gasworks waste on site. Organic contamination (mineral oil, polyaromatic hydrocarbons (PAHs) and benzene, toluene, ethylbenzene and xylenes (BTEX)) is widespread. Elevated levels of total cyanide and elemental sulfur mark inorganic contamination. The emplacement of the PRB system at this site involved the demolition of remaining subsurface foundations, the installation of a slurry wall along the western, northern

and part of the eastern perimeter of the site as well as the installation of the subsurface
reactor in the central northern boundary of the site (Fig. 15.1).

The phased installation of the PRB elements allows us to focus this chapter on the
hydraulic response of the natural groundwater flow system to the demolition works
carried out on the site that involved dewatering of the site, and on the hydraulic
response to the installation of the slurry wall. In addition to this, we qualitatively
analyse the relationship between head fluctuations and variations in precipitation.
We show how numerical groundwater flow modelling provides valuable information
with regard to the prediction of the hydraulic response of natural groundwater flow
to the installation of man-made subsurface structures.

15.2. Geology and topographic setting

The site is at an urban–rural interface: the areas to the south, west and east are
predominately residential and disused industrial areas; to the north are wetlands.
The site generally dips gently from approximately 19.5 to 17.5 m AOD in a north-
western direction (Fig. 15.1) with a distinctive drop in elevation of 1.5–2.0 m
(indicated by shading in the diagram) along the northern border of the site. The
River Bann flows in a south–north direction, meandering approximately 300 m
to the west and 400 m to the north of the site. Regional groundwater flow is
in a north-westerly direction. The regional geology of the area is dominated by
graben-style normal faulting within Tertiary basalts that allowed the accumulation
of the Lough Neagh Group of clays within the Bann valley. These are overlain by

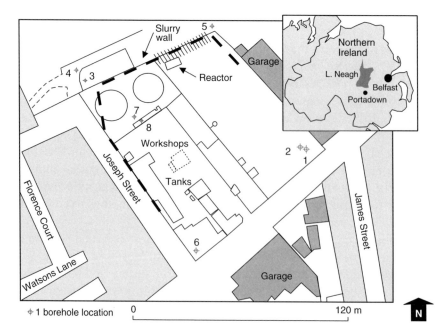

*Fig. 15.1. Plan view of research site with indicated borehole locations, slurry wall and
reactor position*

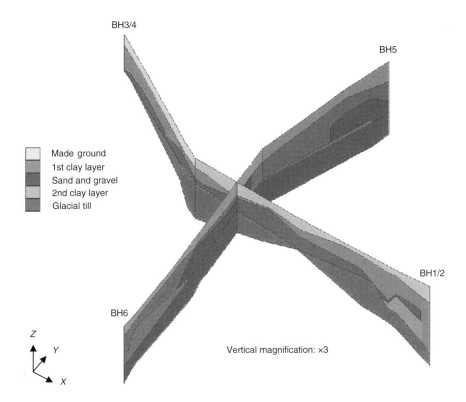

Fig. 15.2. Fence diagram of site lithologies

glacial tills and locally occurring glacial sands, gravels and lake deposits. Flood plain alluvium occurs within the vicinity of the River Bann.

Forty-two trial pits and eight boreholes provided information on the site-specific subsurface lithologies (Fig. 15.1) and allowed us to develop a detailed three-dimensional geological model of the site (Fig. 15.2). Five alluvial/periglacial stratigraphic units can be distinguished at the site, overlying an impervious glacial till aquitard. These are, given with approximate thicknesses, man-made ground (1–2 m) overlying laminated clays (0.5–2.0 m), which overlie silty sands and gravels (0.5–1.5 m). These in turn overlie silty clays (1.0–2.5 m), which rest on sands and gravels (1.5–3.5 m). This classification is highly simplified due to the fact that borehole and trial pit logs show a degree of lateral heterogeneity caused by discontinuous deposition and grading within the above units. The lower sediments generally infill a hypothesized post-glacial kettle hole situated to the north of the site. A basin-like feature in the glacial till controls the pinching out of the lower stratigraphic units at boreholes 3 and 4, with the thickening of the same units at borehole 5 in a northerly direction. The relatively slow melting of an area of permafrost after glacial retreat most likely produced this basin structure. The resulting depression once the permafrost had gone became the kettle hole, which was filled in by postglacial sediments (now the area of wetlands). The upper stratigraphic units dominated by the laminated clays represent the alluvial flood plain of the River Bann.

15.3. Installation phases

Prior to the installation of the PRB underground structures, the remaining shallow subsurface foundations of the former workshops and buildings on the site were removed in the period March to April 2001. The demolition of the subsurface gasholder tanks to the north of the site required dewatering in order to facilitate the excavation of foundations at depth. To control the groundwater flow through the reactor and allow effective risk management of contaminated groundwater across the site, a 600 mm wide slurry wall (Na-bentonite–cement–fly-ash mix) was installed during April and May 2001. The cement–bentonite slurry wall was installed by the slurry trenching method as outlined by Day *et al.* (1999). The slurry wall was installed along the western, northern and part of the eastern boundaries of the site (Fig. 15.1). The base of the slurry wall was keyed into impermeable glacial till at a depth of 8–14 m below ground level. Once installed, the slurry wall was designed to function at a hydraulic conductivity of 1×10^{-9} m/s (UK Institution of Civil Engineers, 1998). The reactor at the northern boundary of the site was turned on 6 months after slurry wall installation, during which mass transport and aquifer response to the slurry wall installation was established.

15.4. Hydrogeology

Groundwater levels at the site have been recorded from nested and multilevel piezometers across the site (Fig. 15.1). Monitoring was carried out prior and during the installation works, and in particular for a 6 month period following the installation of the slurry wall to assess its effect on the natural groundwater flow system. Figure 15.3 shows the recorded hydraulic head data from specific piezometers across the site, up- and downstream of the slurry wall. Figure 15.3 also depicts the daily precipitation record at the meteorological station in Hillsborough approximately 20 km east of the site. Flow across the site was shown to be predominantly in a south-east/north-west direction (Fig. 15.4).

The earthworks at the site began on 20 March 2001. The head data recorded prior to that showed only minor fluctuations and did not correlate closely to changes in precipitation, i.e. changes in potential groundwater recharge, on a short timescale. The major earthworks included the demolition of the subsurface foundations of the two gasholder tanks to a depth of ~6 m below the ground surface. These were at the north-western corner of the site beside borehole 7. This demolition necessitated dewatering at this location, which is translated into a marked decrease in measured head across the northern part of the site. At boreholes 1, 2 and 6 on the southern perimeter of the site, the influence of the dewatering was less apparent.

The slurry wall was installed during the second excavation phase from 20 April to 9 May 2001. Installation of the slurry wall started along the eastern leg, and finally along the northern and western perimeter of the site (Fig. 15.1). The immediate effect of

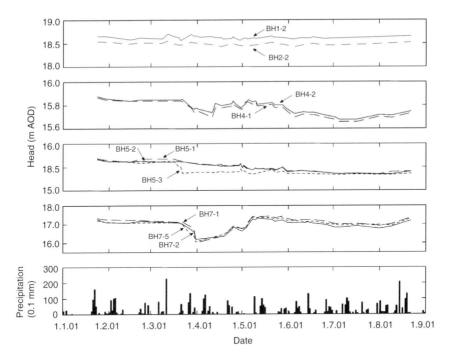

Fig. 15.3. Recorded hydraulic heads from individual piezometers and daily precipitation record from the meteorological station at Hillsborough

the cessation of dewatering and subsequent installation of the slurry wall was a pronounced recovery of the measured heads at the multilevel piezometers at borehole 7 upstream of the slurry wall. The water table rose above the pre-dewatering levels in borehole 7 immediately after slurry wall completion. However, groundwater levels at borehole 7 began to decrease 20 days after slurry wall installation, and within 50 days all multilevel heads were below pre-dewatering levels and falling. The installation of the slurry walls had no immediate effect on boreholes 1 and 2.

Close to the major dewatering location, at borehole 4, downstream of the slurry wall, the cessation of dewatering led to a temporary partial recovery of the observed heads, but the subsequent installation of the slurry wall led to a continuous decline in head until July 2001.

Borehole 5 containeds nested piezometers (5-1, 5-2 and 5-3). At piezometer 5-3, screened in the made ground, an immediate response of the observed heads to the dewatering was noted. At borehole 5 the piezometers installed in the deeper aquifer units (5-1, 5-2) showed a dampened reaction to the dewatering. The installation of the slurry wall upstream of this borehole led to a continuous decline in heads until August 2001. Of particular interest at this location was the lack of recovery of the water table from the perched aquifer within the made ground as a direct response to the cessation of the dewatering. To the north and the east of this borehole location, the topography drops markedly, and the construction of the slurry wall hindered

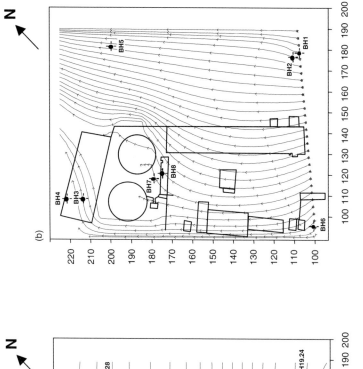

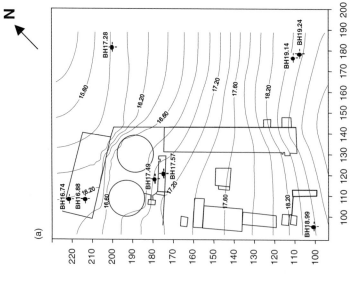

Fig. 15.4. Modelled (a) equipotential heads and (b) particle tracking before excavation of foundations

lateral recharge to this perched aquifer within the made ground. The record of similar head values by piezometers 5-1, 5-2 and 5-3 in July 2001 indicated that the earlier head equilibrium, as seen in February 2001, had been established following a transition phase due to delayed leakage from the lower aquifers.

The declining head data during June–July at borehole 7, upstream of the slurry wall, was probably linked to a low number of intense precipitation events during the summer period concurrent with increased temperatures (evaporation). A general observed increase of heads across the site in August 2001 was attributed to increased precipitation and recharge, and would thus indicate a more rapid head response to changes in potential recharge than observed before removal of subsurface structures.

15.5. Numerical modelling

Numerical modelling of groundwater flow was undertaken using the widely available Visual Modflow and Processing Modflow packages. During earlier investigation work, interference slug tests were carried out on boreholes 1 and 2. This provided data on the hydraulic conductivity and storativity of the silty sand and gravel aquifer. These data were used as fixed parameters during automated calibration, using WSPEST, to estimate the hydraulic conductivities of the remaining lithologies and average annual recharge across the site.

The conceptual groundwater model of the site follows the earlier description of the lithologies encountered. A degree of lateral variation across the site allows the upper and lower sand layers to become continuous around the area of borehole 7. This lateral variation also causes the lower silty sands and gravels to pinch out around boreholes 3 and 4. This geological feature is also responsible for a local variation from expected regional groundwater flow. Regional groundwater flow would be expected to follow a predominately north-westerly direction. The pinching out of the lower stratigraphic units causes a localized groundwater flow more to the north. The presence of the foundations of the gasholding tanks further controls the groundwater flow (Fig. 15.4). Dewatering during the demolition period (20 March to 20 April 2001) was simulated by a series of pumping wells encompassing the area of the gasholding tanks. The volume of water pumped and removed was calculated, and this volume removal was simulated during the demolition period. The results of calculated and modelled heads versus observed heads for the drawdown are shown in Fig. 15.5.

The emplacement of the impermeable slurry wall and expected rebound of the drawdown from the demolition phase were simulated. The presence of the slurry wall and removal of foundations had an effect on the local groundwater flow. The foundations of the gasholding tanks had previously controlled local groundwater flow around the north-western corner of the site, which had retarded flow times across the site in the south and south-western area as the groundwater mounded and flowed around the gasholders. The slurry wall now controls the majority of groundwater flow on site, and all groundwater flow near the source of contamination. Simulations using a completely impervious (1×10^{-12} m/s) slurry wall (Fig. 15.7) resulted in greater modelled than observed rebound of heads in the multilevel borehole 7. Simulations using a heterogeneous or 'leaky' wall provided a closer approximation to the observed field results (Fig. 15.6). The degree of reversed drawdown or 'mounding' of groundwater at the leaky slurry wall is also less than that of the impervious wall. The degree of leakage is such that the modelled hydraulic conductivity of the wall is in the range 1×10^{-7} to 1×10^{-8} m/s, i.e. above the design specification of 1×10^{-9} m/s.

15.6. Discussion

The installation of the slurry wall significantly affected the head distribution across the site as expected, in particular at the northern site boundary. Head data around boreholes 1 and 2 showed no significant correlation to the works carried out on the site. The most pronounced and rapid hydraulic responses were observed at borehole 7. This was due to the backfilling of the excavated area with demolition rubble. Borehole 5, due to its topographic position, led to a distinct behaviour of the head data from the perched uppermost aquifer. This can be attributed to the fact that the made ground around that borehole is predominately loosely compacted fill raised above the natural ground level. The backfilled demolition waste from the gasholder excavation together with the loosely compacted fill around borehole 5 caused an area of higher hydraulic conductivity at the northern border of the site. This is the area that the reactor occupies. This area of higher hydraulic conductivity across the northern side of the site within the slurry wall will have the beneficial effect of homogenizing contaminants in the groundwater before they enter the reactor.

Figure 15.4 shows no dramatic changes in the general groundwater flow direction; however, declining head data downstream of the slurry wall indicated groundwater flow diverging around the slurry wall. The reactor had not been turned on during this study, and a predominant preferential flowpath around the slurry wall occurred as expected. The Fig. 15.7 model results confirmed that the areas on site containing the contamination were effectively contained within the slurry wall, and that clean water entering the site was effectively by-passed around to the east of the site. Qualitative comparison of the head data with the precipitation record from the Hillsborough

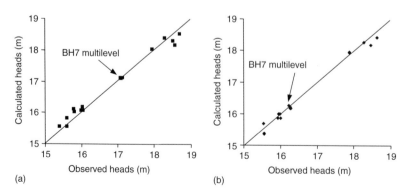

Fig. 15.5. Comparison of BH7 drawdown between (a) normal groundwater flow and (b) dewatering during removal of gasholding foundations

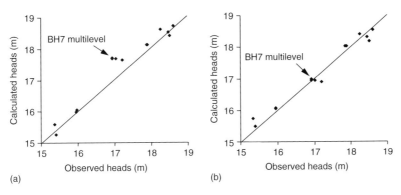

Fig. 15.6. Comparison between simulations of (a) an impervious and (b) a 'leaky' slurry wall. The 'leaky' simulation closely resembles the field data

meteorological station indicated a general lack of direct response of the hydraulic heads to precipitation events. Better comparisons may be possible with more local meteorological stations; however, no data are available at this time. The general increase in head across the site in August 2001 might well be attributed to increased intense rainfall events. The removal of building foundations at the site likely contributed to a more rapid response of groundwater head to potential recharge events. Further monitoring data without the imprint of building foundations will continue to clarify this relationship.

Modelling results prior to the installation of the PRB showed that the gasholder foundations controlled the movement of groundwater at the site, and that a reasonable drawdown during the dewatering phase was effectively simulated. The overall performance of the slurry wall cannot yet be accurately simulated as it cannot yet be determined whether the slurry wall has a slightly higher permeability than expected or if there is a degree of seasonal variation affecting groundwater flow. The overall hydraulic conductivity of the slurry wall may be more complicated than expected: Philip (2001) found a degree of heterogeneity ranging from 1×10^{-7} to 1×10^{-10} m/s within cement–bentonite slurry walls. It is interesting to note that model calibration was met when a hydraulic conductivity of between 1×10^{-7} and 1×10^{-8} m/s was used. This poses questions as to the reliability of and testing methods for slurry walls within the subsurface. New well installations upgradient and downgradient will be used to verify the effective hydraulic conductivity of the wall and to further model the responses of the system to seasonal changes in water level. The slurry wall was completed at the start of the summer of 2001, and all observed data to date were collected during the summer months. A single modelled annual recharge estimate (used in the model) may be too vague to accurately simulate groundwater flow during a rapid phase of subsurface development. Further refinement of the recharge parameter estimate into monthly stress periods, together with precipitation data collected closer to the site, is ongoing. Monitoring of observed head data during the 2002 will provide further calibration and verification of the model of this PRB system

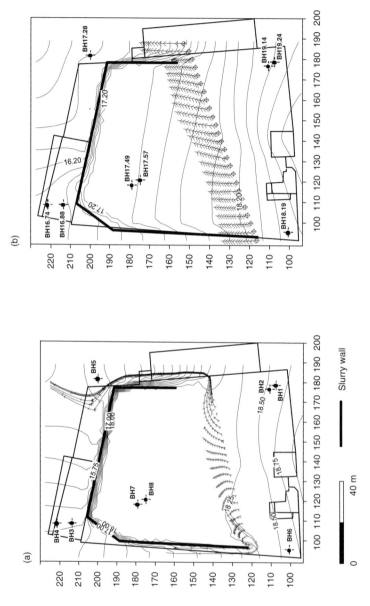

Fig. 15.7. Comparison of modelled equipotential heads and particle tracking of (a) an impervious slurry wall and (b) a 'leaky' slurry wall

15.7. Conclusion

The long-term monitoring of hydraulic heads at the Portadown research site prior to the installation of the slurry wall and reactor provided an observed behaviour for the natural groundwater flow system. Monitoring during and after the installation provided an understanding of the impact of the PRB system on the intermediate scale regional hydrogeology. It also provided an initial assessment of the hydraulic performance of the slurry wall. Discrepancies between the observed and modelled head recovery upstream of the slurry wall indicated a lower hydraulic conductivity than expected. This discrepancy could also originate from temporal variations in groundwater recharge, not accounted for in the present numerical model. Continuous monitoring allowed us to gain confidence in the validity of the numerical groundwater flow model, previously calibrated to non-works-affected conditions. Continued monitoring of the slurry wall performance is ongoing to determine the dominant mechanisms of groundwater–wall interaction.

15.8. Acknowledgements

The authors would like to acknowledge QUB, DEL, DOE(NI)EHS and EPSRC grant GR/M89768 for the support of this work. They would also like to thank K. Walsh, M. Craig and T. Needham for work on this site.

15.9. References

BENNER, S. G., BLOWES, D. W. and PTACEK, C. J. (1997). Full-scale porous reactive wall for the prevention of acid mine drainage. *Ground Water Monitoring and Remediation* **17**(4), 99–107.

BENNER, S. G., BLOWES, D. W., GOULD, W. D., PTACEK, C. J. and HERBERT Jr. R. B. (1999). Geochemistry of a permeable reactive barrier for metals and acid mine drainage. *Environmental Science and Technology* **33**(16), 2793–2799.

BENNNER, S. G., BLOWES, D. W. and MOLSON, J. W. H. (2000). Modelling preferential flow in reactive barriers: Implications for performance and design. *Ground Water* **39**(3), 371–379.

DAY, S. R., O'HANNESIN, S. F. and MARSDEN, L. (1999). Geotechnical techniques for the construction of reactive barriers. *Journal of Hazardous Materials* **B67**, 285–297.

GILLHAM, R. W. and O'HANNESIN, S. F. (1994). Enhanced degradation of halogenated aliphatics by zero-valent iron. *Ground Water* **32**(6), 958–967.

PHILIP, L. K. (2001). An investigation into contaminant transport processes through single-phase cement–bentonite slurry walls. *Engineering Geology* **60**, 209 –221.

THOMAS, A. O., DRURY, D. M., NORRIS, G., O'HANNESIN, S. F. and VOGAN, J. L. (1995). The *in situ* treatment of trichloroethene-contaminated groundwater using a reactive barrier – results of laboratory feasibility studies and preliminary design considerations. In: W. J. van den Brink, R. Bosman and F. Arendt (eds), *Contaminated Soils '95*, pp. 1083–1091. Kluwer, Dordrecht.

TRATNYEK, P. G., JOHNSON, T. L., SCHERER, M. M. and EYKHOOLT, G. R. (1997). Remediating ground water with zero-valent metals: chemical considerations in barrier design. *Ground Water Monitoring and Remediation* **17**(4), 108–114.

UK INSTITUTION OF CIVIL ENGINEERS (1998). *Specification for the Construction of Slurry Trench Cut Off Walls as Barriers to Pollutant Migration*. Thomas Telford, London.

16. A finite-volume model for the hydrodynamics of flow in combined groundwater zone and permeable reactive barriers

D. B. Das
Hydrology and Ecology Section, Faculty of Civil Engineering and Geosciences, Delft University of Technology, PO Box 5048, NL-2600 Delft GA, The Netherlands

V. Nassehi
Chemical Engineering Department, Loughborough University, Loughborough LE11 3TU, Leicestershire, UK

16.1. Introduction

Permeable reactive barriers (PRBs) are used in the subsurface for emplacement of reactive materials to intercept contaminant plumes, to provide preferential flow paths through the reactive zones and for *in situ* transformation of the contaminant(s) into environmentally acceptable forms to attain remediation concentration goals at the discharge of the barrier. They constitute an emerging technology for groundwater treatment and offer many advantages over the traditional methods of *ex situ* treatment. Most current research concerning PRBs is, however, dedicated to examining the kinetics of the geochemical reactions in the barriers, and very little is being done on the hydrodynamics of the groundwater flow through PRBs. Further, in many circumstances the groundwater flow regime may involve free-flow zones (a section which is devoid of porous materials) such as large underground cracks/fissures, etc., adjacent to the PRBs. Alternatively, it may be necessary to construct a PRB near an underground free-flow region for site remediation, because such flow conditions have considerable impact on contaminant transport loads, as indicated in many studies concerning preferential flow channels.

The main purpose of this chapter is to present a methodology which may be adopted to predict movement of contaminated groundwater through combined free-flow zones and PRBs when the flow is not affected by porosity variation. To model overall fluid mobility efficiently in the combined domains, free- and porous-flow zones must be studied in conjunction with each other. In general, subdomains in the coupled flow system can be distinguished by an interfacial surface, which represents a transition zone for fluid mobility and contaminant transport. Flow models for such combined areas require description of not only the fluid dynamic characteristics in the individual domains but also the mass and momentum transfer behaviour across the interfaces. Our aim is to describe a general methodology which can be applied to study groundwater hydrodynamics in PRBs in cases where it involves adjacent free domains, specifically at the interface between the permeable barrier and the groundwater zone. The present study should also help in the accurate evaluation of a number of important factors such as the proposed locations for the PRBs with respect to the groundwater hydrology, direction of groundwater flow within the barriers, materials for the PRBs, suitable barrier configurations, etc.

16.1.1. Modelling approaches

The impact of the combined flow on the overall transport behaviour depends on many distinguishing features, such as the dimensions of the pathways, its behaviour in combination with the surroundings and the characteristics of the porous material, e.g. its porosity and permeability. The number of permeable interfaces between the free- and the porous-flow domains and the aspect ratios of the subdomains also influence the fluid dynamics. To model fluid flow in such domains, two approaches are usually adopted: firstly, formulations based on an assumption of continuum domains; and, secondly, formulations based on discrete pathways.

In the former case, the porous section is treated as a pseudo-fluid layer, and the whole domain is considered to be a single domain. As such, one suitably formulated equation of motion in conjunction with other equations for continuity of mass and pollutant species balance is solved. The mathematical model recognizes the free- and the porous-flow domains based on a spatially varying permeability. Such a single-domain approach is usually preferred in systems where the flow transition from free- to porous-flow sections is not distinct and the structural properties of the permeable domain change progressively. Application of this approach is most commonly found in metal solidification problems involving mushy zones.

However, if the permeability of the porous media is relatively high and constant, so that the interface between adjacent free and porous media forces a transition from free to porous flow, or vice versa, the whole domain must be viewed as a combination of adjacent flow fields rather than a continuous single domain. In such transport phenomena, the second approach should be utilized, where the appropriate equations of motion describing the flow in different subdomains are used.

Mathematical models based on the multi-domain approach for most combined flows are well established for artificial flow systems (Nassehi and Petera, 1994; Gartling et al., 1996; Gobin et al., 1998; Nassehi, 1998; Chen et al., 1999). However, extensions of these formulations to composite groundwater flow have been uncertain. The main difficulty in this task is the realistic representation of flow behaviour at the interfacial surface, which stems from two main sources. The first is the inherent irregularity in the size and shape of the domains. Secondly, randomness can be related to a generalized lack of knowledge about the processes involved and the impossibility of an exhaustive analytical description. In the case of combined free and porous flow in the subsurface, such as through PRBs, not only can the interface be expected to have a random shape and size but there is also a complete lack of physical evidence for the mass and momentum transport behaviour across the interface.

16.1.2. The finite-volume method

In order to preserve the compatibility of the underground fluid dynamic characteristics at the interfaces, numerical schemes are usually used, as they can deal with such problems. It is well known that finite-element schemes can readily cope with curved and complex problem domains (Nassehi and Petera, 1994; Gartling *et al.*, 1996; Nassehi, 1998). However, realistic modelling of the underground flow processes requires three-dimensional computations, which become excessively expensive if finite-element methods are used. Other numerical techniques that can, apparently, resolve such difficulties due to their inherent mathematical strength are, for example, the finite-volume (Patankar, 1980; Versteeg and Malalasekera, 1995) and the spectral (Canuto *et al.*, 1988; Guo, 1998) methods. These methods should, therefore, be adopted when representing such underground fluid flow (Das and Nassehi, 2001; Das *et al.*, 2001, 2002).

A two-domain mathematical model that can be used for combining three-dimensional zones of permeable domains and free-flow channels will be described. The permeable domain is assumed to be a continuum medium; and the average macroscopic flow properties, such as velocity and pressure, are evaluated on representative elementary volumes. The present study deals with such a combined flow system for which there is a complete lack of experimental/field data, particularly for interfacial flow properties, to compare them directly with predicted values. The finite-volume method is, therefore, adopted in the present work for discretizing the governing flow equations to algebraic forms and solving them, as it can conserve fluid materials at each grid cell (unlike other numerical methods).

Though, in general, the computational costs of using the spectral method are less than those of the finite-volume method for the same spatial scale, due to the above advantage the present mathematical formulation is based on the standard finite-volume technique instead of the spectral method. In doing so, the physical region of combined flow is truncated to a computational domain where boundary conditions necessary for solving the problem are imposed. The results presented here represent the hydrodynamic conditions for a physical water flow process. In particular, the influence of the aspect ratios of the combined domain on the flow, based on different thicknesses of the porous layer, is analysed. The model developed here provides, in effect, a hydrodynamic model for predicting contaminant mobility in combined domains of free and porous flow in the subsurface. It is envisaged that this treatment of subsurface water flow could make predictions for water quality in the PRBs more accurate and realistic.

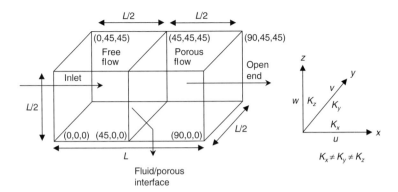

Fig. 16.1. Representative three-dimensional domain of combined free and porous flows for modelling groundwater flow through a PRB

16.2. Formulation of the mathematical model

As mentioned before, the present work is based on a two-domain flow model for simultaneously simulating the hydrodynamics of combined water flow in the subsurface. A representative three-dimensional domain of combined free and porous flow for modelling groundwater movement through a permeable reactive barrier is shown in Fig. 16.1. The non-dimensional forms of Navier–Stokes equations (equation (1)) are adopted as the governing equations for momentum transfer in the free-flowing fluid. The flow in the porous domain is represented by the dimensionless Darcy equation in its transient form (equation (2)). The validity of these equations in the respective cases is well known. Constant properties of the fluid are also defined. The governing equations for motion, as adopted in this work, are

$$\text{Re}\,\nabla P_f^o = -\rho^o\left(\frac{\partial v_f^o}{\partial t^o} + \text{Re}(v_f^o \cdot \nabla v_f^o)\right) + \mu^o \nabla^2 v_f^o \tag{1}$$

$$\text{Re}\,\nabla P_p^o = -\rho^o\frac{\partial v_p^o}{\partial t^o} - \frac{1}{\text{Da}}\frac{\mu^o}{\boldsymbol{K}^o}v_p^o \tag{2}$$

where the superscript 'o' indicates dimensionless terms and the subscripts 'f' and 'p' refer to the free- and the porous-flow domains, respectively. P and v are the pressure and velocity fields, respectively, while ρ and μ are the constant density and viscosity of the fluid, respectively. $\boldsymbol{K}$ is the second-order tensor representation of the isotropic permeability in the porous medium. Re and Da are the Reynolds and Darcy numbers, respectively.

The equations of continuity for conservation of mass in both free and permeable domains are

$$\nabla \cdot v_{f,p}^o = 0 \tag{3}$$

$$e \equiv f, p \tag{4}$$

The three components of the non-dimensional velocity field in the longitudinal (x), lateral (y) and vertical (z) directions are designated u, v and w.

As the matching interfacial condition for linking the equations of motion, the modified form of the Beavers and Joseph (1967) formulation in its dimensionless variant is used (Das *et al.*, 2002). For a Darcian flow region, Beavers and Joseph proposed an empirical slip–flow boundary condition at the interface, describing the proportionality between the shear rate at the interface and the slip velocity through a dimensionless slip coefficient that depends only on the structural properties of the interface. The validity of the formulation has been verified through both theoretical modelling studies (Saffman, 1971) and experiments (Beavers *et al.*, 1970). The applicability of the formulation to different combined flow regimes has been tested by, among others, Salinger *et al.* (1993, 1994), Gobin *et al.* (1998) and Das *et al.* (2002). As adopted in this work, the interfacial boundary conditions are (Das and Nassehi, 2000; Das, 2002; Das *et al.*, 2002):

Longitudinal velocity component, u:

$$u_f^o = u_p^o \tag{5}$$

Lateral velocity component, v:

$$\left(\frac{\partial v^o}{\partial x^o} + \frac{\partial u^o}{\partial y^o}\right)_f = \frac{\gamma_{yz}}{\sqrt{\mathrm{Da}\, K_{yy}^o}}(v_f^o - v_p^o) \tag{6}$$

Transverse velocity component, w:

$$\left(\frac{\partial w^o}{\partial x^o} + \frac{\partial u^o}{\partial z^o}\right)_f = \frac{\gamma_{yz}}{\sqrt{\mathrm{Da}\, K_{zz}^o}}(w_f^o - w_p^o) \tag{7}$$

Pressure, P:

$$\left(-P^o + 2\mu^o \frac{1}{\mathrm{Re}}\frac{\partial u^o}{\partial x^o}\right)_f = -P_p^o \tag{8}$$

γ_{yz} is a slip coefficient at the interfacial surface characterized only by its structural properties. K_{yy} and K_{zz} are respectively the components of the permeability tensor in the lateral (y) and the transverse (z) directions.

The boundary conditions imposed for the solution of the hydrodynamic equations in the present work consist of inlet and initial boundary conditions for velocity and pressure. At the exit, a 'no boundary condition' or 'open boundary condition' is imposed to handle the stress-free section of the domains. As mentioned before, the governing equations are discretized and reduced to algebraic forms using the finite-volume method. The temporal discretization of the equations for predicting the transient flow behaviour is based on the explicit method. Detailed descriptions of the type of grid used in this study and the derivation of the working equations are described elsewhere (Das and Nassehi, 2001; Das *et al.*, 2002).

16.3. Numerical results and discussion

As mentioned before, combined flow systems in the subsurface are influenced by a large number of factors and, as such, general analyses of the problem are intractable. In the present framework, the fluid dynamics of combined flow for different aspect ratios of the domains is considered. Through this analysis, an attempt is thus made to investigate the implications of having different length scales of the domain in the subsurface. A detailed analysis of the associated flow in the subsurface for a global aspect ratio (x:y:z :: 1:1:1) has been presented by Das (2002) and Das *et al.* (2002), where the hydrodynamics of the characteristic flow at different sections of the domains, i.e. free flow, porous flow and the interface, was investigated. The investigation revealed that water in the free-flow section has a unidirectional path, in general. On the other hand, water moves in from the open portion of the porous domain and reverses its direction at the free–porous flow interface. However, the front of velocity reversal and the centre of circulation move away from the inter-face with time. After a certain interval the flow in the porous domain becomes unidirectional, and moves towards the exit of the domain. Figure 16.2 presents typical profiles of free and porous flow for an aspect ratio of x:y:z :: 1:1:1 at time level 25. The longitudinal (x) coordinate of the interface is '45', and it lies on the y–z plane. The flow reversal occurs because of a complex pressure differential across the interface specifically, plus the combined flow domain in general (Das *et al.*, 2002). The Reynolds and Darcy numbers used in the simulation are 1.0 and 1×10^{-14}, respectively. While the Reynolds number indicates a typical creeping flow regime in the porous section, the Darcy number represents a sample filling material in the permeable barrier with low permeability and very fine average grain diameter.

However, with changes in the aspect ratios of the domain in the subsurface, different patterns of velocity reversal may be observed (Das, 2002). In Fig. 16.3, typical velocity fields in the porous section, for an aspect ratio of x_p:y:z :: 0.5:1:1 for the porous domain, are shown at two time levels, $t = 25$ and $t = 45$. In effect, this presents a case where the thickness of the porous domain has been reduced to half. Velocity profiles then correspond to a large extent to the profiles presented in Fig. 16.2 for the porous section. The aspect ratio of the free-flow section is x_f:y:z :: 1:1:1 in this instance, and the velocity profile therein is the same as for the previous case. The interfacial velocity across the interface is the same for both aspect ratios above, and is presented in magnified form in Fig. 16.4 for two time levels, $t = 25$ and $t = 45$. The velocity fields in this work are predicted at the cell nodes. Therefore, the velocity fields are not presented at the interface (x coordinate '45'). As evident, the front of flow reversal is observed near the interface at time level 25, but moves away at time level 45.

With reduced thickness of the domain in the lateral direction (y), the front of velocity reversal becomes much less mobile. The amount of water flowing towards the interface increases with time, which may make the interfacial velocity unstable. Such trends for an aspect ratio of $x_{f,p}$:y:z :: 1:0.5:1 at time levels 25 and 45 are shown in Fig. 16.5. Comparison between the velocity profiles in Figs 16.4 and 16.5 indicates clearly that the front has moved away from the interface in the former case but not in the latter case. Also, the velocity distribution in Fig. 16.5(b) is of particular interest,

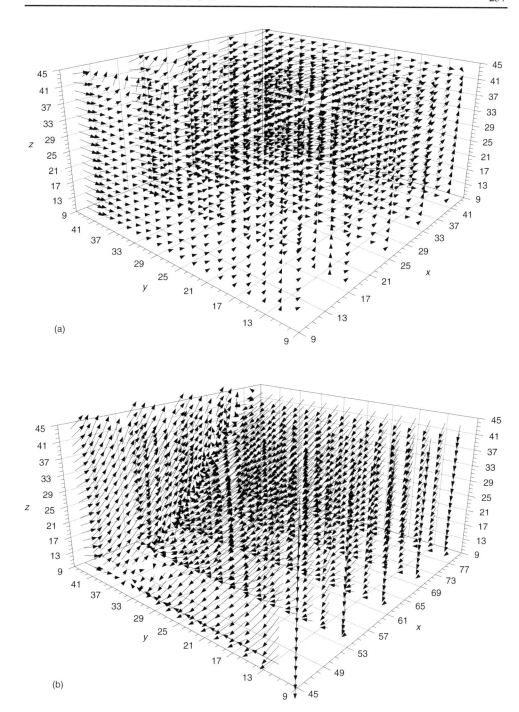

Fig. 16.2. Individual velocity profiles in subdomains of combined free- and porous-flow regions in the subsurface.

(a) Free-flow section, x_f:y:z :: 1:1:1, t = 25. (b) Porous-flow section, x_p:y:z :: 1:1:1, t = 25

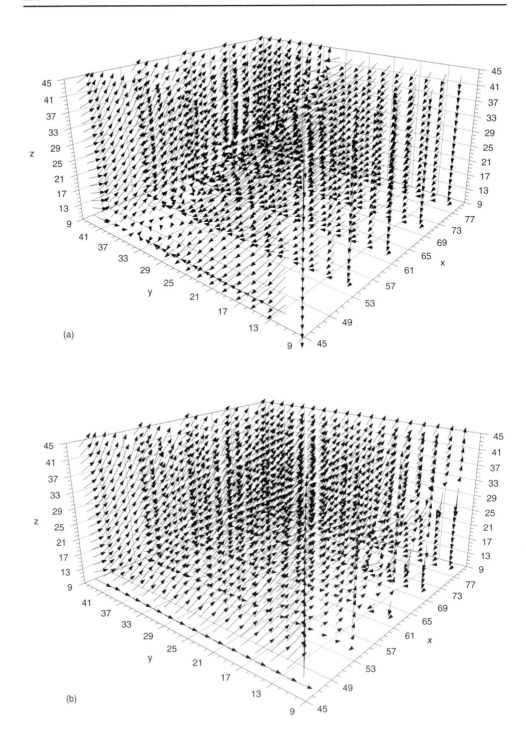

Fig. 16.3. Velocity profiles in the porous-flow region in the subsurface.
(a) Porous-flow section, x_p*:y:z :: 0.5:1:1,* t = 25. *(b) Porous-flow section,* x_p*:y:z :: 0.5:1:1,*
t = 45

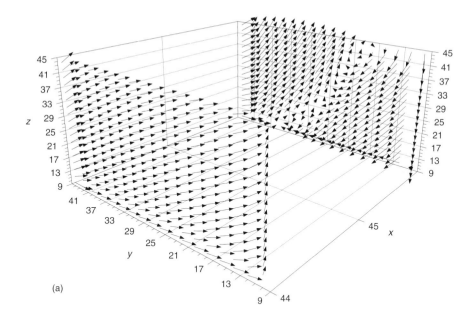

(a)

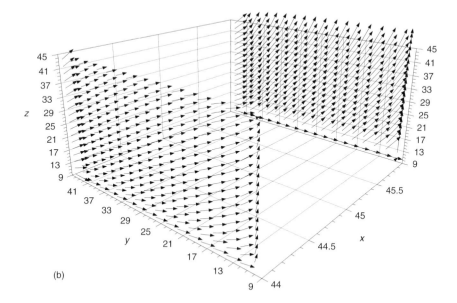

(b)

Fig. 16.4. Velocity profiles across the free–porous interface in the subsurface.
  (a)(i) x_f:y:z :: *1:1:1*, x_p:y:z :: *1:1:1*, t = *25;* *(ii)* x_f:y:z :: *1:1:1*, x_p:y:z :: *0.5:1:1*, t = *25*.
  (b)(i) x_f:y:z :: *1:1:1;* x_p:y:z :: *1:1:1*, t = *45;* *(ii)* x_f:y:z :: *1:1:1;* x_p:y:z :: *0.5:1:1*, t = *45*

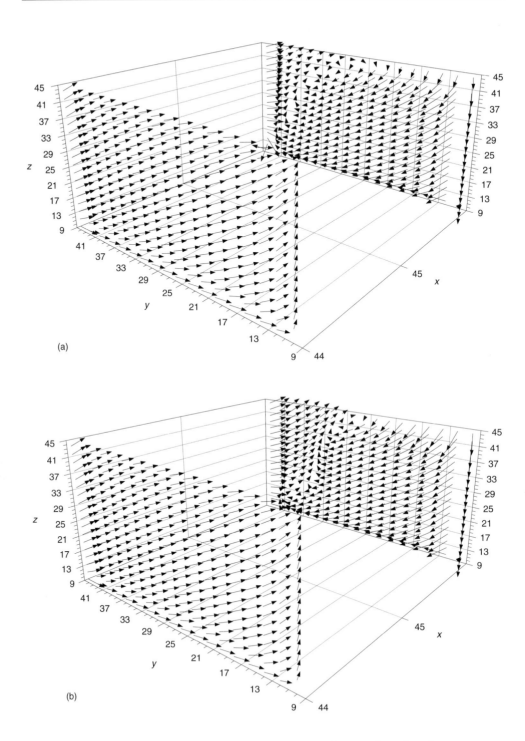

Fig. 16.5. Velocity profiles across the free-porous interface in the subsurface.
(a) $x_{f,p}$:y:z :: 1:0.5:1, t = 25. *(b)* $x_{f,p}$:y:z :: 1:0.5:1, t = 45

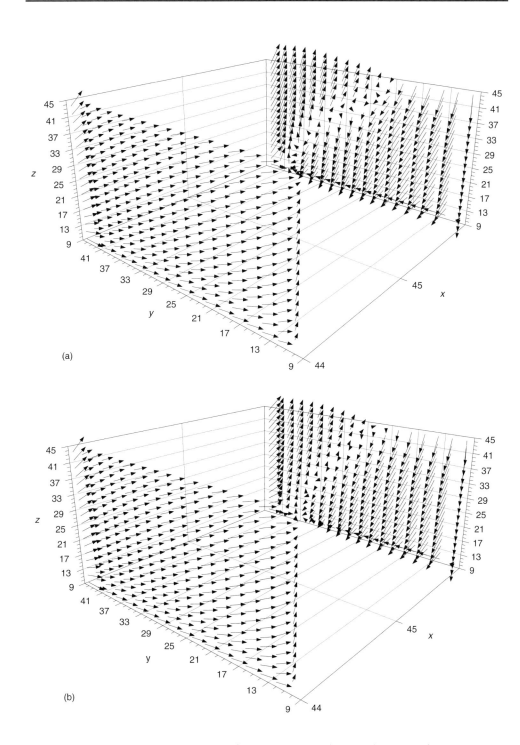

Fig. 16.6. Velocity profiles across the free-porous interface in the subsurface.
(a) $x_{f,p}$:y:z :: *1:1:0.5,* t = *25. (b)* $x_{f,p}$:y:z :: *1:1:0.5,* t = *45*

as the fluid at the interface begins to be unstable. Hence, the possibility arises for water in the porous section to move to the free-flow section due to the greater pressure on the porous side of the interface. Similar observations are also made if the transverse length of the domain is minimized. Figure 16.6 presents two typical profiles to illustrate the phenomena for an aspect ratio of $x_{\mathrm{f,p}}{:}y{:}z :: 1{:}1{:}0.5$ at time levels 25 and 45.

16.4. Conclusions

Simulated three-dimensional velocity profiles for combined free- and porous-flow through PRBs have been presented in this chapter. Due to the complex pressure distributions in the free–porous flow interface, the fluid reverses its direction. The aspect ratios of the domains play a significant role in determining the locations of the centre of flow circulation and the front of flow reversal. In many cases, the fronts of flow reversal move away with time from the interfacial surface. But there may not be a universal physical significance to such flow phenomena, as with different aspect ratios the pattern may change. It can, therefore, be concluded that the scales of the domain determine the flow reversal and other important factors such as the rate of fluid circulation. This, in turn, necessitates that each independent case of combined water flow through the subsurface is investigated based on specific problem domains for the management of underground water quality.

16.5. Acknowledgement

B.G. Technology, UK, is acknowledged for providing the necessary funds to carry out this work.

16.6. References

BEAVERS, G. S. and JOSEPH, D. D. (1967). Boundary conditions at naturally permeable wall. *Journal of Fluid Mechanics* **30**(1), 197–207.

BEAVERS, G. S., SPARROW, E. M. and MAGNUSON, R. A. (1970). Experiments on coupled parallel flows in a channel and bounding porous media. *Journal of Basic Engineering, Transactions of the ASME* **92D**, 843–848.

CANUTO, C., HUSSAINI, M. Y., QUARTERONI, A. and ZANG, T. A. (1988). *Spectral Methods in Fluid Dynamics*. Springer-Verlag, New York.

CHEN, Q.-S., PRASAD, V. and CHATTERJEE, A. (1999). Modelling of fluid flow and heat transfer in a hydrothermal crystal growth system: use of fluid-superposed porous layer theory. *Journal of Heat Transfer, Transaction of the ASME* **121**, 1049–1058.

DAS, D. B. (2002). Hydrodynamic modelling for groundwater flow through permeable reactive barriers. *Hydrological Processes* (in press).

DAS, D. B. and NASSEHI, V. (2000). Computational methods in the modelling and simulation of pollutants' mobility in contaminated land – a review of the physical flow processes. In: S. G. Pandalai (ed.), *Recent Advances in Chemical Engineering*, Vol. 4, pp. 117–136. Transworld Research Network, Trivandrum.

DAS, D. B. and NASSEHI, V. (2001). LANDFLOW: A 3D finite volume model of combined free and porous flow of water in contaminated land sites. *Water Science and Technology* **43**(7), 55–64.

DAS, D. B., KAFAI, A. and NASSEHI, V. (2001). Application of spectral expansions to modelling of underground water flow. 3rd International Conference on Future Groundwater at Risk, 2001, Lisbon, Portugal.

DAS, D. B., NASSEHI, V. and WAKEMAN, R. J. (2002). A finite volume model for the hydrodynamics of combined free and porous flow in sub-surface regions. *Advances in Environmental Research* **7**(1), 49–72.

GARTLING, D. K., HICKOX, C. E. and GIVLER, R. C. (1996). Simulation of coupled viscous and porous flow problems. *Computational Fluid Dynamics* **7**, 23–48.

GOBIN, D., GOYEAU, B. and SONGBE, J.-P. (1998). Double diffusive natural convection in a composite fluid-porous layer. *Journal of Heat Transfer, Transactions of the ASME* **120**, 234–242.

GUO, B.-Y. (1998). *Spectral Methods and their Applications*. World Scientific, Singapore.

NASSEHI, V. (1998). Modelling of combined Navier–Stokes and Darcy flows in crossflow membrane filtration. *Chemical Engineering Science* **53**(6), 1253–1265.

NASSEHI, V. and PETERA, J. (1994). A new least-squares finite element model for combined Navier–Stokes and Darcy flows in geometrically complicated domains with solid and porous boundaries. *International Journal of Numerical Methods in Engineering* **37**, 1609–1620.

PATANKAR, S. V. (1980). *Numerical Heat Transfer and Fluid Flow*. Hemisphere, Washington DC.

SAFFMAN, P. G. (1971). On the boundary condition at the surface of a porous material. *Studies in Applied Mathematics* **50**, 93–101.

SALINGER, A. G., ARIS, R. and DERBY, J. J. (1993). Modelling the spontaneous ignition of coal stockpiles. *AIChE Journal* **40**(6), 991–1004.

SALINGER, A. G., ARIS, R. and DERBY, J. J. (1994). Finite element formulations for large-scale coupled flows in adjacent porous and open fluid domains. *International Journal for Numerical Methods in Fluids* **18**, 1185–1209.

VERSTEEG, H. K. and MALALASEKERA, W. (1995). *An Introduction to Computational Fluid Dynamics: the Finite Volume Method*. Addison Wesley Longman, Harlow.

Part VII
Active and passive methods – a comparison

17. Technical and economic comparison between funnel-and-gate and pump-and-treat systems: an example for contaminant removal through sorption

P. Bayer, C. Bürger, M. Finkel and G. Teutsch
University of Tübingen, Centre for Applied Geoscience, Sigwartstrasse 10, D-72076 Tübingen, Germany

17.1. Introduction and background

Over the past 20 years, source zone treatment has been general practice in the field of remediation. Several technologies and strategies were developed to either excavate, isolate or flush (decontaminate) *in situ* the zones where contaminants had penetrated into the subsurface. Practitioners, however, are faced with the fact that the source zone is not sufficiently identifiable and only partially accessible. Therefore, in most cases, source decontamination has proved to be quite costly and often not very successful. Thus, management of the plume has become a valid (and often the only feasible) option. Lacking alternatives, simple pump-and-treat concepts have mainly been used to treat or to contain the contaminant plume through downgradient pumping wells. Unfortunately, field experience has shown that the long-term persistence of source zones puts operation times within the range of decades at least (e.g. US National Research Council, 1994).

Consequently, the *pump-and-treat* approach appears to be a cost-intensive and, under many circumstances, suboptimal solution. One should bear in mind, though, that the impact of any payment in future will decrease with increasing operation time (US Government, 1992; Ross *et al.*, 1995). This is due to the economic necessity to discount future payments (i.e. convert to net present value) if the net interest rate exceeds the rate of inflation.

Reactive barriers (funnel-and-gate systems or continuous walls) were invented as an inexpensive long-term operational option. They are, nowadays, considered a valid alternative to pump-and-treat systems for these cases (Starr and Cherry, 1994; US Environmental Protection Agency, 1995). The idea is that the higher initial investment costs for constructing a reactive barrier will be compensated in the long run by the lower operational costs (passive technology). However, this is not always true, and will depend on factors such as the type of contamination and the availability of adequate 'passive' technologies, the contaminant concentrations, the width and depth of the contamination, the hydraulic conditions at the site, etc.

In this chapter, total remediation costs for a funnel-and-gate system are compared with the costs for both a conventional pump-and-treat system and a pump-and-treat system with additional hydraulic barriers. This barrier-supported pump-and-treat design, hereafter called 'innovative pump and treat', minimizes the pumping rate required for plume containment (Bayer, 1999). Based on a simplified hypothetical contaminated site, cost-optimal design alternatives are calculated for the three remediation options by means of a design optimization framework that combines groundwater flow modelling, hydrochemical treatment modelling, economic modelling and mathematical optimization. The section on their comparative assessment focuses on the following decision influencing factors: type of contaminant, contaminant concentration and interest rate.

17.2. Design optimization framework

The overall strategy can be divided into three main steps. In the first step the hydraulic performance of various design alternatives is evaluated. For that, a groundwater site model is used to predict the effect of various design alternatives for the three remediation options: (a) pump and treat, (b) innovative pump and treat (with additional hydraulic barriers) and (c) funnel and gate. Particles are placed along the borders of the modelled source area and tracked downstream by simulation to verify successful plume containment for each design alternative. In the second step the treatment costs are calculated based on the flow rate of the water, which is either extracted from the subsurface (pump and treat) or flows naturally through the gate (reactive barrier). In the third step the total remediation costs are calculated on a net present-value basis (dynamic cost analysis) for each design alternative.

The main advantage of this three-step strategy is that the influence of uncertain cost-driving parameters (contaminant concentration or external factors like interest rate) can be analysed independently, and hence in a fast and transparent way. Sensitivity analyses are also carried out, which yield ranges rather than discrete values for total costs, reflecting the uncertainty of some input parameters.

17.3. Case study

A hypothetical heterogeneous confined aquifer (thickness 10 m) is generated in a two-dimensional hydraulic computer model. The model comprises 100×100 m square cells with sides of 10×10 m. The hydraulic conductivity distribution is shown in Fig. 17.1. A characteristic high conductivity zone (size 250×250 m), which runs from north-west to south, is located in the central part of the contaminated area. A regional hydraulic gradient of 0.001 is established through 'constant-head' boundary conditions, leading to a mean groundwater flow in the north–south

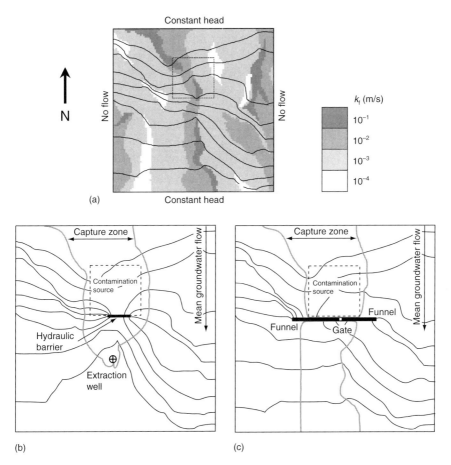

Fig. 17.1. Hydraulic conductivity field of the case study site. The flow field is indicated by potential lines (spacing 0.1 m head difference). (a) The 'natural' flow field; the dotted square represents the contamination source. (b) The flow field under the influence of the hydraulically optimal innovative pump-and-treat system. (c) The flow field under the influence of the optimal funnel-and-gate system

direction. 'No flow' boundary conditions are assumed on the eastern and western domain boundaries.

The groundwater flow conditions, as well as the contaminant emission from the source area, are assumed to be constant over time (steady state). The assumption of constant contaminant emissions (mass per time and cross-sectional area) is in accordance with the fact that a long-term concentration tailing is observed during pump-and-treat operation at most sites where the (organic) contaminant source is not entirely removed (Keely, 1989). In general, initially high contaminant concentrations tend to decrease rapidly to lower, but still above normal, levels and then remain constant, sometimes over decades. Figure 17.1 also depicts the flow fields for the case study site, assuming the installation of an innovative pump-and-treat system (Fig. 17.1b) and a funnel-and-gate system (Fig. 17.1c).

17.4. Hydraulic performance evaluation

17.4.1. Conventional pump-and-treat systems

For simplicity, it is assumed that only one well is used in the conventional pump-and-treat approach presented here. In order to find the optimal position of the extraction well downgradient of the source – defined by the minimum pumping rate required to guarantee plume containment – an optimization method based on 'evolution strategies' (Schwefel, 1995) is used. In our simple example, the optimal position for the abstraction well is at a distance of about 20 m downstream from the contamination area, slightly outside the central high conductive zone (see Fig. 17.1). The optimal pumping rate was found to be 22.1 l/s. These results were confirmed by calculating the minimum containment pumping rates for all other possible well locations. It should be noted that using the 'evolution strategies' approach computation time is reduced by a factor of three compared with a 'brute force' approach.

17.4.2. Innovative pump-and-treat systems

The task here is to find the optimal configuration of the barrier (length and location) that minimizes the total remediation costs (reduction of treatment costs due to a lower rate of extracted contaminated water, minus additional investment for barriers). For barrier lengths of up to 750 m (enclosing 75% of the contaminated area) the minimal pumping rate (with the same well location used for the conventional pump-and-treat approach) necessary for containment has been determined systematically.

Figure 17.2 shows the functional dependence of the extraction rate reduction, compared with conventional pump and treat, for varying barrier lengths, with

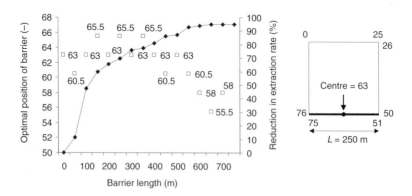

Fig. 17.2. Variation of the maximal reduction in extraction rate with respect to the conventional pump-and-treat system. For each barrier length the reduction value is representative of the optimal barrier location. The location (centre) is expressed in relative values, starting from the north-west corner of the source area (position = 0) clockwise

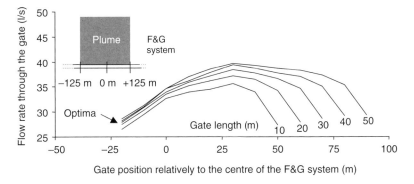

Fig. 17.3. Volume flow rate as a function of gate position and gate length (gate position is specified by the distance of the centre of the gate from the central axis through the source zone in the direction of the mean flow)

their corresponding locations. The values shown represent the maximal reduction (optimal location) for a specific barrier length. Surprisingly, the optimal location varies little, and the best results are always obtained with the barrier located on the downgradient side of the contamination source.

In this example, the additional hydraulic barrier could lead to a reduction in the pumping rate of about 94% if optimally positioned. As expected, the optimal locations (around position 63) coincide with locations which cut off the high conductivity channel. Obviously, longer barriers lead to larger reductions in the extraction rate, but the relative increase (ratio of gain in reduction to elongation of barrier) diminishes rapidly for longer barrier lengths. The optimal (in terms of a maximum reduction of costs) barrier configuration was found for a barrier length of 100 m, with a barrier centre position of 45 m (see Fig. 17.1b).

17.4.3. Funnel-and-gate systems

Again for simplicity, only single-gate systems are considered here. These are aligned perpendicularly to the regional hydraulic gradient for a fixed distance of 10 m downstream of the contamination source boundary. The hydraulic effect of the funnel-and-gate system was systematically analysed for different designs. The total system length was varied from 370 to 610 m, and the gate lengths considered ranged between 10 and 50 m, with variable gate positions. Figure 17.3 shows the flow rate through the gate as a function of gate position and gate length. Only system configurations achieving full capture were taken into account. The optimal configuration in terms of a minimal volume flow rate through the gate was found to be a total funnel length of 520 m and gate length of 10 m. These results revealed that only a few gate positions within the high conductivity zone were suitable for total plume containment – the optimal one with respect to the minimum flow rate is shown in Fig. 17.1c. It should be noted that the number of suitable gate positions increases with total system and gate length.

17.5. Evaluation of remediation costs

17.5.1. Conceptual model of sorptive removal – assumptions and input parameters

Based on the cost analyses of on-site treatment units for sorptive removal of contaminants on to granular activated carbon (GAC) (Bayer, 1999) as well as of 'sorptive' funnel-and-gate systems (Teutsch et al., 1997), cost functions were derived to quantify total remediation costs. These functions relate treatment costs directly to various cost-controlling factors such as contaminant concentration, flow rate, unit GAC costs, discount rate and electric current price.

A key factor for the cost calculation is the determination of the refill period or service life-time of the on-site (pump and treat) or in situ (permeable barrier) GAC reactor. In this study, transport through the GAC reactor is described by an approximation of the analytical Rosen formula (Rosen, 1954) published by Crittenden et al. (1986). This model considers the effects of sorption kinetics of a particular contaminant and GAC type. It is part of a hydrochemical modelling and cost calculation routine, which provides necessary refill periods to meet the remediation target.

For this study the reactor volume is determined by the flow rate and the assumption of a desired contact time of 1 h. Consequently, it is assumed that GAC containers come in the corresponding sizes. Therefore, container costs (CC) are directly dependent on the flow rate. A fixed charge of 1000 euros and an additional 250 euros per tonne of GAC are assumed for the actual exchange work and added to the costs for GAC. Unit costs (euros per mass) for GAC as well as for the installation of the funnels and barriers, etc. are taken from a database built with typical prices (free of tax) for Germany. To avoid step-functions resulting from staggered prices (e.g. price dependence on purchase quantity), mathematical regression is used in some cases to achieve continuous cost functions. Since a thorough description of the cost analysis approach is beyond the scope of this study, we refer the reader to Teutsch et al. (1997) and Bayer (1999) for a more detailed description.

Box 17.1. Input parameters used for modelling reactive transport through the gate/reactor

Contaminant:		Freundlich coefficient (mg/l)	Freundlich exponent (g/kg:mg/l)
Trichloroethene	(TCE)	1100	28
Benzene	(B)	1780	14.3
p-Xylene	(X)	175	85
Acenaphthene	(ACE)	9.26	205.8
Apparent diffusion coefficient for all contaminants			1×10^{-13} m^2/s
GAC parameters:			
Effective porosity		0.5	
Aggregate density		1.2 mg/cm^3	

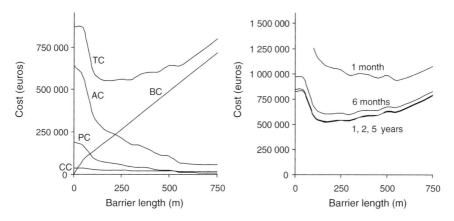

Fig. 17.4. Total costs for TCE as a function of barrier length and container life-time

Cost comparisons for different remediation alternatives cannot be valid in general, as simplifying assumptions always have to be made (US National Research Council, 1997). In this study, expenditures that are site-specific, such as costs for site investigation and planning as well as for monitoring and labour, are not taken into account. Operation time is set to 30 years, as suggested by Freeze and McWorther (1997). The discounting of future costs to net present values is calculated on the basis of a current inflation rate of 2.5% and a current interest rate of 5%. More details on discounting procedures are given by the US National Research Council (1997).

In order to assess the influence of the contaminant type on the design and costs of funnel-and-gate and pump-and-treat systems, remediation costs were calculated and compared for several organic contaminants with different physicochemical properties: trichloroethene (TCE) as a representative of the chlorinated solvents group, benzene (B) and xylene (X) for BTEX, and the polycyclic hydrocarbon acenaphthene (ACE). The following standard input parameters were used: contaminant concentration in groundwater, 1000 mg/m^3; drinking water standard to be reached at the reactor outlet, 10 mg/m^3. F300 (Chemviron) was chosen as the GAC adsorbent. Box 17.1 shows the parameters used to model sorptive removal in order to determine the necessary reactor volumes. It is worth noting that we do not address contaminant mixtures but instead compare single contaminants assuming that, in any case, one contaminant can be determined that is critical to the dimension of the reactor volume, and that effects due to competitive sorption can be neglected.

17.5.2. Results for pump-and-treat systems

Figure 17.4 shows the total costs for TCE as a function of barrier length and desired container life-time (1 month to 5 years). In this way, the optimal pump-and-treat design (barrier length and container size) can be determined directly from the chart. A container life-time of 2 years appears to result in the lowest total costs. This result is not unexpected as, with a longer container life-time (corresponding to larger container volume), sorption processes gradually reach equilibrium conditions, therefore diminishing the negative effect of sorption kinetics (Finkel *et al.*, 1998). This simply means that a larger percentage of the total sorption capacity is being used. Yet, increasing the container life-time considerably (5 years) will not lead to a larger cost reduction as the initial investment (including GAC costs) increases nearly linearly with the container volume but the cost reduction of future investments by discounting shows an exponential behaviour.

In the following investigations a container life-time of 1 year is assumed. The results for all contaminants considered are given in Fig. 17.5. Individual cost elements associated with certain parts of the remediation system, i.e. barrier costs (BC), pumping costs (PC), container costs (CC) and GAC costs (AC), are shown together with the total costs (TC). Obviously, PC and BC are independent of the type of contaminant. The contaminant-specific costs comprise AC and CC.

It is noteworthy that for all contaminants considered, a benefit can be achieved by using additional hydraulic barriers. Furthermore, Fig. 17.5 reveals that the optimal

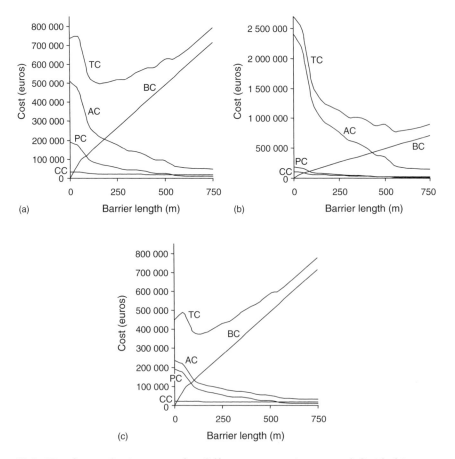

Fig. 17.5. Total remediation costs for different contaminants, subdivided into costs of the individual elements of funnel-and-gate systems, in relation to the length of the additional hydraulic barrier: (a) ACE, (b) B, and (c) X

Table 17.1. Total cost minima for conventional and innovative pump-and-treat approaches for different contaminants

Contaminant	Cost, conventional (euros)	Cost, innovative (euros)	Savings (%)	Barrier length (m)
TCE	868 240	551 448	36	200
X	736 312	499 469	32	150
B	2 705 241	782 912	71	550
ACE	449 397	380 827	15	150

pump-and-treat design is highly dependent on the type of contaminant. While rather short additional hydraulic barriers (100–150 m) are favourable for comparatively strong sorbing contaminants (here, ACE and X), relatively long barriers (500 m) are needed for cost-optimal remediation of less sorbing contaminants such as B.

This relationship between optimal design (and costs) and contaminant type (properties) is further substantiated in Table 17.1. The lowest costs are estimated for ACE, the highest for B. Looking at the potential savings of the innovative pump-and-treat approach compared with the conventional system, the order is reversed. Potential savings for ACE are negligible. Due to its comparatively high sorptivity, unit treatment costs (euros/(m^3 s)) of ACE are relatively low, and, therefore, a reduction in pumping rate is not as cost-relevant compared with B, which possesses a comparatively low sorptivity. In the case of B this means, recalling Fig. 17.2, that even if the pumping rate can be reduced only slightly by extending the length of the additional hydraulic barrier, this investment pays off by lowering the treatment costs. Therefore, innovative pump-and-treat systems yield significant cost savings, especially for low sorbing contaminants (as here, for B up to 54%).

Since sorption isotherms of the contaminants addressed here are non-linear, it is to be expected that not only the type but also the concentration of the contaminant in groundwater will influence the dimension of the reactors and, hence, the remediation cost. Figure 17.6 shows how costs develop if the contaminant concentration changes (results for B and TCE). The effects observed are, in a qualitative sense, quite similar for both contaminants, although B and TCE have different sorption properties: (a) the innovative pump-and-treat approach outperforms the conventional pump-and-treat system at all concentration levels, (b) barrier lengths, adjusted according to optimal cost reduction, increase for higher concentration levels, and, (c) savings from installation of additional barriers increase with increasing concentrations. However, absolute values for cost savings are considerably higher for B compared with TCE.

17.5.3. Results for funnel-and-gate systems

As described above, different designs of funnel-and-gate-systems have been analysed to find the optimal design with respect to the hydraulic performance of a system. For any given gate length, one optimal design has been specified that guarantees complete capture with a minimum flow rate through the gate. When searching for an optimal design of a funnel-and-gate system with respect to total costs, a second design parameter, the most favourable reactor volume, also has to be optimized.

All in all, five relationships exercise an influence on the optimal funnel-and-gate configuration:

(1) residence time increases with increasing reactor volume, as the flow rate is virtually constant for the cases investigated (a change in the gate thickness induces only minor changes in the flow rate through the gate);
(2) sorption processes gradually reach equilibrium conditions with increasing residence time, diminishing the negative effect of sorption kinetics (as has already been mentioned within the context of pump-and-treat systems);
(3) for a given reactor volume, systems with a large gate length and a small thickness have more favourable economics than systems with a narrow gate with a large thickness, due to the fact that for a large gate length, less funnel costs have to be paid (since the total length of the system is constant);
(4) the number of refills within the remediation time decreases with increasing reactor volume;
(5) investment costs for gate construction increase with increasing gate volume.

As the cross-sectional area of the gate is defined by the aquifer thickness and the gate length, the gate/reactor thickness becomes the main design parameter for the reactor volume. Therefore, the sensitivity of the total costs to the gate thickness has been analysed for each of the five hydraulically optimized designs (for given gate lengths of 10, 20, ..., 50 m).

To illustrate the complex system of interacting effects, Fig. 17.7 and Table 17.2 show how gate thickness influences both overall cost minimization and the optimal design, and how this influence again depends on the specific characteristics of the problem considered (e.g. type of contaminant, values assumed for relevant parameters such as the interest rate). An example is shown in Fig. 17.7a, with TCE as

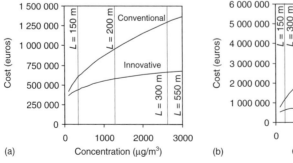
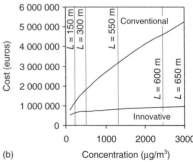

Fig. 17.6. Total costs for conventional and innovative pump-and-treat systems as a function of contaminant concentration: (a) TCE and (b) length (vertical lines show barrier lengths L adjusted for cost-optimal design)

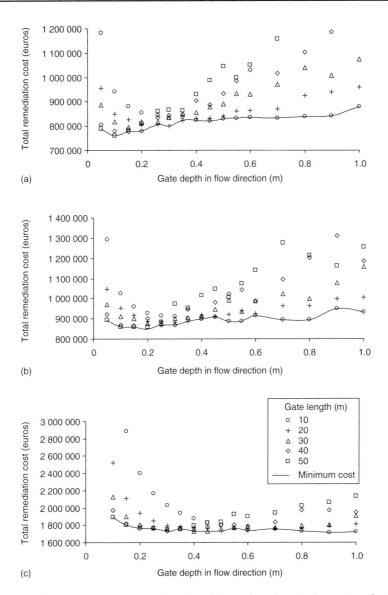

Fig. 17.7. Total remediation costs and optimal funnel-and-gate design in relation to the thickness of the gate/reactor for different contaminants and interest rates: (a) TCE, interest rate 5%, (b) TCE, interest rate 2.5%, (c) length, interest rate 5% (symbols indicate optimal solutions for each of the five hydraulically optimized designs)

the contaminant (standard input parameters as specified in Section 17.5.1). In this case, systems with a large cross-sectional gate area are optimal for comparatively small gate thicknesses, whereas systems with smaller gates become favourable when the reactor thickness is increased. Interestingly, the cost-optimal design is found for the largest cross-sectional area and a reactor thickness of 0.1 m, which obviously is not a realistic option. Figure 17.7b shows the result for the same investigation assuming that the interest rate is only 2.5% (instead of 5%). Hence, since inflation is assumed to be 2.5%, operational costs are not discounted. As a logical consequence, the total costs increase. The point of interest is that, compared with Fig. 17.7a, system optimization tends to result in designs with smaller gate thicknesses and larger gate depths (best option for gate thickness: 0.2 m). The same tendency can be observed if B is the contaminant (Fig. 17.7c). In this case, the system with a 10 m long gate with a thickness of 1 m is cost-optimal. In order to completely understand the behaviour of these system effects, further investigations concerning the influence of individual parameters, and their relative importance in given situations, need to be undertaken.

17.5.4. Comparison of the three remediation options

As illustrated in detail above, total remediation costs are, in general, influenced by various parameters. Furthermore, the sensitivity of total costs to these parameters differs considerably depending on which of the three remediation options – (1) conventional pump and treat, (2) innovative pump and treat, and (3) funnel and gate – is considered. Consequently, when comparing these options, it is not possible to address all the issues that may be of interest within the framework of this investigation. For the same reason, it is not possible to rank the options considered, in a conclusive way – this will always be a question of the specific situation addressed, e.g. of the values assumed for the cost-relevant parameters. The following discussion is therefore limited to some key issues.

Table 17.2. *Parameters of cost-optimal funnel-and-gate systems depending on the gate thickness chosen*

Gate thickness (m)	Funnel-and-gate length (m)	GAC mass (tonnes)	No. of refills	Total cost (euros)
0.05	410	12.5	23	738 510
0.10	410	25.0	10	729 420
0.15	410	37.5	7	738 313
0.20	410	30.0	9	751 446
0.25	410	25.0	10	757 521
0.30	410	30.0	8	760 446
0.35	450	17.5	15	774 087
0.40	450	20.0	13	774 647
0.45	450	22.5	11	785 940
0.50	450	25.0	10	772 862
0.55	450	27.5	9	775 474
0.60	450	30.0	8	799 374
0.70	450	35.0	7	786 052
0.80	450	40.0	6	793 081
0.90	450	45.0	5	831 259
1.00	450	50.0	5	824 603

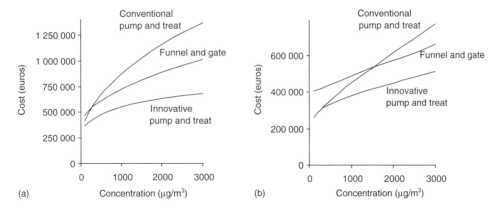

Fig. 17.8. Comparison of the total costs of the three remediation options as a function of contaminant concentration: (a) TCE and (b) ACE

17.5.4.1. Sensitivity of costs to contaminant type and concentration

It is clear that remediation costs principally increase with increasing contaminant concentration and decreasing sorptivity. However, due to different volume rates of groundwater to be treated, the sensitivity of the three remediation options is different. This is shown in Fig. 17.8 where total remediation costs for TCE and ACE are compared. For both contaminants the innovative pump-and-treat approach is least sensitive to changes in contaminant concentration. Furthermore, it outperforms both of the other options at all concentrations of TCE and ACE. Since the conventional pump-and-treat approach is, by contrast, the economically worst option in most cases, there is a clear evidence for the large economic benefit that can be obtained by means of additional hydraulic barriers. Only if the contamination at a site comprises strong sorbing compounds at low concentrations is there no pay-off by using the innovative pump-and-treat approach. This is illustrated in Fig. 17.8b for ACE concentrations below 400 mg/m^3.

17.5.4.2. Sensitivity of costs to interest rates

Making an economic comparison between pump-and-treat systems (commonly associated with high operational costs) and funnel-and-gate systems (which are considered a technology where costs are dominated by initial investment), one would expect that interest rates will primarily impact the costs of the pump-and-treat system. However, as Fig. 17.9 reveals, for the sorptive removal considered here, this is true only for the conventional pump-and-treat approach. The innovative pump-and-treat system is less sensitive to interest rates than the funnel-and-gate system. Similarly, minimization of the pumping rate leads to a reduction in the sensitivity of the innovative pump-and-treat system to interest rates as a parameter controlling the operational cost. The comparatively high sensitivity of the funnel-and-gate system to interest rates is caused by the periodic need to replace the GAC reactor material. Conversely, in the case of a one-time zero-valent iron-filings filling, the total remediation costs are more or less independent of interest rates (Bayer *et al.*, 2001).

It is noteworthy that interest rates are relevant to the cost-optimized design of an innovative pump-and-treat system: if the assumed interest rate decreases, the optimal design tends to favour longer or additional barriers (Fig. 17.10).

17.5.4.3. Further remarks

It should be mentioned that adequate positioning of the additional hydraulic barriers requires that the site characterization stage delineates a high-conductivity channel crossing the contaminated area. If not, the site groundwater model and hence the remediation optimization cannot take this into consideration. This problem was addressed by Wagner and Gorelick (1989) some 10 years ago. Another important fact to mention is that the funnel-and-gate system was optimized only with respect to the gate position and length. For instance, neither the position of the whole system relative to the source area nor multiple gate options have been investigated within the framework of this study. Furthermore, additional costs needed for additional water treatment are not taken into account.

The results presented for pump-and-treat systems are based on an assumption that the water leaving the on-site treatment system can be discharged at no cost into the local sewer system or some surface water stream. However, depending on the respective country or city regulations, a considerable discharge fee for treated water of up to 3 euros/m^3 might need to be taken into account. These costs would have to be added to the unit pumping and treatment costs. As has been shown by Bayer *et al.* (2001), such additional costs may influence the final decision on a technology more than anything else – in general towards (passive, non-pumped) technologies such as the funnel-and-gate system, to which discharge fees may not apply.

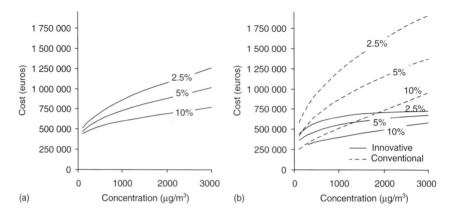

Fig. 17.9. Dependency of the total costs of the three remediation options on net interest rate as a function of contaminant concentration (example for TCE): (a) funnel-and-gate system; (b) conventional and innovative pump-and-treat system

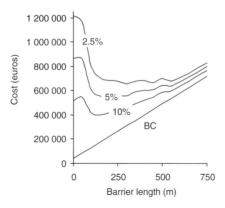

Fig. 17.10. Influence of interest rate on total costs and optimized design of an innovative pump-and-treat system

17.6. Summary and conclusions

A comprehensive analysis and cost optimization approach has been presented in this chapter for the long-term management of organic contaminant plumes in groundwater. This approach is intended to support decision-makers in comparing remediation design alternatives in a convenient and transparent way. Using a hypothetical example, it is shown that there is no simple answer to the question 'What is the optimal cost solution?' The uncertainty in some of the cost-controlling parameters, and the fact that the options compared in this study show different sensitivities to these parameters, need to be addressed. Consequently, a unique ranking of solutions with respect to costs is not possible. In addition, external factors such as specific site restrictions and public acceptance, will eventually drive a final decision on the 'optimal' remediation option.

Further research will focus on a better way to deal with these uncertainties and on a better insight into the complex system of controlling parameters. Limited knowledge about the contaminant source and the spatial distribution of hydraulic conductivity will have to be considered by means of geostatistical approaches, eventually leading to cost – reliability distributions for each remediation alternative. Expected remediation costs will then be related to their reliability, including the probability that the alternative proposed will actually meet the remediation target.

17.7. Acknowledgement

Support for this project has been provided by the German Federal Ministry of Education and Research (BMBF).

17.8. References

BAYER, P. (1999). Kostenanalyse innovativer Pump-and-Treat Konzepte [Cost analysis of innovative pump-and-treat systems]. Diploma thesis, Centre for Applied Geoscience, University of Tübingen.

BAYER, P., MORIO, M., BÜRGER, C., SEIF, B., FINKEL, M. and TEUTSCH, G. (2001). Funnel-and-gate vs. innovative pump-and-treat systems: a comparative economic assessment. Ground-water Quality Conference 2001, Sheffield.

CRITTENDEN, J. C., HUTZLER, N. J., GEYER, D., ORAVITZ, J. L. and FRIEDMAN, G. (1986). Transport of organic compounds with saturated groundwater flow: model development and parameter sensitivity. *Water Resources Research* **22**(3), 271–284.

FINKEL, M., LIEDL, R. and TEUTSCH, G. (1998). A modelling study on the efficiency of groundwater treatment walls in heterogeneous aquifers. *Proceedings of the GQ '98 Conference, University of Tübingen. IAHS Publication*, No. 250, pp. 467–474.

FREEZE, R. A. and MCWORTHER, D. B. (1997). A framework for assessing risk reduction due to DNAPL mass removal from low-permeability soils. *Ground Water* **35**(1), 111–123.

KEELY, J. F. (1989). *Performance Evaluation of Pump and Treat Remediations*, EPA/540/4–89/005, US EPA, ORD. R. S. Kerr Environmental Research Laboratory, Ada, Oklahoma.

ROSEN, J. B. (1954). General numerical solution for solid diffusion in fixed beds. *Industrial and Engineering Chemistry* **46**(8), 1590–1594.

ROSS, S. A., WESTERFIELD, R. W. and JORDAN, B. D. (1995). *Fundamentals of Corporate Finance*, 3rd edn. Irwin.

SCHWEFEL, H.-P. (1995). *Evolution and Optimum Seeking*. Wiley, New York.

STARR, R. C. and CHERRY, J. A. (1994). *In-situ* remediation of contaminated ground water: the funnel-and-gate system. *Ground Water* **32**(3), 465–476.

TEUTSCH, G., TOLKSDORFF, J. and SCHAD, H. (1997). The design of *in-situ* reactive wall systems. A combined hydraulic–geochemical–economic simulation study. *Journal of Land Contamination and Reclamation* **5**(3), 125–130.

US ENVIRONMENTAL PROTECTION AGENCY (1995). In situ *Remediation Technology Status Report: Treatment Walls*. US Environmental Protection Agency, Washington DC.

US GOVERNMENT (1992). *Guidelines and Discount Rates for Benefit–Cost Analysis of Federal Programs. OMB Circular A-94*. US Government Printing Office, Washington DC.

US NATIONAL RESEARCH COUNCIL (1994). *Alternatives for Groundwater Clean-up*. National Academy Press, Washington DC.

US NATIONAL RESEARCH COUNCIL (1997). *Innovations in Ground Water and Soil Cleanup: From Concept to Commercialisation*. National Academic Press, Washington DC.

WAGNER, B. J. and GORELICK, S. M. (1989). Reliable aquifer remediation in the presence of spatial variable hydraulic conductivity: from data to design. *Water Resources Research* **25**(10), 2211–2225.

18. Engineering and operation of groundwater treatment systems: pump and treat versus permeable reactive barriers

E. Beitinger
URS Deutschland GmbH, Office Frankfurt, Heinrich-Hertz-Strasse 3, D-63303 Dreieich, Germany

18.1. Evaluation of best available techniques (BATs)

A wide range of 'old fashioned' technologies as well as innovative approaches are available for the remediation of contaminated groundwater (Gavaskar *et al.*, 1998; US Environmental Protection Agency, 1998) (Tables 18.1 and 18.2). The primary objective of each preliminary investigation and design is to evaluate the BAT, including cost considerations. The BAT must also be the most economic method (BATNEEC, BAT not entailing excessive cost) to achieve defined target levels according to risk assessment results.

The BAT might be assessed by common site-specific evaluation processes as described in Section 18.2 below. There is no existing standardized procedure for BAT evaluation for groundwater remediation technologies. However, in Europe the BAT concept is defined in Article 2 of the Integrated Pollution Prevention and Control (IPPC) Directive of 1996 (EU, 1996). In essence, the IPPC Directive deals with the minimization of pollution from various point sources throughout the EU. Even though the concept is applicable only to industrial installations covered by Annex I of the Directive, it may be a helpful guide for selection purposes in general.

Permits for industrial plants must take into account the entire environmental performance of the plant, i.e.:

* emissions to air, water and land;
* generation of waste;
* use of raw materials;
* energy efficiency;
* noise;
* prevention of accidents;
* risk management, etc.

So-called BREFs (BAT reference documents) will be completed by the end of 2004 for some 30 industrial sectors. Eight have now finalized, and can be downloaded from the BREF website of the IPPC Bureau (http://eippcb.jrc.es/pages/FActivities.htm).

Table 18.3 details an example for a screening matrix of treatment technologies by the US Environmental Protection Agency which can be used for a first overview.

18.2. Selection criteria and evaluation process

To determine the BAT(s), all appropriate and applicable groundwater treatment technologies and/or combinations need to be evaluated using selection criteria and an evaluation process which is either qualitative, quantitative or semi-quantitative. The process itself should be fully described to allow a clear understanding of how the best choice was developed.

Table 18.1. Well-developed technologies for groundwater remediation

Remediation technology	Modified/added technologies
Pump and treat	Air stripping Granulated activated carbon (GAC) Ion exchange Precipitation/coagulation/flocculation Separation (filtration/reverse osmosis) Ultraviolet oxidation Bioreactors
In situ biodegradation	Nitrate enhancement Oxygen enhancement Bioslurping
Circulating wells	
Phase extraction	Free phase extraction (liquid) Vapour extraction Dual-phase extraction
Natural attenuation	Enhanced Not enhanced

Table 18.2. Innovative technologies

Remediation technology	Examples
Hot water or steam flushing/stripping	Hydrofracturing
Permeable reactive barriers	Zero-valent iron Granular activated carbon Reactive zones
Surfactant-enhanced recovery	Tensides
ORC/HRC	Oxygen/hydrogen-releasing compounds

Table 18.3. Treatment technologies screening matrix

	Development status	Treatment train (excluding off-gas treatment)	Residual produced	O&M or capital-intensive	Availability	System reliability/maintainability	Clean-up time	Overall cost	Non-halogenated VOCs	Halogenated VOCs	Non-halogenated SVOCs	Halogenated SVOCs	Fuels	Inorganics	Radionuclides	Explosives
SOIL, SEDIMENT, SLUDGE																
A. In situ biological treatment																
1. Bioventing	F	N	N	N				+	+				+	−		−
2. Enhanced bioremediation	F	N	N	O&M					+	+			+			+
Aerobic																
Anaerobic																
3. Land treatment	F	N	N	N					+	+			+	−		
4. Natural attenuation	F	N	N	O&M			−		+							
5. Phytoremediation	P	N	N	N	−	o	−	+	o	o	o	o	+	−	−	+
Enhanced rhizosphere biodegradation																
Phytoaccumulation																
Phytodegradation																
Phytostabilization																
B. In situ physical/chemical treatment																
6. Electrokinetic separation	F	Y	L	O&M	+	o	o	o	o	o	o	o	−	+	o	−
7. Fracturing	F	Y	N	O&M	−	o	o	o	o	o	o	o	o	o	−	o
Blast enhanced																
Lasagna process																
Pneumatic fracturing																

Table 18.3 (Contd).

	Development status	Treatment train (excluding off-gas treatment)	Residual produced	O&M or capital-intensive	Availability	System reliability/ maintainability	Clean-up time	Overall cost	Non-halogenated VOCs	Halogenated VOCs	Non-halogenated SVOCs	Halogenated SVOCs	Fuels	Inorganics	Radionuclides	Explosives
8. Soil flushing Cosolvents enhancement	F	N	L	O&M	+	O	–	–	+	+	O	O	O	+	–	–
9. Soil vapour extraction	F	N	L	O&M	+	+	O	+	+	+	O	O	+	–	–	–
10. Solidification/stabilization *In situ* vitrification	F	N	S	B	O	O	O	O	+	+	+	+	–	+	+	–
C. *In situ* thermal treatment																
11. Thermally enhanced soil vapour extraction Electrical resistance heating Radio frequency/ electromagnetic heating Hot air injection	P	N	N	N	–	+	–	+	O	O	+	+	+	–	–	–
D. *Ex situ* biological treatment (assuming excavation)																
12. Biopiles	F	N	N	N	+	+	O	+	+	+			+		–	+
13. Composting	F	N	N	N	+	+	O	+	+	+			+		–	+
14. Fungal biodegradation White rot fungus	F	N	N	O&M	–	–	–	+						–	–	+
15. Landfarming	F	N	N	N	+	+	–	+	+	+			+		–	
16. Slurry phase biological treatment	F	N	N	B	O	O	O	O	+	+			+		–	+

E. *Ex situ* physical/chemical treatment (assuming excavation)

Item																			
17. Chemical extraction	F	Y	L	B	o	o	o	-	o	o	o	o	+	o	o	+	o	+
Acid extraction																		
Solvent extraction																		
18. Chemical reduction/oxidation	F	Y	S	N	+	+	+	+	o	+	+	o	o	o	o	+	-	+
19. Dehalogenation	F	N	V	B	-	-	-	-	-	-	o	o	-	-	-	-	-	-
Base-catalysed decomposition																		
Glycolate/alkaline																		
polyethylene glycol (A/PEG)																		
20. Separation	F	Y	S	O&M	+	o	o	+	o	o	o	o	+	+	+	+	-	+
Gravity separation																		
Magnetic separation																		
Sieving/physical separation																		
21. Soil washing	F	Y	SL	B	+	o	+	+	o	o	o	o	+	+	+	+	-	+
22. Soil vapour extraction	F	N	L	N	+	+	+	o	+	+	+	+	o	o	-	-	-	-
23. Solar detoxification	P	N	N	Cap.	o	o	o	o	+	+	+	+	+	+	-	-	-	-
24. Solidification/stabilization	F	N	S	Cap.	+	+	+	+	o	-	-	o	o	-	+	+	+	-
Bituminization																		
Emulsified asphalt																		
Modified sulfur cement																		
Polyethylene extrusion																		
Pozzolan/Portland cement																		
Radioactive waste solidification																		
Sludge stabilization																		
Soluble phosphates																		
Vitrification/molten glass																		

F. *Ex situ* thermal treatment (assuming excavation)

Item																			
25. Hot gas decontamination	P	N	N	B	o	+	+	+	+	-	-	-	-	-	-	-	-	-
26. Incineration	F	N	LSV	B	+	o	+	+	+	+	+	+	+	+	+	+	-	+
Circulating bed combustor																		
Fluidized bed																		
Infrared combustion																		
Rotary kiln																		

Table 18.3 (Contd).

	Development status	Treatment train (excluding off-gas treatment)	Residual produced	O&M or capital-intensive	Availability	System reliability/ maintainability	Clean-up time	Overall cost	Non-halogenated VOCs	Halogenated VOCs	Non-halogenated SVOCs	Halogenated SVOCs	Fuels	Inorganics	Radionuclides	Explosives
27. Open burn/open detonation	F	N	S	B	+	+	+	+	–	–	–	–	–	–	–	+
28. Pyrolysis	F	N	LS	B	–	–	+	–	o	o	+	+	o	–	–	–
Fluidized bed																
Molten-salt destruction																
Rotary kiln																
29. Thermal desorption	F	Y	LS	B	+	o	+	o	+	+	+	+	+	–	–	+
High temperature																
Low temperature																
G. Containment																
30. Landfill cap	NA	N	NA	N	+	+	–	+	o	o	o	o	o	o	–	o
Asphalt/concrete cap																
RCRA Subtitle C cap																
RCRA Subtitle D cap																
31. Landfill cap enhancements	NA	N	NA	N	+	+	–	+	o	o	o	o	o	o	–	o
Water harvesting																
Vegetative cover																
H. Other treatment																
32. Excavation, retrieval, and off-site disposal	NA	N	NA	N	+	+	+	–	o	o	o	o	o	o	–	o

GROUND WATER, SURFACE WATER, LEACHATE

I. *In situ* biological treatment

No.	Technology					
33.	Co-metabolic treatment	P	N	N	O&M	–
34.	Enhanced biodegradation	F	N	N	O&M	+
	Nitrate enhancement					
	Oxygen enhancement with air sparging or hydrogen peroxide					
35.	Natural attenuation	F	N	N	O&M	+
36.	Phytoremediation	P	N	N	N	o
	Enhanced rhizosphere biodegradation					
	Hydraulic control					
	Phytodegradation					
	Phytovolatilization					

J. *In situ* physical/chemical treatment

No.	Technology					
37.	Aeration	F	Y	V	N	+
38.	Air sparging	F	Y	V	N	+
39.	Bioslurping	F	Y	L V	N	+
40.	Directional wells (enhancement)	F	N	NA	Cap.	–
41.	Dual-phase extraction	F	Y	L V	O&M	+
42.	Fluid/vapour extraction	F	Y	L V	O&M	+
43.	Hot water or steam flushing/stripping	P	Y	L V	Cap.	+
44.	Hydrofracturing	P	Y	N	N	–
45.	In-well air stripping circulating wells	P	Y	L V	Cap.	+
46.	Passive/reactive treatment walls	F	N	S	Cap.	+
	Funnel and gate					
	Iron treatment wall					

K. *Ex situ* biological treatment

No.	Technology					
47.	Bioreactors	F	N	S	Cap.	+

Table 18.3 (Contd).

	Development status	Treatment train (excluding off-gas treatment)	Residual produced	O&M or capital-intensive	Availability	System reliability/ maintainability	Clean-up time	Overall cost	Non-halogenated VOCs	Halogenated VOCs	Non-halogenated SVOCs	Halogenated SVOCs	Fuels	Inorganics	Radionuclides	Explosives
48. Constructed wetlands	F	N	S	Cap.	–			o	o	o	o		o	+	–	+
L. Ex situ physical/chemical treatment (assuming pumping)																
49. Adsorption/absorption	P	N	S	I	o	I	I	–	o	o	o	o	–	+		–
Activated alumina																
Forager sponge																
Lignin adsorption/sorptive clays																
Synthetic resins																
50. Air stripping	F	N	L, V	O&M	+	+	o	+	+	+	o	o	o	–	–	–
51. Granulated activated carbon (GAC)/liquid phase carbon adsorption	F	N	S	O&M	+	+	+		+	+	+	+	o	–	–	–
52. Ion exchange	F	Y	S	N	+	+	o	+	–	–	–	–	–	+	o	–
53. Precipitation/coagulation/ flocculation	F	Y	S	N	+	+	o	+	–	–	–	–	–	+	o	1
Coagulants and flocculation																
54. Separation	F	Y	S	B	+	+	+	–	+	+	+	+	+			–
Distillation																
Filtration/ultrafiltration/ microfiltration																
Freeze crystallization																
Membrane pervaporation																
Reverse osmosis																

Technology															
55. Sprinkler irrigation Trickling filter	F	Y	S L	N	+	o	o	+	+	+	+	+	−	−	−
56. Ultraviolet oxidation Ultraviolet photolysis	F	N	N	B	+	−	NA	o	+	+	+	+	−	−	+
M. Containment															
57. Deep-well injection	F	N	N	N	+	o	NA	+	o	o	o	o	o	o	o
58. Groundwater pumping Surfactant enhanced recovery Drawdown pumping	F	N	L	B	+	+	NA	−	o	o	o	o	o	−	o
59. Slurry walls	F	N	N	Cap.	+	+	+	+	o	o	o	o	o	−	o
N. Air emissions/off-gas treatment															
60. Biofiltration	F	NA	S L	N	o	+	+	+	+	+	+	+	−	NA	−
61. High-energy destruction High-energy corona Tuneable hybrid plasma reactor	P	NA	N	I	−	−	NA	o	+	+	+	+	o	NA	o
62. Membrane separation	P	NA	N	I	−	−	NA	o	+	o	o	o	−	NA	o
63. Oxidation Catalytic oxidation Internal combustion engine oxidation Thermal or ultraviolet oxidation	F	NA	N	N	+	+	NA	+	+	+	+	+	−	NA	o
64. Vapour phase carbon adsorption VOC recovery and recycle	F	NA	S	N	+	+	NA	+	+	+	+	+	o	NA	+

Source: US Environmental Protection Agency website (Website: http://www.frtr.gov/matrix2/section3/table3_2.html)

+, better; o, average; −, worse; Y, yes; N, no; B, Both; Cap, capital; F, full; I, inadequate; L, liquid; NA, not applicable; O&M, operation and maintenance; P, pilot; S, solid; SVOC, semivolatile organic compound; V, vapour; VOC, volatile organic compound

Selection criteria may be sorted into major groups, such as:

- development status;
- availability;
- efficiency;
- emission control;
- residues produced;
- operation and maintenance/performance;
- clean-up time;
- capital and operation costs;
- risks;
- compliance with local laws and policies;
- public acceptance.

Up to 20 or 30 selection criteria may be applicable, and should be chosen for specific project purposes.

Each single criteria can be rated by a benefit factor ($+1$, $+2$) or a negative effects factor (-1, -2). In this qualitative evaluation, '0' signifies little apparent positive or negative effect.

Groups of selection criteria ratings or each single criteria rating can be weighted additionally by multiplication with an 'importance weight factor'.

The sum of the selection criteria, multiplied by the 'importance weight factor' will result in a total qualification number for the selection of the best remediation alternative.

'Non-feasible' technologies should be excluded in a pre-evaluation process considering all potential criteria for exclusion.

To exclude non-feasible technologies, and to make the best choice of applicable remediation technologies, a wide range of site-specific data, covering hydrogeology, geochemistry and hydrochemistry, contaminant distribution and contaminant migration must be considered.

18.3. Data gathering and data gaps

On most contaminated sites, data gathering is focused on the identification of contaminants and the actual delineation of plumes and/or free phase volumes in the unsaturated and saturated zones below the spills. To determine potential remediation alternatives, additional data must be evaluated in respect of hydrogeology (Table 18.4), hydrochemistry (Table 18.5) and migration (Tables 18.6 and 18.7) of dissolved and undissolved chemicals. This includes those data which identify potential problems with treatment technologies, such as precipitation of iron and manganese, precipitation of calcium and magnesium, and bio-clogging (Beitinger *et al.*, 1998).

The general objective of data gathering is to fully understand all criteria which could either exclude or determine treatment alternatives. A so-called conceptual site model will need to be developed to enable an understanding of the distribution, migration, adsorption, degradation, convection, diffusion and other transport and retention mechanisms, as well as of chemical reactions and physical behaviour.

Bore log evaluation, groundwater monitoring wells, extensive pump tests and geophysical investigations are the common methods of choice for the gathering of reliable hydrogeological data. Heterogeneous underground structures in the unsaturated and saturated zones must be carefully surveyed.

Geochemical and hydrochemical data are of major importance for evaluation of migration and degradation processes *in situ* as well as treatment technologies *in situ* and above ground (the latter assuming *pumping*). Groundwater samples are

Table 18.4. Hydrogeological data

	Data	Remarks
1	Hydrogeology	General description of geology, aquifers/aquitards, anomalies
2	Depth to groundwater table	(m)
3	Aquifer thickness	(m)
4	Groundwater flow direction	
5	Hydraulic permeability	k_f (m/s)
6	Groundwater gradient	J
7	Transmissivity	T
8	Aquifer confinement	Pressure
9	Surface water bodies	Description, distance
10	Weather conditions	Precipitation rates, wind factor
11	Surface conditions	Surface covers, plants, asphalt

Table 18.5. Geochemical and hydrochemical data

	Data	Remarks
12	pH of soil and water	
13	Electrical conductivity, TDS (salinity of water)	TDS: total dissolved solids
14	Redox potential	
15	Oxygen content	
16	Temperature	
17	Iron	
18	Manganese	
19	Calcium	Hardness
20	Magnesium	Hardness
21	Carbon dioxide	Precipitation of Ca, Mg
22	Sulfate, sulfide	Potential inhibitors
23	Nitrogen, TKN	Nitrate, nitrite, ammonium. TKN: total Kjeldahl nitrogen
24	Other chemical compounds	'Background levels', metals
25	BOD, COD	Biological and chemical oxygen demands

Table 18.6. Contaminant distribution

	Data	Remarks
26	Identification of contaminants	Types of pollutants
27	Delineation of plume	Area, depth, concentration in soil and groundwater
28	Plume activity	Increasing/decreasing, time factor
29	Free phase spreading	LNAPL/DNAPL (light and dense non-aqueous phase liquids)
30	Residual saturation (Sr)	Unsaturated zone
31	DOC	Dissolved organic content
32	TOC	Total organic content (including suspended particulate matter in water)
33	Potential receptors	Identification, distance, sensitivity
34	Age of pollutants	Ageing/degradation processes
35	Migration with time	
36	Spill location/contaminant sources	Points of emission

Table 18.7. Contaminant migration

	Data	Remarks
37	Density	LNAPL/DNAPL
38	Liquid viscosity	
39	Interfacial tension with water	
40	Solubility	
41	Vapour pressure	
42	Henry's law constant	
43	Partitioning coefficient	K_d
44	Organic content in the soil	f_{oc}
45	Octanol–water partition coefficient	K_{ow}
46	Organic carbon partitioning coefficient	K_{oc}
47	Ion exchange capacity	Clay fractions
48	Biodegradability	
49	Grain size distribution	
50	Bulk density of aquifer material	
51	Air permeability in soil	Soil vapour
52	Soil porosity	
53	Water content	
54	Soil heterogeneity	

collected by pumping when the pH, electrical conductivity, redox potential, oxygen content and temperature show constant values.

Actual spreading of contaminants as well as monitoring data over time will help us to understand emission rates, spill migration velocities and retardation, as well as degradation processes. Screening methods may help identify all pollutants of concern.

It is important to understand the behaviour of contaminants in the subsurface environment to properly design and successfully implement a remediation system.

18.4. Conceptual design/dimensioning criteria

Based on the results of the selection process and the identification of one or several remaining BATNEECs, a conceptual design is drawn up using appropriate dimensioning criteria. An overview of potential dimensioning criteria is shown in Table 18.8.

The conceptual design report should also include the following information:

- emissions and emission control;
- discharge or re-infiltration of treated groundwater;
- power consumption (electricity, fuels, etc.);
- waste streams generated and waste disposal;
- input of materials such as GAC or lime;
- remediation target levels;
- overall efficiency;
- maintenance (manpower, parts);
- monitoring;
- cost estimation;
- health and safety issues.

Table 18.8. Dimensioning criteria (example)

Dimensioning criteria	Pump-and-treat system with GAC filter	Permeable reactive barrier with GAC filling
Aquifer thickness	(m)	(m)
Hydraulic permeability	$>10^{-4}$ m/s	No restriction
Well capacity (all wells)	(m³/h)	–
Width of funnel and gate	–	(m)
Hydraulic gradient		
Seepage velocity (V_a)	–	(m/year)
Groundwater throughflow	–	(m³/day)
Contaminant mass flow	(g/h)	(g/day)
Iron content	Pretreatment: de-ironing	No restriction
Hardness	Pretreatment: softening	No restriction
Carbon dioxide	Pretreatment	No restriction
DOC	(g/h)	(g/day)
TOC	(g/h)	(g/day)
Loading capacity (GAC)	0.1–3%	Potentially higher
Filter height	2 m	>0.3 m
Retention time	10–60 min	6 h to several days
Surface load	<15 m/h	≪15 m/h

Potential risks and additional data needs have to be identified. Treatability studies (laboratory scale, bench scale, full size) may help to reduce risk and costs due to site-specific uncertainties or the use of innovative technologies (long-term efficiency, unknown potential for failure or unidentified chemical reactions, etc.).

18.5. Treatability studies/final design

To evaluate the feasibility of conceptual groundwater treatment plants, bench scale treatment facilities can be installed and operated for several months at the site by using existing monitoring wells of at least 100 mm diameter, or by installing new wells to be used as future remediation wells. As an example, a bench scale test for the planned installation of a permeable reactive barrier is described below.

A pilot groundwater treatment plant was installed at a former industrial site in Essen, Germany, where organic solvents had been stored and processed in a small chemical plant for several decades (Bütow and Beitinger, 1998). Leakage and handling losses caused significant soil and groundwater contamination, mainly by BTEX compounds (benzene, toluene, ethylbenzene and xylol) and CHCs (chlorinated hydrocarbons). The contaminated aquifer has low hydraulic conductivity and is only 2–3 m thick. The aquifer is covered by 4–11 m thick silty and clayey layers (loess). During investigations and the conceptual remediation design stage, it was determined that the site was a suitable candidate for the installation of adsorbent walls, since conventional remediation and contamination control measures could not be applied in a cost-efficient manner.

Subsequently, Woodward Clyde International (WCI) and the Institut für Wassergefährdende Stoffe (Institute for Water-contaminating Substances, IWS) reported on various alternatives for the installation of an adsorbent wall following a feasibility study. The study also established the data necessary to arrive at the dimensions of the adsorbent wall. The feasibility study recommended that pilot tests be conducted on the site for this purpose. The objective of the pilot tests was to obtain precise information on the adsorption potential for the contaminants at the site, the type and quantity of the required adsorbent material, the functioning of filters at different flow speeds, and the long-term effectiveness as well as the attendant risks, if any, of installing an adsorbent wall.

Conducting the pilot tests involved the following principal tasks:

- Selecting a suitable adsorbent for the tests, depending on water quality and the relevant contaminant concentrations at the site;
- Structural design and planning of the pilot plant;
- Operation and sampling from the pilot plant as well as carrying out laboratory analyses;
- Assessment of the pilot tests.

Planning, conducting and assessing the pilot tests required 8 months.

Several subtasks had to be carried out for the tests to be successful and representative, namely designing the pilot plant, determining the operating parameters of the

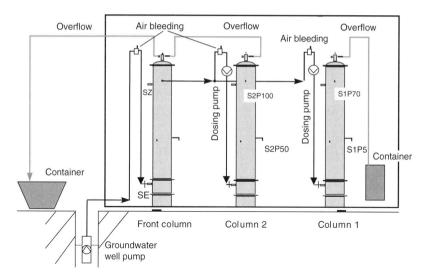

Fig. 18.1. Sketch of the pilot plant

pilot plant, and selecting a suitable site for the installation of the pilot plant (Fig. 18.1).

The design of the pilot plant had to take into account several peripheral conditions:

- continuous supply of groundwater to the pilot plant directly at the site;
- tests to determine actual site conditions (column 1, scale 1:1);
- tests to extrapolate durability of 30 years (column 2);
- regular periodic sampling of the experimental columns at various depths of the activated carbon bed to obtain characteristic curves.

The final design of the pilot plant is shown in Fig. 18.1.

The column dimensions were as follows: $h = 1.90$ m, $d_i = 0.25$ m and $A = 0.05$ m².

The pilot plant was fed with groundwater which was pumped directly from the aquifer into the front column. Two dosing pumps located behind a gravel bed in the front column fed groundwater into columns 1 and 2.

The front column contained a 40 cm gravel filter (size: 2–3.15 mm). The gravel filter served to hold back sediments as well as to eliminate iron and manganese. Sampling points were located on the front column at SE (in front of the gravel filter) and SZ (behind the gravel filter).

Column 1 contained:

- a 45 cm gravel filter (size: 2–3.15 mm);
- 5 cm of activated carbon (Norit ROW 0.08 supra);
- a 5 cm gravel filter (gravel size: 2–3.15 mm);
- 65 cm of activated carbon (Norit ROW 0.08 supra).

The thickness of the activated carbon bed in column 1 corresponded to the recommended thickness of the activated carbon bed of the adsorbent wall in the feasibility study. Sampling points were located on column 1 at S1P5 (sampling point behind 5 cm of carbon) and S1P70 (sampling point behind 70 cm of carbon).

Column 2 contained 100 cm of activated carbon (Norit ROW 0.08 supra). Sampling points were located on column 2 at S2P50 (behind 50 cm of activated carbon) and S2P100 (behind 100 cm of activated carbon). The treated water was led via an overflow into a trough located outside of the container.

The operating parameters of the pilot plant were selected on the basis of the following criteria, among others:

- hydrogeological site conditions;
- suitable topographical site conditions for installation of the pilot plant;
- simulation of various performance periods;
- likely results on termination of the pilot tests.

Following the recommendations made in the feasibility study, activated carbon was selected as the adsorbent. For the preliminary tests, two products made by Norit (GAC 830 and ROW 0.8 supra), were recommended; these are particularly well suited as adsorbents for BTEX and volatile CHCs. For both the products, isotherms were plotted using water from the site, in order to obtain adsorption results derived from actual site samples as far as possible.

A comparison of the adsorption isotherm plots clearly showed Norit ROW 0.8 supra to be the more effective product in this situation.

As shown in Fig. 18.1, the plant had a total of seven sampling points. Samples from the points marked SE and SZ characterized the conditions at the inlet to the pilot plant. SZ was located behind the gravel filter, ahead of the fork leading to columns 1 and 2. Samples from all points were taken at 12 day intervals during the entire period of operation.

Groundwater analyses were based on the contamination at the site; their scope was determined by the feasibility study to install an adsorbent wall. The analyses covered field parameters, general parameters and parameters to quantify BTEX and volatile CHC contamination.

For each sample, the following field parameters were established:

- temperature;
- pH;
- conductivity;
- oxygen content;
- redox potential.

The values determined were expected to provide information on the extent of contaminant breakdown within the plant and the consequent changes to the ground-water milieu.

The general analysed parameters included global parameters for total organic compounds as well as parameters for iron and manganese. A global parameter for all organic compounds was used in order to study whether it could serve as a substitute for analyses of individual substances. Moreover, these global parameters were used to check whether the results of individual analyses were plausible. Iron and manganese contents were determined in order to check whether precipitation of these substances would clog the adsorbent wall.

Separate analyses were carried out for BTEX and volatile CHCs. The number of parameters analysed (16) was deliberately large, so as to cover important decomposition products such as vinyl chloride.

Analyses of the activated carbon were carried out by Norit prior to the tests, in order to select the right grade of activated carbon. The activated carbon used was also analysed after the plant was disassembled. The tests conducted included thermogravimetric analyses to determine contaminant content in the activated carbon columns and microscopic analyses to determine microbial activity as well as iron and manganese build-up.

In order to study whether the activated carbon used could be regenerated, three samples were taken from the lower section of column 2 and tested for heavy metal and organic compounds. Contaminant concentrations in the activated carbon used should be highest at this sampling point of column 2 – a hypothesis confirmed by the thermogravimetric analyses. The individual parameters (eight for heavy metals, four for BTEX and 12 for volatile CHCs) and global parameters studied indicate that the conditions for regeneration had been met. This has significant cost-saving potential for the adsorbent wall since the depleted activated carbon would not need to be disposed of.

Water samples collected at 11 day intervals were tested for numerous parameters; in total, over 1600 individual results were obtained for water samples taken during operation of the pilot plant.

The pilot tests confirmed the findings of the feasibility study, to the effect that the site is suitable for the installation of an adsorbent wall. The following statements can be made with respect to the tests:

- The pilot tests showed significant contaminant retention in the activated carbon, much higher, in fact, than that assessed in the feasibility study. Contaminant breakthrough for toluene and trichloroethylene was determined at sampling point S2P50 (i.e. after flow through 50 cm), in column 2, only at the end of the 5 month pilot test operation. By this time, throughput had reached 600 times the bed volume.
- The pilot tests indicated that the durability of the wall, given a 70 cm thick activated carbon layer, would be much longer than the 30 years estimated in the feasibility study. The thickness of the carbon layer should therefore be reduced when the wall is installed.
- The DOC concentrations established during the pilot tests could be traced almost entirely to the contaminants detected at the site. It is therefore to be expected that the adsorbing potential of the activated carbon will not be impaired by natural organic compounds, such as humin.
- Data pertaining to the contaminant breakthrough suggested that the depletion of the adsorbing capacity of the activated carbon was accompanied by a sharp peak in the concentration of volatile substances. A suitable monitoring system should therefore be set up when the adsorbent wall is installed.
- The finding that the activated carbon could be regenerated after disassembling the plant suggests economic operation of the adsorbent wall.

- Laboratory analyses of the water and activated carbon samples indicate that iron and manganese precipitation would be insignificant and would not clog the addsorbent wall.
- Microbial activity could not be detected in the gravel filter nor in the activated carbon; it may be concluded that under the given site conditions the build-up of a bacterial film does not pose a risk.
- Preliminary laboratory tests to determine the choice of activated carbon as well as pilot tests must be carried out in all cases prior to setting up an adsorbent wall, given the variance in site conditions.

18.6. Operation/maintenance/monitoring/risks

Efficient operation and maintenance, monitoring and minimization of potential risks during the operation period are the main driving forces for the completion of a successful and cost-effective remediation scheme. One of the major problems with regard to groundwater treatment systems is the time needed to achieve given remediation target levels. A comparison of 28 groundwater remediation schemes for CHCs in the USA (US Environmental Protection Agency, 1999) showed that only two out of 28 schemes achieved their expected results within the given time period. Pump-and-treat systems must be operated for several years or decades and need careful maintenance during operation. Pumps need inspection, anti-freezing devices need to be installed during winter operation, filters and strippers may be blocked by iron, hardeners or bio-clogging. Materials such as GAC must also be exchanged periodically.

For monitoring purposes, monitoring wells must be identified with defined parameters for chemical analysis. A monitoring plan will help to calculate costs, and needs to be accepted by regulatory authorities to prove the successful operation of the treatment technology. Costs for monitoring also depend on the technology ultimately selected and installed, as well as the frequency at which samples are taken and analysed.

18.7. Costs

The cost of a remediation scheme derives from two main sources: capital costs for buying or leasing the necessary installations and operation costs (Beitinger, 1998). Table 18.9 gives common cost parameters that need to be considered for cost estimation purposes.

In several cases a net present value calculation may allow future costs to be reduced by the discount rate for the period between the spending date and the date of the accrual (if the accrual is tax free and without costs increasing year by year).

The net present value calculation will result in lower total costs for treatment plants with low investment and higher operational costs (e.g. pump-and-treat systems operating for several years or decades), in comparison to permeable reactive barriers, which have high immediate capital costs for installation but generate low operating costs (Table 18.10).

Table 18.9. Cost parameters

Capital expenditure	Operational and maintenance costs
Purchase cost (including taxes)	Personnel costs
Mobilization/installation	Energy costs (electricity, fuels)
Start-up costs	Consumable materials (GAC, lime, etc.)
Interest rate	Maintenance
	Monitoring (sampling, chemical analysis, reporting)
	Discharge costs
	Residuals, waste disposal costs
	Years of operation
	Fees and taxes

Table 18.10. Cost comparison (example)

Dimensioning and cost criteria	Pump-and-treat system	Permeable reactive barrier
Aquifer	10 m	10 m
Hydraulic permeability	10^{-4} m/s	10^{-4} m/s
Hydraulic gradient	0.001	0.001
Pore volume	0.2	0.2
Groundwater capacity	20 m³/h	20 m³/h
Installations	2 wells each	3 gates (10 × 15 m),
	10 m³/h	tunnel length: 270 m
	($R \approx 159$ m, $s \approx 4$–5 m)	
Contamination (CHCs)	1 mg/l	1 mg/l
GAC loading	1% CHC/weight	1% CHC/weight
GAC volume	2 × 10 m³	150 m³
Loading time	~6 month	~3.5 years
Investment	0.25 million euros	1 million euros
Depreciation time	5 years	20 years
Depreciation costs	50 000 euros/year	50 000 euros/year
Interest rate	5%	
Interest costs	12 500 euros/year	50 000 euros/year
Total capital costs	**62 500 euros/year**	**100 000 euros/year**
Electricity (10 W/h)	13 140 euros/year	–
Operation (manpower)	10 000 euros/year	–
GAC	40 000 euros/year	40 000 euros/year
Repair/maintenance	30 000 euros/year	10 000 euros/year
Discharge of groundwater (0.50 euro/m³)	43 800 euros/year	–
Monitoring	10 000 euros/year	10 000 euros/year
Total operational costs	**146 940 euros/year**	**60 000 euros/year**
Total costs	209 440 euros/year	160 000 euros/year
Specific costs per treated volume (175 200 m³/year)	1.30 euros/m³	0.65 euro/m³

18.8. References

BEITINGER, E. (1998). Permeable Treatment Walls – *Design, Construction and Cost. NATO/CCMS Pilot Study Phase III*, EPA 542-R-98-003. US Environmental Protection Agency, Washington DC.

BEITINGER, E., TARNOWSKI, F., GEHRKE, M. and BURMEIER, H. (1998). Permeable treatment walls for *in-situ* groundwater remediation – how to avoid precipitation and bio-clogging. Consoil '98, Edinburgh.

BÜTOW, E. and BEITINGER, E. (1998). Pilotversuche zur Ermittlung der Standzeit einer Aktivkohle-Adsorberwand. ITVA-Symposium Altlasten, Kassel.

EC (1996). Council Directive 96/61/EC of 24 September 1996 concerning integrated pollution prevention and control. *EC Official Journal* **L257**, 0026–0040. Website: http://www.europa.eu.int/comm/environment/ippc.

GAVASKAR, A. R., GUPTA, N., SASS, B. M., JANOSY, R. J. and O'SULLIVAN, D. (1998). *Permeable Barriers for Groundwater Remediation*. Batelle Press, Columbus, Ohio.

US ENVIRONMENTAL PROTECTION AGENCY (1998). *Permeable Reactive Barrier Technologies for Contaminant Remediation*, EPA/600/R-98/125. US Environmental Protection Agency, Washington DC.

US Environmental Protection Agency (1999). *Groundwater Cleanup: Overview of Operating Experience at 28 Sites*, EPA 542-12-99-006. US Environmental Protection Agency, Washington DC.

19. Discussion: status, directions and R&D issues

C. McDonald
School of Civil Engineering, University of Leeds, Leeds LS2 9JT, UK

19.1. Introduction

When setting off on this 'state-of-the-art' review, the editors took the intentionally bold challenge of comparing permeable reactive barriers with pump-and-treat systems, with natural attenuation as a baseline reference. Our purpose was to provoke a range of contributions from those with real experience in developing, implementing or evaluating the alternative technologies – and in that we were not disappointed.

We were conscious of thereby taking a risk of provoking 'turf wars' between vested interests in the various approaches. But we believe it to be a measure of the maturity of the business of groundwater remediation that, in such an interdisciplinary and international field, we got no such reaction. Rather, we found a real thirst to assess objectively the pros and cons of what has been achieved to date and to share thoughts on priorities and directions for the future.

A large measure of agreement emerged that all three approaches – natural attenuation, pump and treat, and reactive barriers – have their distinct roles to play, and find applications for which they are the *best practical environmental option (BPEO), best available technique not entailing excessive cost (BATNEEC)*, or *best available technique (BAT) for IPPC Directive purposes*. Indeed, what was enlightening was the number of sites where more than one approach was being incorporated, either separately or as hybrids, and both as trial or pilot projects and as serious remediation solutions (Remediation Technologies Development Forum (RTDF), 2002).

Let us take an overview of some of the applications, findings and concerns emerging. As well as referring to the chapters appearing in this volume, we also draw upon discussions at the European Science Foundation GPoll Workshop '*In-situ Reactive Barriers versus Pump-and-Treat*', held at BAM, Berlin, in October 2001, and wish to acknowledge the many thoughtful contributions from that occasion.

19.2. Setting the scene

19.2.1. Is there a level playing field?

Our starting point herein was a status review of groundwater remediation using active and passive processes (see Chapter 1). Our purpose now is to reassess the prior state-of-the-art outlined there, in light of all the experiences reported in later chapters.

At the outset we must caution that *there is in practice no level playing field between active and passive processes*. On an international basis, at least, pump and treat is a relatively long-established approach, relying heavily on technically proven water treatment methods. Reactive barriers are comparatively novel. It is hardly surprising, then, that both the opening review (see Chapter 1) and many other chapters give disproportionate emphasis to reporting and explaining the hopes for and experiences with those newer and lesser-known passive technologies.

19.2.2. Where do our findings come from?

Another wise preliminary must be to recognize the strengths and limitations of our data, which has tended to come from three types of study.

Firstly, there is review information from wider sources, not only that mentioned above (Simon *et al.*, 2002), but that incorporated as background in various chapters, or as contributions by participants at the GPoll workshop. However, even from the USA, where there are already said to be some 60–80 barrier installations, experience is limited by the fact that most of them (maybe 90%) use zero-valent iron as their reactive agent.

Secondly, there are the individual test situations reported in more detail here, such as:

- *Hydrocarbons.* A pilot plant used to test solutions for a former industrial site suffering benzene, toluene, ethylbenzene and xylene (BTEX) and chlorinated hydrocarbon (CHC) contamination, resulting in selection of activated carbon as an adsorbent, plus many significant recommendations for the design and operation of a full-scale facility (see Chapter 18).

  A model project on land heavily contaminated with complex chlorinated aliphatic and aromatic hydrocarbons (see Chapter 10). The aim was to develop, test and optimize new technologies for permeable reactive barriers under control- led *in situ* conditions. The underground reactor facility comprised 20 reactors in five shafts, testing physicochemical, microbial and combined techniques.

- *Metals.* The relocation of heap leaching wastes at a former uranium mine, the exposure of which leads to rapid increase in dissolution of the uranium (see Chapter 14). The aims of this project were research and investigation of full-scale use of a lime (calcium oxide)-based reactive barrier.

  The management of mine flooding waters from the underground workings of another closed uranium mine complex (see Chapter 13). Hydraulic channelling or caging, initial flushing/sweeping, immobilization, various pump and treat or collect and treat approaches, construction of reactive barriers, *in situ* treatment methods, natural or enhanced precipitation or sorption, and passive water treatment systems, were all seen as possible options for an optimized flooding strategy.

- *Mixed contaminants.* The collection of groundwater polluted by chromate (Cr^{VI}) and hydrophobic organic compounds (see Chapter 2). In the context of a research project, part of the site drainage system was connected with a reactor shaft containing zero-valent iron in the form of steel wool mixed with calcite. This was to be operated as a subsidiary part of the pump-and-treat system already installed as a final site remedy, which involved a separate on-site treatment unit where the contaminants are removed by conventional methods (ion exchange and activated carbon) to meet regulations.

Thirdly, there is the important ongoing German RUBIN network on permeable reactive barriers:

- *Twelve reactive barrier projects, of which nine deal with actual installations* (see Chapter 3). The evaluation is to be over about 4 years, with the aim of delivering comprehensive data on all aspects of both pilot and full-scale case studies, leading to a reliable assessment of benefits and drawbacks, applicability and viability. Current R & D needs are addressed, and projects tailored for solving technical, economic, administrative and other issues relating to permeable reactive barrier implementation.

While we are thus drawing upon a wealth of experience in diverse real-world situations, it has to be borne in mind that *many are only pilot scale projects*, that *the numbers are still limited*, and that *often only preliminary or interim results are available*.

19.2.3. Treating what with what?

Although most of the early focus with reactive barriers has been on using zero-valent iron or activated carbon, which had a background of application to industrial wastewaters, the range of active agents shown to be worth consideration is now much wider (see Chapter 1).

The basic list of target contaminants in Europe may be little different from that in North America, but due to population density and settlement patterns, and the complex political and industrial history, the priorities for and challenges of treatment can be more diverse.

Uranium, for instance, is a heavy metal of great concern for ground and surface waters, especially in central and eastern Europe, mainly due to its chemical toxicity. This is true for some naturally occurring groundwaters, but in particular for areas where uranium mining and milling took place (or still does), near nuclear energy sites or military facilities, and heaps where coal and fly ash have been dumped (see Chapter 12). Possible approaches to remediation of uranium contamination, discussed in this book, are:

- *Redox reactions.* Uranium in its hexavalent oxidation state is very mobile, and only reduction to tetravalent uranium seems to offer a sustainable means of removing uranium from water. This may be done through *strong reductants such as* Fe^0 *and* CH_4 or by *biodegradation of organic matter* (see Chapter 12).
- *Precipitation.* The use of calcium oxide (lime)-based reactive barriers for removing uranium has already been mentioned (see Chapter 14). A primary advantage is in their low cost compared with other materials. Their main disadvantage is their relatively short life-time. This type of reactor should therefore be used when the target water contamination is expected to last for only a short period. This is the case when uranium-containing wastes are relocated: a rapid increase of the uranium concentration is observed just after relocation, and the main task is to attenuate this increased uranium con-centration during the following 2–3 year period.
- *Ion exchange and adsorption.* These are more widely applicable to both *ex situ* and *in situ* treatment, and for metals, halogenated organic compounds or radio-nuclides (see Chapter 6), e.g.:
  - zeolites for soluble organics, inorganic cations and anions;
  - goethite for cations and oxyanions, such as arsenic.

  Sorption is a fast reaction, and can also play a key first step in the natural attenuation of uranium from water, followed by the slower steps of precipitation and co-precipitation (see Chapter 12).

  Although good experience exists for the use of natural zeolites for drinking water preparation and for removal of heavy metals from industrial wastewaters, there is little knowledge of their utilization in permeable reactive barriers.

  To be suitable for use as a reactive agent, zeolites must meet the following conditions (see Chapter 11):
  - high sorption capacity and selectivity for the target contaminants;
  - fast reaction kinetics and high hydraulic permeability.

  Laboratory batch and column tests have now shown that zeolites, and especially sodium-activated zeolites, are indeed potential sorbents for barrier fillings (see Chapter 11).

Reflecting on the treatment toolkit which all that illustrates, our key observation is that *in principle such agents and processes are applicable, to a greater or lesser extent, above and below ground, and whether the contaminant is pumped or free flow. They do not appear to us to be intrinsically tied to either active or passive treatment systems.* Indeed, some are thought to have significant roles to play in natural attenuation.

19.2.4. Do they work and how do they compare?

Investigation and monitoring of the natural attenuation of organic contaminants in the USA has shown some encouraging outcomes. By contrast, an evaluation of 28 groundwater remedies for CHCs in the USA showed that only two achieved their expected results within the given time period (see Chapter 18). Bearing in mind our caution over the lack of a level playing field, and sticking to the broad picture for a moment, how have active and passive approaches worked in practice?

Over the past 20 years *source zone treatment and/or removal* has been a first recourse in the field of remediation. Several technologies and strategies were developed to either excavate, isolate or flush (decontaminate) *in situ* the zones where contaminants had penetrated into the subsurface (see Chapter 17). Where that is feasible, it usually remains a vital part of any approach.

Practitioners are, however, often faced with the fact that a source zone is not sufficiently identifiable and only partially accessible. *Plume management* is a valid alternative where technical, economic or ecological factors prevent or limit source removal (see Chapter 17).

The most common plume management approach to date has been through pump and treat: extraction via downgradient pumping wells for treatment in a surface plant. It is relatively cheap to install and safe to operate. However, operational and maintenance costs are generally high, and field experience has found contaminant sources to persist much longer than originally anticipated, sometimes potentially for decades (see Chapter 17).

Pump-and-treat systems need careful maintenance during operation. Pumps require inspection, anti-freezing devices need to be installed during winter operation, and filters and strippers may be blocked by iron, hardeners or bio-clogging. Treatment materials such as granular activated carbon must also be exchanged periodically (see Chapter 18).

Consequently, pump-and-treat approaches can appear cost-intensive over the longer term and, under many circumstances, a suboptimal solution. One should bear in mind, though, that the impact of future payments decreases with increasing operation time, due to the economic necessity to discount future payments (i.e. convert to net present value) (see Chapter 17).

Reactive barriers (funnel-and-gate systems or continuous walls) were thus invented as an inexpensive long-term operational option. The idea is that the higher initial investment costs for constructing a reactive barrier will be compensated in the long run by lower operational costs. However, this turns out not always to be true (see Chapter 17).

In short, there can be no general rule that either active or passive methods perform better or are more cost-effective.

19.2.5. How fundamental are the distinctions anyway?

Without wishing to imply that they are representative, it is significant that none of the specific test sites identified above (for hydrocarbons, metals or mixed contaminants) incorporated a conventional *in situ* reactive barrier, either continuous wall or funnel-and-gate. While this was partly due to their site situations and experimental status, in our view it also points to a much richer variety and choice of practical solutions than presupposed by the simple dichotomies of active versus passive or pump and treat versus the reactive barrier:

- *Model site incorporating 20 reactors in five underground shafts* (see Chapter 10). This was one example which relied on a pumped system with vertical flow. Placing the reactors underground was seen to offer the right temperature and other conditions to simulate an *in situ* barrier. But it was accepted that an equivalent system could have been constructed above ground, where it might instead have offered greater temperature control and flexibility. Constructing the reactors in shafts allowed the reactive materials to be exchanged more easily. But that is also a consideration for some non-experimental installations.
- *Mine flooding water case* (see Chapter 13). This is informative in two respects. Firstly, it is noteworthy for the whole battery of approaches – immobilization, *in situ* chemical treatment, pump and treat or collect and treat, reactive barriers, natural or enhanced precipitation or sorption, reed bed treatment – which are being investigated and implemented, to deal with different areas of the uranium mine field at different stages of prevention and clean-up. Secondly, even the full-scale permeable barriers contemplated would not, it seems, be typical walls or gates, but plugs in existing mineworkings, or horizontal surface level trickle beds.
- *Uranium heap leach relocation* (see Chapter 14). This is the only example described of a full-scale solution having been implemented, which indeed took the form of a horizontal barrier. Lime was spread on the surface and mixed with some portion of the wastes being relocated, using agricultural techniques. (There were actually a series of such reactive barriers, built overlapping each other low in the tip, for stability reasons.)
- *Soil mixing*. There are other solutions, not covered in these chapters. For instance, reagents can be placed in the ground by soil mixing, e.g. with organophilic or hydrophobic clays in a granular form. There are already examples in the USA and UK, and there will be many more. Installation is relatively cheap, and the outcomes correspondingly uncertain. But there seems to us no basic reason why it may not in due course be given as full a technological underpinning as reactive barriers or pump and treat. There are also parallels with the prospects foreseen for enhanced natural attenuation e.g. by addition of waste gypsum or complexed iron in the subsurface.

Permeable reactive walls or barriers can be seen as just one subset of (ground)water treatment reactors. They have close relatives in various types of reactive zone. There are analogies to natural processes and some repository designs (see Chapter 4). The issues of whether they are situated above or below ground, are pumped or passive, arc permanent or replaceable, are all important to the performance and economics of individual installations, but should be dealt with as matters of design rather than as issues of great principle. *The common feature that they are all reactors, relying on fundamentally similar processes, is what should drive their technologies forward to mutual advantage.*

19.3. Setting about remediation

19.3.1. What contaminants have we got here?

We now turn from the big issues we have discussed so far to ask what we can learn about the 'nitty gritty' of specifying reactors of whatever type. And although it is a field in its own right, we would be remiss not to begin with a brief word about site investigation and characterization.

For this we would commend the US Environmental Protection Agency's 'adaptive sampling' approach to site investigation, as in their recent published guidelines (Robbat, 2001). That is the way to minimize mobilization costs and maximize useful information. It does not obviate the need for detailed investigation, but concentrates that progressively on the most relevant source areas and species, in the context of the greatest risks.

As for the detailed investigation, a knowledge of the total amount of heavy metals in soil is not enough to assess environmental impact and decide on the suitability of polluted soils for potential use or upon a reclamation strategy. It is important to have a quantitative understanding of the distribution of heavy metal chemical forms and species over the solid phase and pore water in soil. In one of the case studies reported here a six-step sequential extraction procedure was used to determine the distribution of lead and zinc among various fractions of soils from 30 locations (see Chapter 9).

Toxicity tests, using bio-indicators (such as fish) to gauge whole toxicity may help economize on multi-species analysis. Some UK sites already use toxicity maps. DNA tests may also help eventually, but the feeling is that they will not completely replace traditional speciation methods.

19.3.2. How should remediation approach be selected?

Having embarked upon site investigation, how does one choose the appropriate remedial technology – the total system as distinct from just the active agent – for a site? Several of our chapters have addressed that centrally and/or illustrated it well:

- *Comprehensive methodology* (see Chapter 18). Groundwater treatment systems for the remediation of contaminated sites have to be developed and evaluated individually for each site. The best available technique (BAT) should be the result of adequate data gathering plus an evaluation process in which all appropriate technical, ecological and economic criteria are considered carefully and objectively.

  In most cases it will be advantageous to develop a reliable conceptual site model and to perform pumping and treatability tests. The conceptual site model should enable an understanding of distribution, migration, adsorption, degradation, convection, diffusion, and all other transport and retention mechanisms, as well as chemical reactions and physical behaviour.

  Selection criteria may be sorted into major groups such as: development status; availability; efficiency; emission control; residuals produced; operation and maintenance/performance; clean-up time; capital and operation cost;

monitoring; risks; compliance with applicable laws and policies; public accept-ance. A ranking and weighting system may be applied to assist in selection of the best option.

Site-specific assessments should be provided for the chosen technology, to encourage better acceptance by relevant authorities and affected neighbours. New technologies must be proven on the best available technique not entailing excessive cost (BATNEEC) principle.

Following further data assembly, a conceptual design is performed using appropriate dimensioning criteria, and addressing issues such as remediation target levels and monitoring. To evaluate the feasibility of the concept-ual design, pilot bench scale treatability studies over some months are then recommended.

- *Practical systems approach* (see Chapter 13). Flooding a mine needs very careful planning, in order to minimize long-term costs. Mine flood waters generally need treatment. Major long-term pumping and treatment costs require the development of flooding strategies and optimization of the flooding process. A systems approach is appropriate for this complex task.

  Various options and measures of a basic and supporting nature typically exist to achieve optimization of the flooding process. The applicability of options and measures should be evaluated, on the basis of existing conditions at the individual mines, by cost–benefit analyses and careful risk management.

  An optimized flooding concept includes not only an understanding and control of the hydraulics but also the evaluation of potentially applicable water treatment processes and possible alternatives such as passive treatment systems, *in situ* (e.g. reactive barriers) or *ex situ* (e.g. wetlands).

  In the case of the Königstein mine, which contained water highly con-taminated with 60 mg/l of uranium, extensive bench scale tests of various materials have shown that a mixture of zero-valent iron and lignite can immobilize the critical contaminants effectively. Two options exist to use these materials, either to fill the control drift to form an *in situ* reactive barrier, or to construct a solid bed reactor at the surface, which would necessitate continued pumping from the flooded drift.

- *Practical scientific approach* (see Chapter 14). Laboratory tests (1 m^3 open tanks):

  - The average uranium concentration in the seepage (throughout the whole experiment) was 16 mg/l without the reactive barrier, but only 0.61 mg/l in the apparatus with the barrier. During the experiment 8.6 g of uranium was dissolved, of which 8.1 g was retarded by the reactive barrier.

  - From a practical viewpoint it is important to know whether the calcium oxide has to be introduced under the wastes or can be placed on top of the pile. The uranium concentration was found to be much less if the calcium oxide barrier was placed underneath. The precipitation of the uranium essentially occurred in the barrier itself.

Large-scale field test (six test cells, totalling 25×7.5 m):

- Seepage from the test cells without barriers contained much more uranium than from those with reactive barriers. The uranium concentration remained below 1 mg/l only provided lime consumption was higher than 3 kg/m^3.

Different though those perspectives may be, in our view they are consistent in talking about *solutions which are site-specific, involve a methodical succession of treatability tests on the actual contaminants, and incorporate realistic and objective consideration of operational, economic and environmental factors* appropriate to each circumstance.

19.4. Combined approaches

19.4.1. How can technologies vary over place and time?

The mine flooding example deserves our further consideration for the pragmatic way it packages a number of distinct technologies:

- *Packaging over time* (see Chapter 13). An initial phase of pump and treat can be the optimal strategy. This is called the washing or sweeping phase. The aim of this phase is to 'scoop off' an initial contaminant load and to avoid long-term dilution of the contaminant source with less contaminated infiltration water.

    For the Wismut field (except the Königstein mine) the optimal flooding strategy is to let the flood level rise to its natural maximum and collect the seepage at hydraulic low points. This seepage collection option promises better use of natural attenuation processes. It eases the change from conventional water treatment systems to passive systems. The efficiency of conventional treatments at a late stage of flooding, when the contaminant levels are relatively low, is much reduced, and passive systems can then be used at a much higher level of efficiency.

    The most difficult case is given when contaminated infiltration waters originating from high levels within a mine (which cannot be flooded or hydraulically separated) continuously discharge their contaminant load into the flooded portion of the mine. This is often the case with old metal mines, where mining historically started in the rich oxidized zone of the ore body above groundwater level, and which consequently provided and provides an acid mine drainage problem.

    As long-term water treatment costs represent a heavy financial burden, the aim is to keep the necessary period of water treatment as short as possible. One strategy to achieve this is to retain as much contamination as practicable within the mine, i.e. to reduce or stop the mobilization of contaminants or to facilitate natural or artificial precipitation or sorption in a geochemically stable form, for example by *in situ* lime precipitation or by sorption within a reactive barrier.

- *Packaging by location* (see Chapter 13). At Königstein, one aquifer is an important groundwater resource for the Dresden region. Its pollutant potential comes mainly from sulfate base metals, iron and radionuclides. There is a

two-phase model for pollutant outputs: an initial short phase of fast contaminant release, followed by a long phase of much slower release, when
contaminant concentrations are no longer expected to represent a major risk to
the environment and specifically to the water supply aquifer.

A concept of controlled flooding is to be followed. Its major element is a
control drift surrounding the deposit at its hydraulically open sides, which allows
the collection of flood waters draining through a shaft into the drift. Initially
these will be pumped to the surface and treated by a conventional high-density
sludge (HDS) lime precipitation process.

Once pollutant concentrations have reached sufficiently low levels, use of the
control drift as such will be abandoned, and natural hydraulic conditions will
develop. The flooded control drift can then be used as a horizontal well, from
which contaminated flood water may be pumped to the surface and treated.
However, investigations are presently being carried out with a view to using the
drift downstream of the deposit as a reactive barrier. At that stage a reactive
barrier could be a more economic and efficient method than conventional pump
and treat. A system of wetland bioremediation in also being developed in case of
leakage of the flooding waters via a fault line.

We consider the principles so well illustrated here – of *controlled release, phased
treatment, and a series of location-specific solutions* – to be no less applicable to
different types of complex groundwater pollution prevention and remediation
problem.

19.4.2. What other 'treatment trains' are there?

The case studies have highlighted some other interesting and promising biochemical
and physical treatment combinations:

*   *Naphthalene sorption by modified zeolites* (see Chapter 11). It was noted that
    naphthalene is itself used for the sorption of metals. So on a site with mixed
    contaminants, it might be possible to fix metals on the surface as a second stage.
*   *Biosorptive flotation* (see Chapter 6). This was examined in the context of
    a pump-and-treat system targeting metal contaminants. The metal ions are
    abstracted on to sorbents, in the fine or ultrafine size range, followed by a
    flotation stage for separation of the metal-loaded particles.

    A wide range of flotation techniques are highly developed in mineral process-
    ing, soil washing, and water and sewage treatment. With collectorless electro-
    flotation, an electric field gradient aids the self-flocculation of suspended
    matter.

    A rock salt solution may be used as the electrolyte, generating hypochlorite to
    oxidize cyanide and hydroxides (of nickel, lead, zinc and copper) to form metal
    precipitates. Sacrificial aluminium alloy electrodes have proved effective.

    Biosorptive flotation seems a viable and effective separation process for the
    removal of metal ions from dilute aqueous solutions, no matter what sort of
    biosorbent is applied. Recovery of metals is even possible from the concentrate.

Coupled reactor systems are already included in the 20-reactor/5-shaft trials, and it is *considered likely that combinations of the various physicochemical and microbial technologies will need to be applied* (Weiss *et al.*, in Chapter 10). We strongly concur with that opinion.

19.5. Targets and agents

19.5.1. How is the active agent selected or improved?

Two of the studies led to valuable guidance on choosing the main classes of active agent:

- *Choice of sorbent* (see Chapter 7). Qualitative and quantitative sorption data are required to predict the behaviour of toxic elements in surface water and groundwater and choose the best decontamination procedures. The scale and location of the contaminated media is one of the selection criteria between pump-and-treat systems and reactive barriers. The nature of the pollutants and contaminated media themselves primarily guide the selection of the sorbent. The next step in selection is the sorption mechanism, including sorption kinetics. Finally, quantitative models and computer codes can be used in the technical development of the decontamination method.
- *Choice of ion exchanger* (see Chapter 2). The research reactor for mixed chromate and organic contamination mentioned earlier highlighted issues even in specifying a proven agent, in that case zero-valent iron. Steel wool had been selected as having the best performance on the basis of laboratory tests. Only the first month's results were available from an experiment intended to last several years. But they raised questions as to whether one was seeing surface oxidation or redox effects. There were doubts, on the basis of findings from elsewhere, as to how long the process would work with chromates. There were suggestions that granulated iron might have produced better results with the hydrocarbons. But even then, there were said to be few sources of granular zero-valent iron, of the quality needed, in Europe. It was essential to analyse the surface properties of commercially available iron materials.

Nor, these days, is it simply a matter of *selection*. Two of the other cases reported covered experiments aimed at *improving the performance* of potential active agents:

- *Covalent surface modification of clays* (see Chapter 11). Surface modification of special clay minerals by cationic surfactants, to produce organophilic zeolites and organoclays with enhanced sorptive properties, is a well-documented process. A different approach is the use of chemical compounds that can be attached by covalent bonds to the mineral surface, leading to high stability of the surface modification.

    Here the use of diphenyldichlorosilane (DPDCS) to enhance natural zeolites and diatomites was tested, after preliminary experiments showed the best surface modification results compared with other chlorosilane types.

The success of the surface modification was documented by measurement of physical and chemical properties of the minerals, which changed dramatically. Aromatic organic contaminants showed a high affinity to the phenyl groups of the surface-modified material. Sorption of naphthalene from water was greatly enhanced by the surface modification, compared with the untreated clays.

The surface modification of the two materials exhibited great stability even under extreme conditions, which would facilitate their application in permeable reactive barriers and waste water treatment plants.

- *Phytoextraction of heavy metals* (see Chapter 9). Phyto-available fractions of heavy metals were increased with complexing ligands or chelates. Using disturbed and undisturbed soil column experiments the effects of a single dose and weekly additions of ethylenediaminetetraacetic acid (EDTA) and ethylenediaminedissuccinic acid (EDDS) were compared for the uptake of lead (and zinc and cadmium) by Chinese cabbage (*Brassica rapa*) and for the leaching of heavy metals through the soil profile. Both chelates applied to the soil increased the concentrations of lead in the leaves of the test plant.

  EDDS is a natural, non-toxic, product used in detergents. It was found to be a promising new chelate for enhanced, environmentally safe phytoextraction of soils contaminated mainly with lead. In contrast to EDTA, EDDS addition caused only minor leaching of lead, and was less toxic to plants and soil micro-organisms.

  It is possible to recover heavy metals when such biomasses are burnt. However, even the highest concentrations of heavy metals in harvestable plant tissues achieved in this study were still far below the concentrations required for efficient phytoextraction procedures.

Even on an essentially empirical level, it is evident that *there is much worth trying and learning about the tailoring of reactive agents* for remedial purposes.

19.5.2. Are the chemical processes understood?

Uranium again provides a good case in point for demonstrating the extent and limitations of our knowledge in relation to the requirements of remediation.

Uranium is a water-soluble element, so long as the general boundary conditions are oxidizing and it is in the uranium(VI) state. Otherwise, uranium(IV) shows an extremely low solubility, while uranium(V) is metastable. The removal of uranium from water thus requires either the reduction of uranium(VI) to uranium(IV), by means of ion exchange, or an effective sorption material (see Chapter 12).

However, ion exchange and sorption are known as processes with limited capacities on the one hand, and fast kinetics on the other, and in jeopardy of desorption due to changes in boundary conditions. Thus, reduction of uranium(VI) to uranium(IV) is assumed to be the only sustainable process to remove uranium from water (see Chapter 12).

Removal of uranium(VI) by zero-valent iron has been suggested as a feasible low-cost technique to control uranium contamination in surface water and groundwater. However, it is still in question under what boundary conditions the reduction of uranium(VI) to uranium(IV) occurs, which is very likely a process being triggered and catalysed by means of micro-organisms (see Chapter 12).

Modelling uranium removal requires reliable thermodynamic and kinetic data. However, data describing the kinetics of mineral precipitation, in particular, are rarely available. Where they are there can be downscaling problems, in that data from the nuclear industry tends to relate to higher concentrations (see Chapter 6). Chemical speciation models based on thermodynamic principles are sensitive to the quality of their databases – default values supplied with the widely available computer programs may not be relied upon (see Chapter 6).

Nor is there much information about the mechanisms of retardation/fixation and the possibility of re-oxidation of uranium(IV) to uranium(VI) during the complex process. Diverging results were obtained from remobilization experiments with iron columns. In some cases, remobilization with HCl was negligible; in other cases it was significant. More detailed studies on this type of binding are therefore in hand (see Chapter 12).

Not unexpectedly, such findings can be seen to support a *clear need for continuing research work on reactive agents*.

19.6. Outcomes

19.6.1. How well do the reactors work?

While reference has already been made to the US study which found that only two out of 28 groundwater remedies for CHCs had achieved their expected results within the given time period (see Chapter 18), that may be as much a comment on human optimism as on technological failure. Another view one hears about the US experience with reactive barriers is that what failures there have been have tended not to be fundamental. They are ascribed to poor site characterization or misunderstood site hydraulics rather than failure of the reactor system itself.

Does the evidence we have confirm such a view? Four of the case studies included here provide specific insights into their remediation performance: one on natural attenuation, one on pump and treat, one trial column reactor with zero-valent iron, plus the lime-based barrier for uranium leachate:

*   *Natural attenuation of BTEX contamination* (see Chapter 8). Biodegradation processes for BTEX were monitored in groundwater at a former military airfield over a 4 year period. Pollution at this site was strongly influenced by a fluctuating groundwater level: the contaminant release rate was reduced but the plume extent expanded during groundwater peaks.

    As with many earlier studies, the general effectiveness of natural bio-degradation processes was considered proven by both direct evidence (*in situ* metabolite formation) and indirect evidence (depletion of electron acceptors and characteristic hydrochemical shifts).
*   *Microbial remediation via pump and treat* (see Chapter 8). After the removal of most of the BTEX contaminant source at this former gasworks site, a

combination of on-site and *in situ* microbial remediation was carried out by pumping contaminated groundwater. The water was treated in bioreactors, and was then re-infiltrated upstream of the former source.

However, in this case the anaerobic oxidation of the aromatic hydrocarbons declined, due to infiltration of oxygen into the aquifer as a result of the remediation activities. The oxygen was used preferentially to oxidize iron monosulfides, which were then precipitated in the aquifer during hydrocarbon degradation by combined sulfate and iron reduction.

During the active remediation phase, the stability of the contaminant plume was thus temporarily disturbed, and concentrations of aromatic hydrocarbons increased in the groundwater downstream from the source area.

- *The mixed contaminant site* (see Chapter 2). This was the project in which an experimental reactor shaft containing zero-valent iron (steel wool mixed with calcite) had been installed within the treatment plant for groundwater polluted by chromate (Cr^{VI}) and hydrophobic organic compounds.

  The plume contained an average concentration of about 5 mg/l of chromate (with maximum values of 80 mg/l) plus an average concentration of 2 mg/l of halogenated aliphatics. It was found that the chromate concentration could be reduced below the target value of 0.2 mg/l.

  However, the degradation of the organic compounds in the reactor system was insufficient, so that the remedial target for such pollutants was not reached. It was thought that passivation of the surface of the steel wool – due to concurrent chromate reduction – plus insufficient residence time had led to this unsatisfactory performance with the hydrocarbons.

19.6.2. How well do the system hydraulics work?

As explained earlier, there has been little research on the hydraulic performance of actual installations, but the practical case study reported here has offered unexpected results:

- *Funnel and gate* (see Chapter 15). Groundwater modelling was used to assess the performance of the bentonite cement containment wall forming the funnel at this former gasworks site.

  Discrepancies between observed and modelled head recovery upstream of the slurry wall indicate a higher hydraulic conductivity than expected. The slurry wall had been designed to function at a hydraulic conductivity of 1×10^{-9} m/s, but the results of flow modelling suggested an effective hydraulic conductivity for the slurry wall of between 1×10^{-7} and 1×10^{-8} m/s.

  This poses questions as regards the reliability of and testing methods for slurry walls. But the discrepancy could also originate from temporal variations in groundwater recharge, not accounted for in the numerical model. Longer-term monitoring will be required to confirm the reason. It is also expected that permeability will reduce in time, due to bicarbonates in the groundwater.

Even if it is confirmed, such a deficiency may not be critical with the type of contamination and funnel-and-gate configuration in question. It should be possible to make the gate more highly conductive to compensate: although the regulators will doubtless seek evidence that funnel leakage is then tolerable, and that the reactor design is still able to provide adequate residence time.

19.7. Monitoring and durability

19.7.1. How are reactor systems monitored?

Monitoring has not been a central issue here, not because of unimportance, but since we felt it deserved and needed books and workshops of its own, and that nobody could yet provide them meaningfully, at least for operational reactive barriers. That presumption has only been reinforced by the work reported here. This section is thus geared more to problem definition than to findings.

It is not even clear what the monitoring priorities should be, not least because there is often much uncertainty about how much contaminant there is to start with. Does one have to look to measure absolute amounts, or are concentrations a sufficient index? Does one focus on contaminants removed or those remaining? What emphasis should be given to understanding flows in the reactor itself? And what emphasis upon flows outside the reactor?

Of course, the easy answer is that we need all those things; but one must be realistic. And the real issue is what are the critical parameters for the cost-effective and environmentally responsible monitoring of a full-scale operational installation, as distinct from those parameters which (validly) pre-occupy the R & D community in uneconomic pilot plants?

Opinions we have heard leave us with unease. There is a view that monitoring should focus on high-risk contaminants within lists 1 and 2; the priority has merit, but as a comment on frequency rather than need (for which it might too easily be interpreted). Some say that the effectiveness of an engineered pump-and-treat system can be more readily and directly measured, calculated and guaranteed than that of a permeable reactive barrier: but even if that is currently true, it defies us why it should necessarily remain so. One of our authors stresses the need to consider flow or concentration heterogeneities when designing monitoring systems for reactive barriers (see Chapter 4); we agree completely, but are no wiser than his call for further work on it. Some practitioners report a planning requirement to monitor aquifers outside the installation itself; but such monitoring may only involve the checking of water levels, which tells one precious little.

We can be more positive about the promise of *in situ* real-time sensors for monitoring groundwater contamination and remediation processes. Even if only covering 'signal' contaminants, these could help significantly to remove uncertainties and provide more confidence to problem holders and regulators. *We look forward to the outcome of the rapid developments underway in the sensing field, and to their integration into viable monitoring protocols for contaminant remediation reactor systems of every type.*

19.7.2. How often and for how long are they monitored?

Although essentially only a subset of the issues just touched upon, much discussion – and expense – already centres on these questions. They are, of course, closely interrelated, in that economic and intelligent monitoring relies on adjusting sampling and measurement frequencies to changes in perceived risk over time.

In the absence of well-founded risk assessments, many regulators are prepared to take a pragmatic approach. The example was given to us of a reactive barrier for list 1 and list 2 contaminants. First-year monitoring was to be on bimonthly basis, with the regulator retaining a possibility of reverting to monthly. There were then to be annual reviews for 4 years, with monitoring frequency still able to go up or down. After 4 years, there was to be the possibility of handing in the waste licence against a long-term monitoring agreement. Or the licence would remain intact until a major 10 year review.

We have some reservations about any protocol which treats monitoring frequencies of all parameters identically, or which jumps straight from monthly or bimonthly to annual monitoring: but such refinements aside, that gives the gist of a responsive approach to the uncertainties endemic even in current best practice.

One has to concede the importance of monitoring commitments to securing confidence and permits. The potential costs of having to monitor indefinitely are a real problem. Operators can often vastly improve their lot by using early monitoring results not simply to fulfil requirements and 'cover backs' but to better understand the site, reactor system and remediation processes, and to convey that under-standing to the regulator. The industry as a whole would benefit if it could develop systems which can be left to run safely and reliably for longer periods, and help ensure that best available technology (BAT) briefs in this field cover industry-wide standards and protocols for monitoring.

19.7.3. For how long are the reactor systems effective?

Given the state of play in permeable reactor barrier development in Europe, it is a little premature to expect definitive answers to this question. Long-term durability is an aspect to be addressed both by the RUBIN network (see Chapter 3) and at the 20-reactor/5-shaft model site (see Chapter 10). It has also been the central theme of the EC-funded PEREBAR project (Perebar, 2000), from which relevant findings are awaited. Perhaps the most useful thing to do meanwhile is recognize two key dimensions of the issue.

The first is whether and how *clogging or soiling of a reactive medium* can be avoided. Typical mechanisms are precipitates which fill pores or coat active surfaces, or catalysts which are deactivated through microbial fouling. Chapter 11 summarizes research under way on electrokinetic techniques for the prevention of barrier clogging. The idea is to apply an electric field upstream of the barrier, to reduce the amount of groundwater constituents that flow into the reactor and might impair its function by coating grains of the reactive medium or clogging with precipitates.

The second point to make is that reactors can be and are being designed so that *the reactive agent (and if appropriate its carrier medium) may be easily recharged or exchanged upon exhaustion* and 'breakthrough' of the contaminant.

19.7.4. Can contaminants be remobilized?

An aspect of performance on which there was more active awareness in the studies reported here was the risk of unintentional remobilization of contaminants from the reactor medium – as distinct from remobilization when a reactor is recharged by controlled flushing. There is particular concern about the long-term potential for re-release of heavy metals, as these are not as amenable as organic compounds to natural attenuation.

A thorough understanding of sorption mechanisms is critical to evaluating the efficiency of immobilization techniques using specific sorbents. For example, if sorption occurs by simple electrostatic interaction, the sorbed species may be remobilized by changes in environmental conditions. If sorption occurs by inner-sphere complexation or by surface precipitation, the sorbed species tend to be stable and not easily remobilized (see Chapter 6).

Zero-valent iron is commonly cited as a uranium-removing reagent for permeable reactive barriers. But while its uranium removal effects are well known, the mechanisms of uranium retardation and fixation are not yet understood in detail. There is neither much information about the kinetics of these reduction processes under different boundary conditions nor on the possibility of re-oxidation of uranium(IV) to uranium(VI) during the complex processes (see Chapter 12).

Similar considerations have exercised the authors of the chapters on the two practical remediation examples for uranium presented here:

* *Uranium mines flooding contamination* (see Chapter 13). With rising levels, oxygen is depleted in the flood waters, which is reflected in decreasing Eh values. This leads some contaminants (e.g. uranium) to precipitate. Others, however (e.g. arsenic and radium), can be more mobile under the more highly reducing conditions, and therefore desorption processes must be taken into consideration.
* *Heap leach uranium waste relocation* (see Chapter 14). In the very long term there would be conversion to calcium carbonate by atmospheric carbon dioxide, and a risk of uranium ions then going back into solution. But the waste is only exposed to air for 5 years or so in this situation. From the trials undertaken, there would still be some calcium hydroxide available throughout that period of exposure.

With our present state of knowledge, plus the uncertainties there will always be about future environmental conditions, it is difficult to be confident about the eventual fate of heavy metals left in any reactor. But greater awareness of the possibilities and mechanisms may help avoid more predictable remobilization incidents meanwhile.

19.7.5. Can contaminants be recovered?

In this vein it is important to distinguish between degrading and accumulative reactions. The latter can be said to accumulate environmental risks. An *in situ* reactor which stores the contaminant or daughter products will, over time, become a form of waste repository. Such a reactor should in principle be subject to very different regulatory controls to those which destroy contaminants or convert them to a harmless form (see Chapter 4).

It might still be argued that accumulating such material within an engineered body at a known location reduces risk. Indeed, it is always a tantalizing thought that enough metals might be concentrated in a convenient form to serve as a raw material for some future process of reworking and recovery due to metal concentration comparable to ores. This is not impossible, as the history of mineral working itself has shown; but it would inevitably involve a chain of processes which would themselves need to both technically feasible and economically viable. And it is a prospect which is difficult enough to realize when its time has come, let alone to plan for decades or centuries in advance.

19.8. Comparative economics

19.8.1. Have the economics of the various systems been investigated?

This is a front on which we begin to see leaps forward. There is a consciousness of the imperative to address cost-effectiveness, if technologies are to be accepted by problem holders and funding agencies; and meaningful results are beginning to trickle through.

Our initial state-of-the-art review included then available cost figures (see Chapter 1). Certain chapters have added to that picture from their own perspective.

An illustrative cost comparison between pump-and-treat systems and reactive barriers (see Chapter 18) gave rise to a discussion of recent trends in the USA and UK, where improved techniques and lower costs for trenching and slurry wall construction are said to be making long continuous barriers more viable. Whether that will ultimately favour longer reactors any more than longer funnels perhaps remains to be settled.

In Chapter 6 on biosorptive flotation, the cost of an electroflotation device plus a sand filter was said to compare favourably with a conventional treatment system using cyanide oxidation/alkaline precipitation with polymer-aided clarification. In the discussion it was suggested that, although conventional precipitation methods tended to be cheap, they were less effective with low concentrations of contaminant.

The 20-reactor/5-shaft model project is setting out to address not only technical but economic (as well as legal and ecological) aspects (see Chapter 10). The RUBIN network is assembling a comprehensive data set, covering both capital and running costs, to aid the more precise calculation of reactive barrier viability (see Chapter 3). But the most impressive piece of work to date by far is an idealized economic model constructed with the benefit of the RUBIN costs already available (see Chapter 17).

19.8.2. How is a proper basis for comparison made?

The focus of Chapter 17 was on systems based on the sorptive removal of hydrocarbon contaminants (trichloroethylene and acenaphthene) by means of *in situ* or on-site reactors using granular activated carbon. Total remediation costs for funnel-and-gate systems were compared with the costs of 'conventional' and 'innovative' pump-and-treat systems. The latter were assumed to incorporate hydraulic barriers in order to minimize the required pumping rate and, hence, their operating costs.

That, in our view, was a key and inspired departure, which goes to the essence of any proper evaluative procedure. For *if one is to treat both active and passive systems simply as reactor variants*, engineered for a cost-effective fit to specific circumstances, *why should a pump-and-treat system not be evaluated with the benefit (and costs) of a funnel or containment arrangement too*? Indeed, does one not have to do so, in order to be comparing like with like? And is it not precisely that containment element which might offer pump and treat most, in helping to overcome its recognized shortcoming of clean groundwater intrusion and contaminant dilution?

Based on a simplified hypothetical contaminated site, cost-optimal design alternatives were calculated for the three remediation options by means of a design optimization framework that combined groundwater flow modelling, hydro-chemical treatment modelling, economic modelling and mathematical optimization. The comparative assessment focused on three key decision-influencing factors: type of contaminant, contaminant concentration, and interest rate. The results were fascinating.

19.8.3. How do their (modelled) costs compare?

It is clear from the discussion in Chapter 17 that remediation costs principally increase with increasing contaminant concentration and decreasing sorptivity. However, due to the different volume rates of groundwater to be treated, the sensitivity of the three remediation options is different.

For the given hydrocarbons, innovative pump and treat (with the hydraulic barriers) was least sensitive to changes in contaminant concentration. Furthermore it outperformed both other options at any assumed concentration. Since the conventional pump-and-treat approach was the economically worst option in most cases, this was taken as clear evidence for the large economic benefit that may be obtained by means of additional barriers.

Making an economic comparison between pump-and-treat systems (high operating costs) and funnel-and-gate systems (high initial investment costs), one might expect that interest rates would primarily impact on the costs of the pumped system. But for the sorptive removal considered in the study of Chapter 17, that was found to be true for conventional pump and treat only. The innovative pump-and-treat system was less sensitive to interest rates than the funnel-and-gate system, although if interest rates decreased, optimal design tended to favour longer or additional barriers. The comparatively high sensitivity of the funnel-and-gate system to interest rates was caused by the periodic replacement of the activated carbon reactor material. Conversely, in the case of a one-time zero-valent iron filling, total remediation costs were more or less independent of interest rates.

Such conclusions raise questions for some of the conventional wisdom about active systems such as pump and treat being more cost-effective for the early stages of remediation, when concentrations are high, with passive systems such as reactive barriers being more suited to the long-term treatment of the low-concentration tail. If they were to be confirmed, *there could be many situations in which the most critical aspect of intervention, both short and long term, is more effective containment*. Given that, the basis for selecting an appropriate reactor system might then be very different.

We must here repeat the authors' caution. It is not possible to rank the options considered in a conclusive way – this will always be a question of the specific situation addressed (see Chapter 17). That must be right. And different idealized assumptions, or model and data refinements, may well produce contrary results in future studies. But we do happen to be familiar with the power of such idealized models to give valuable insights in other fields. We know they can be highly influential to decisions and design; and rightly so.

19.9. Investment, risk and acceptability

19.9.1. How much remediation is affordable?

The ability to compare remediation options more objectively and systematically, using such methodologies, does not itself resolve all the decision issues. Even if we also had better defined and costed 'do nothing' options – i.e. an evaluation of the losses and risks of leaving sites untreated, except via natural attenuation – there are, as in any investment situation, certain thus-far unquantifiable factors.

Some of these are of a sociopolitical nature. The payback on any investment depends upon one's framework for comparison. Under certain central and eastern European administrations, for example, the overriding consideration is the cost of the penalties if the owner fails to remediate. Under others, there may be a desire to use 'best available technology'; but what is effectively 'available' is as highly fluid as the local economic and security situations.

In the USA, we are told, the perception is that industry is willing to switch from pump-and-treat systems to more capital-intensive approaches if they can see a payback within 10 years. In the EU, apparently, some data indicate that the acceptable payback period needs to be half that. In the UK, pump and treat has been relatively little used; so there is a tendency to compare with the cheaper 'dig-and-dump' approach, paybacks become more distant, and more expensive new technology is even less attainable.

Such uncertainties are mirrored in doubts over time preferences and discount rates. The idealized economic analyses reported above (see Chapter 17) used discounted costs, which is the recognized way to do these things. If one accepts the perspective and results of discounting, the longer-term costs (and benefits) of monitoring, etc., over a time frame of decades, become negligible in net present value terms.

But that is tantamount to saying we do not care about the repercussions that far down the line; let us pass on those problems (and opportunities) to our children – which is hardly consistent with notions of precautionary principles and sustainability which governments espouse today. Nor can it be said to be simply a philosophical issue. Discounting is actually being used to guide decisions. And discount rates could be adjusted to reflect different time preferences for 'sustainability' investments, if there was the evidence and will to do so. It is a conundrum not unique to ground-water remediation.

Meanwhile, we may in some sense be fortunate in often still being practically restricted to dealing with those sites where the payback from some form of action is clear to all: sites for which remediation will release more than enough development or after-use value to cover its cost; sites which pose such a high and evident risk to human health or the ecology that no cautious or responsible government can ignore their condition; or sites for which prevention and clean-up costs are so modest that it makes obvious sense to tackle them now. Perhaps frames of reference will be clearer by the time all the more marginal sites have to be tackled.

19.9.2. How can risk be reduced?

There are sometimes suggestions that active systems or pump-and-treat systems are intrinsically less risky than passive systems or reactive barriers – or vice versa. For instance, it is said that pump-and-treat systems are more easily adjustable and repairable than permeable reactive barriers. That may be so, but begs the question as to whether there could be more need for adjustment or repair in an active than a passive set-up. *We have yet to see persuasive evidence that either is generally more risky than the other, or that either is necessarily more capable of guaranteeing cheaper, faster or more thorough clean-up*.

Improved evaluative methods may themselves help minimize risk. Limited knowledge about the contaminant source and the spatial distribution of hydraulic conductivity will have to be handled by means of geostatistical approaches. Further research should focus on better ways to deal with uncertainties and to manage complex reactor design and operating parameters. Such refinements would eventually lead to cost–reliability distributions for each remediation alternative. Expected remediation costs could then be related to the probability that the alternative solutions proposed would actually meet remediation targets (see Chapter 17).

One way of accounting for the risk that they will not do so is to provide for a fall-back solution in one's evaluation, at cost multiplied by the estimated probability in having to resort to it. The cost may, for example, be the cost of retaining or providing a stand-by pump-and-treat system – as in some of the options mentioned for the uranium mine flooding case (see Chapter 13). Or it might be the cost of removing to an engineered landfill an accumulating reactor which has become a heavy metals repository.

In the USA, apparently, if an approved innovative process fails, the Environmental Protection Agency is willing to pay the costs for back-up treatment, i.e. the
government underwrites that part of the risk.

In Germany, by contrast, we understand that the government will pay only for
monitoring, not for remediation. A reason given for the shorter payback horizon
demanded in Europe is that here the site owner or consultant has to carry such risk
of failure. In other words, the short payback period is a way of internalizing an
allowance for the likely cost of the fall-back solution in their evaluation of the
scheme.

Ultimately, however, the safeguard is to actually provide the back-up, i.e. to build
redundancy into the installation for safety reasons. And in situations of scarce
resources there even comes a time when one has to decide whether to put money
into some such practical back-up solution on the ground, or into yet more
sophisticated models and evaluation reports at some consultancy or university
hundreds of kilometres away!

19.9.3. How can acceptance be gained for solutions?

Acceptability of remediation technologies is perceived to be a major problem:
acceptability to site owners and developers, to private financiers and public funding
agencies, to the local community, and, above all, to the regulators. There is
considered to be a huge amount of inertia in many European administrations, and
an unwillingness to have the courage of convictions, and try new remediation
approaches on their merits.

We are not in a position to either deny or validate such claims. What we would say
is that innovation is not always beneficial, and that regulators have an obvious
responsibility to ask difficult questions as to whether a proposal works. And one of
the best ways of persuading them is evidence from other cases, such as those
presented in this book, plus site-specific data and feasibility assessments.

*A handicap in Europe is the absence of widely accepted quality standards and quality
management schemes, performance assessment procedures and criteria, or training and
certification programmes and manuals, on remedial technologies.* The US Environmental Protection Agency and industry joined together in a forum to progress such
matters some years ago, and we are led to believe that it crossed a watershed in
advancing mutual understanding and collaboration by all sides in the business. To
establish parallel mechanisms here would seem to be a natural way forward.

19.10. Conclusion

We arrive at this end point more impressed by the similarities between passive and
active remediation systems than by their differences. They are broadly able to target
the same contaminants. At the heart of each are reactors which can call upon a range
of fundamentally similar agents and processes. Both require a thorough practical
understanding of groundwater hydrology and of engineering measures to control
it. The specification of either requires a similar raft of site and laboratory

investigations and technological design skills. Their economic and risk assessments rely on similar data and models, produce competitive results, and (at least at present) leave us with equivalent uncertainties. Specialist subsidiary elements (such as excavation or pumping) apart, professionals best able to advise on one approach are quite likely to be among those also best able to advise on the other.

The approaches will show significant scientific, engineering and cost differences in particular applications, but no more than, say, between steel and concrete structures. As with steel and concrete structures, both can often do the job, and the choice between them may depend more on matters such as fluctuating prices or local conditions than on their intrinsic capabilities. As with steel and concrete, which can happily coexist in the same city, building or room, we would expect to see combinations, hybrids and successions of pumped and barrier approaches in the remediation of any complex area of industrial contamination. As with steel and concrete applications, a whole series of guidelines, standards and quality initiatives are needed to give proper assurance to interested parties.

Research and development effort on groundwater remediation which departs from a recognition of such similarities is in our view much more likely to be productive than that hung up on issues as to whether the approaches are new or old, active or passive, or operate above or below ground. Regulations which do likewise would do best for the environment and could save a lot of wasted money.

19.11. References

PEREBAR (2000). Long-term performance of permeable reactive barriers used for the remediation of contaminated groundwater. European Research Project in the 5th Framework Programme. Website: http://www.perebar.bam.de.

REMEDIATION TECHNOLOGIES DEVELOPMENT FORUM (RTDF) (2002). Permeable reactive barriers installation profiles. Website: http://www.rtdf.org/public/permbarr/prbsumms/ default.cfm.

ROBBAT, A. (2001). *A Dynamic Site Investigation Adaptive Sampling and Analysis Program for Operable Unit 1 at Hanscom Air Force Base Bedford, Massachusetts*, EPA X991631-01. US Environmental Protection Agency, Washington DC.

Index

acenaphthene (ACE) 133, 271
accelerated testing 32
active agent 311–12
adaptive sampling 307
aliphatics 35–37
n-alkanes 128–34
ammonium acetate method 142
apatites, sorption mechanisms 123
aromatic hydrocarbons, biodegradation
    under aerobic conditions 131
    under anaerobic conditions 131–4
array of wells 23, 25
arsenic 54, 75
atomic absorption spectroscopy (AAS) 142–3

batch reactors 76
bayerite 124
Beerwalde, flooding of uranium mines 216
benzene, biodegradation 131–2
benzene, toluene, ethylbenzene and xylene
    see BTEX
Bernau, PRB project in 50–2
best available technique not entailing
    excessive costs (BATNEEC) 281, 293,
    301, 308
best available techniques (BATs) 281–2, 301,
    307, 316, 320
best practical environmental option (BPEO)
    301
biobarriers 24–6
biochemical oxygen demand (BOD) 77
biodegradable polymer slurries 21, 85
biodegradation 127–34, 163
biofilm 24
bio-fouling 8
bio-indicators 307
biomass 13, 24
biopolymer slurry technique 85–97
biopolymer trenching 26
bioremediation 23
biosorption 104–10
biosorptive flotation 102, 106, 310

Bitterfeld, SAFIRA test site 158–69
bored-pile cut-off walls 18
Brassica rapa 142, 312
BREFs (BAT reference documents) 282
BTEX 7, 25, 237, 302
BTEX-contaminated sites, monitored
    natural attenuation (MNA) at 128–34,
    313
bubble aeration 102

calcium-oxide-based reactive barriers 221–33
cadmium, chelate-induced phytoextraction
    146–51
calcium diuranate 223
cationic surfactant 10, 24
caisson 21, 86
centre of circulation 254
chabazite 178
chlorinated ethenes 44
chlorinated hydrocarbon (CHC)
    contamination 302
chlorinated volatile organic compounds
    (cVOCs) 44
chlorosilane surface-modified natural
    minerals 183–8
Christian Albrecht University, Kiel, PRB
    project 70
chromium 9–10, 35–42, 101–3
    chromate 9–10, 35–37
    sorption on zeolites 10
chromium-contaminated groundwater in
    subsurface Fe^0 reactor systems 35–41
    field site and hydrogeological setting 35–6
    groundwater remediation concept 36–9
clays, covalent surface modification 311–12
clinoptilolite 178, 179, 181, 184–6, 188
clogging of reactive medium 316
colligends 103
colloids, fixation on 121
complexing agent 8, 114, 116
composite cut-off wall 17
constant capacitance model 120

constant-volume batch reactor 77
contaminant
    measurement of 318
    mixed 303, 314
    plume 23
    remobilization 317
    types 307
contamination, definition 73
continuous stirred-tank reactor (CSTR) 105
contractor method 17, 26
controlled release 310
corrected selectivity coefficient 117
cost(s)
    contaminant type and concentration 277
    investment 266–74
    interest rates 277–8
    of PRBs 298–9
    operational 15, 266–76, 298–9
    pump-and-treat systems cf PRBs 13–14
    remediation using PRB 4
    RUBIN 318
    sorbents 107
    sorption, remediation costs 270–8
cut-off wall construction 16–18
    bored-pile cut-off walls 18
    composite cut-off wall 17
    frozen walls 18
    injection walls 18
    jet grouting 18
    sheet-pile walls 17–18
    single-phase diaphragm wall 16–17
    technology 20–1
    thin walls 17
    twin-phase diaphragm wall 16, 17
cyclohexanediaminetetraacetic acid (CDTA)
    146

Darcy equation 252
Darcy number 254
Denkendorf, PRB project in 53
decommissioning of uranium mines 209–18
deep soil mixing 21
degradation 4–8, 23, 35, 87–9, 129–34, 163–8
desorption 2, 7
diaphragm wall 16–7
diatomite (kieselguhr) 184, 186
cis-dichloroethene (cis-DCE) 35, 44
1,2-dichloroethene (1,2-DCE) 86, 161
cis-1,2-dichloroethene (DCE) 95
trans-dichloroethylene (trans-dichloroethene,
    DCE) 35, 161
diethylenetriaminepentaacetic acid (DTPA)
    146

diffusion 80, 121–123
dig-and-dump approach 320
dioxins 102
diphenyldichlorosilane (DPDCS) 185,186,
    188, 311
dispersed-air flotation 102, 106
dispersion 80
dissolution 2, 9, 12
dissolved-air flotation 103, 107
dodecylamine 106
drain and gate 45, 49
Dresden, PRB project in 54–5
Dresden-Gittersee 54
    flooding of uranium mine 213
drilling and deep soil mixing 21
Dubinin-Radushkevitch model 115–16

Eberhard Karl University, Tübingen PRB
    project 70
economics 318–20
Edenkoben, PRB project in 55–9
electrode reactions 174–5
electroflotation (electrolytic flotation) 103,
    107, 108–10
electrokinetic soil remediation 172
electrokinetic techniques 172–87
electrolysis of water 174–5
electrolytic flotation 103, 107, 108–10
electromigration 172, 173–4
electro-osmosis 172–3
electrophoresis 172, 174
electroremediation 175–7
elemental iron 4, 5, 9, 11, 35
ethylbenzene, biodegradation 131–2
ethylene
        glycol-bis-(2-aminoethyl)-tetraacetic
        acid (EGTA) 146
ethylenediaminedi(o-hydroxyphenylacetic)-
        acid (EDDHA) 146
ethylenediaminetetraacetic acid (EDTA)
        146, 147, 148–9, 150, 152, 312
equations of continuity 252
evolution strategies approach 268

Faraday constant 6, 120
ferric oxyhydroxide 13
ferrous ion 4–5
finite-volume model for flow hydrodynamics
    249–60
    modelling approaches 250–1
    finite-volume method 251

formulation of the mathematical model
    252–3
numerical results 254–60
flocculation 9
flotation 103–4
flooding of uranium mines 209–18, 309
fluidized beds 103
forensic environmental geochemistry 128
free energy 6
Freundlich equation 179–80
Freundlich exponents 167, 270
Freundlich model 116
Freundlich sorption isotherms 7, 167, 186
froth flotation 103
frozen walls 18
funnel-and-gate systems 4, 7, 15, 20, 23–5, 86,
    265–79, 314–15
    hydraulic performance 269, 314–15
    remediation costs 273–6

Gaines–Thomas equation 117
gamma-alumina 124
gas chromatography 128
gas oil contamination 128
Gittersee 54
    flooding of uranium mines 213
goethite 6, 102, 304
Gouy--Chapman model 120
granular activated carbon (GAC) reactor 270
granulated iron 22
'green rust' 6
guar gum 26, 85–89

halogenated hydrocarbons 4–6
heap leach uranium waste relocation 221–33,
    317
heavy metal 8–9, 139–52
    analysis 142–3
    chelate addition, effect on soil
        micro-organisms 151
    chelate-induced phytoextraction of lead,
        zinc and cadmium 146–51
    column phytoextraction experiments
        disturbed soil profile 142
        undisturbed soil profile 142
    leaching 149–50
    materials and methods 140–3
    phospholipid analysis 143
    precipitation 8–9, 11–13
    sequential extraction 142, 143–5
    soil samples and analysis 140–2

sorption mechanisms on inorganic solids
    113–24
uptake by plants 147–9
heptadecane 128
hexadecyltrimethylammonium (HDTMA)
    10, 183
high-density sludge (HDS) precipitation
    process 310
high-pressure injection (HPI) 18
high-pressure jet 15, 21–22
hydraulic conductivity 237–47
hydraulic fracturing 24
hydrocarbons test situations 320
hydrocyclone 22, 103
hydroxyapatite 9, 11
    main sorption mechanisms 123
hydroxyethylenediaminetriacetic acid
    (HEDTA) 146
hydroxypyromorphite 9
hyperaccumulating plants 140
hyperaccumulation 146
hyperaccumulators 140

indene 133
injected systems 23–4
injection walls 18
inner-sphere surface complexes 120
innovative pump and treat 266, 268–9
investment costs 266–74
investment, risk and acceptability 320–3
ion exchange 4, 10, 13, 26, 37
    choice of 311
ion exchange/adsorption 101, 304
ion exchange model 117–18
ion flotation 103
iron/sand mixture 26, 85–97
isoprenoid hydrocarbons 128, 129

jet grouting (roddingjet; high-pressure
    injection; HPI) 18, 21–2
jet pump 21–2

kaolinite 13
Karlsruhe East gasworks site 20
kieselguhr (diatomite) 184, 186
Kjeldahl digestion method 142
Klebsiella 25
Königstein, uranium removal 198–203, 308,
    309
    flooding of mines 213, 216–18, 219
KOP Superfund site 103

Langmuir adsorption isotherm 7
Langmuir model 7, 8, 12, 115, 116
lasagna technique 176
leachate 221–33
lead 9
    chelate-induced phytoextraction 146–51
    plant uptake 144–5
Lichtenberg open pit, flooding of uranium
        mines 214
long-term performance 6, 26, 27
lysimeter 224–6

maximum contamination level (MCL) 194
Melich method 142
metals, test situations 320–1
methyl-mercury 121
methyl-t-butyl ether (MTBE) 178
microbial fouling 316
mixed contaminants 303, 314
modelling 237–47, 249–60
monitored natural attenuation (MNA)
        128–34
    analytical monitoring 128–9
monitoring 18, 73, 83–5, 90–7, 128–34,
        209–17, 237–47, 291–4, 298
monochlorobenzene 161
Monod equation 197
Monod rate 198
monosite 1-pK model 120
monosite 2-pK model 118–19, 120–1
mucoid phenotype 25

Na-clinoptilolite 179, 181
naphthalene 133
naphthalene sorption by modified zeolites 310
natural attenuation 196–8, 200–1
    enhanced 203
natural reactive treatment zones (natural
        RTZ) 74–5
Navier–Stokes equations 252
nitrilotriacetic acid (NTA) 133, 146
Nordhorn, PRB project in 59–60

octadecane 128
Offenbach, PRB project in 60–3
operational costs 15, 266–76, 298–9
organic compounds, sorption of 7–8, 186–8
organic pollutants 4–8
organo-zeolite 10
outer-sphere surface complexes 120
oxidation state 9, 11, 12
oxygen releasing compound (ORC) 26, 168

packaging
    by location 309–10
    over time 309
passive technology 266
PCBs 102
perchloroethene (PCE) see
        tetrachloroethene
PEREBAR project 27, 176, 316
permeability 6–8, 17–20, 24–25
permeability tensor 253
permeable reactive barriers (PRBs) 1, 2
    basics 4–13
    contamination, location of 73–4
    cost comparison between pump-and-treat
            systems and 13–14
    design 73, 74
    dispersion and diffusion 80, 121–123
    engineering 14–26
    monitoring 18, 73, 83, 85, 90–97
    natural 74–5
    pump and treat vs PRBs 281–99
    reaction mechanisms 76
        first-order reactions 76
        second-order reactions 76
    reactor recycle (recycling) 78–80
    reaction time 75
    short circuiting/by-passing 80–2
    technologies
        development in Germany 45–6
        global status 44–5
        reactor types 77
petroleum contamination 128
phased treatment 310
phospholipid fatty acids (PLFAs) 143
PHREEQC 199, 201, 203
phytane 128, 129
phytoextraction of heavy metals 139–52, 312
    chelate-induced 146–51
    column phytoextraction experiments
        disturbed soil profile 142
        undisturbed soil profile 142
phytotoxicity 150–1
pilgrim's pace method 17
1-pK multisite complexation model ('Music')
        120
plug-flow reactors 37
    degree of reaction in 77–8
plume management 265, 305
p-nitrophenol (PNP) 177
Pöhla, flooding of uranium mines 213
point of zero charge (PZC) 119
polyaromatic hydrocarbons (PAHs) 7, 237

porosity 7
Portadown, PRB at 237–47
  geology and topographic setting 238–40
  hydrogeology 240–3
  installation phases 240
  numerical modelling 243–4
precipitate flotation 103–4
precipitation 4, 6, 8–9, 11–3, 26, 102–4,
    106–10
  of uranium 11–3, 223, 225, 229, 304
presumptive remedies 101
pristane 128, 129
Pseudomonas 25
pump-and-treat systems 102, 265–79
  basics 1–2
  conventional 268
  costs 271–3, 298–9
  cf PRBs 281–99
    conceptual design/dimensioning criteria
      293–4
    cost comparison 13–14
    data gathering and data gaps 290–3
    evaluation process 282–90
    microbial remediation vs 313–14
    operation/maintenance/monitoring
      /risks 298
    selection criteria 282–90
    treatability studies/final design 294–8
  innovative 266, 268–9
pyrite 203, 221, 222

reaction 76–8
  first-order 76
  second-order 76
reactive barrier construction 19–26, 283
reactive thin walls 21
reactive treatment zones (RTZs) *see*
      permeable reactive barriers (PRBs)
reactor recycle (recycling) 78–80
reactors
  degree of reaction in 77–8
  monitoring 315–16
  types 76–7
rebound 2, 243
redox potential measurements 94–5
redox reactions 4, 6, 134, 304
reducing agent 4, 5
repository design 73, 74
residence time 7, 20–3, 40, 75–83
retardation 7, 13
retardation factor 7
Reynolds number 254

Rheine, PRB project in 63–8
risk reduction 321–2
roddingjet 18
Ronneburg, flooding of uranium mines 213,
      214, 215, 216
Rosen formula 270
RUBIN 303
  costs 318
  mission, goals and structure 43–4, 46–9

SAFIRA 46
  pilot facility 161–3
  project 158–9
  reactor technologies 163–8
    activated carbon filtration 163–7
    catalysis 167
    coupled redox reactors 167–8
    microbiology 167
  structure of 158
  test site at Bitterfeld 158–61
    geology and hydrogeology 158–60
    groundwater contamination 160–1
saturated hydrocarbons, biodegradation of
    under aerobic conditions 129–30
    under anaerobic conditions 130–1
saturation index 197
scanning electron microscopy (SEM) 90
Schlema, flooding of uranium mines at 213
schoepite 195–6
sequential extraction 140–3
sheet-pile walls 17–18
siderite 6
Silent Piler 21
silanization 184–6
single-phase diaphragm wall 16–17
slurry pump 21–22
slurry wall 237–47
soil mixing 306
Soil Protection Act and Ordinance
      (Germany) 1
soiling of reactive medium 316
Somersworth Landfill Superfund site, PRB
      at 85–97
  cored material testing 88–90
  groundwater wells and monitoring 90–5
  hydraulic testing 95–7
  installation and development 87–8
  site description and characteristics 86
sorbent
  choice of 311
  column studies 181–2
  innovative 178–88

quantitative description of sorption
        processes 179–80
    reaction kinetics 180–1
sorption 101–10, 178–188, 270–4
    case study 266–7
    complexing agents 8, 114, 116
    design optimization framework 266
    desorption 2, 7
    distribution coefficients 7, 13, 114–15
    influence of other factors 123–4
    hydraulic performance evaluation 268–9
    isotherms 115–16, 179
    kinetics 121–2
    mechanisms on inorganic solids 113–24
    models 116–21
        ion exchange model 117–18
        surface complexation models 118–21
    remediation costs 270–8
        contaminant type and concentration 277
        interest rates 277–8
    speciation in solution 121
sorptive flotation 102, 106–8
source zone treatment/removal 305
speciation 139–150
steel wool 37
Stern model 120
stirred-tank reactors 37
strontium, sorption on polyantimonic acid 118
surface complexation models 118–21

tailing 2
tetrachloroethene 5, 6, 44, 86, 95, 178
thin walls 17
toxicity characteristic leaching procedure
        (TCLP) 150
toxicity maps 307
toxicity tests 307
tremie tube/pipe 17, 26, 85, 87
trichloroethylene (trichloroethene, TCE) 35,
        44, 86, 95, 161, 177, 271
trisodium (S,S)-ethylene diamine disuccinate
        (EDDS) 147, 148–9, 150–1, 151–2, 312
twin-phase diaphragm wall 16, 17

University of Applied Science of North-East
        Lower Saxony, Suderburg, PRB
        project 69
uraninite 11–12
uranium and uranium mining 11–13, 231
    concentration 226–8
    decommissioning by flooding 209–19, 306,
        317

flooding strategies 212–18
    mine remediation 210–12, 303–4
    pump and treat vs collect and treat 212
    leachate composition 223
    leachate dry content 227
    leachate pH 225, 226, 232
    leachate specific electrical conductivity
        226, 227
    leachate volume 225
    sulfate--uranium correlation 228
    leaching from wastes 221–2
    migration, calcium oxide-based reactive
        barriers to attenuate 221–33
        field test results 231–2
        laboratory experiments 223–33
        large-scale field test 229–31
        open-air experiments 224–32
        process steps 223
        reactive barrier arrangement 228–9
        reactive barriers in practice 232–3
    removal from water 193–206
        natural attenuation processes 196–8
        water chemistry 195–6
    sources 193–4
    toxicity 194
uranium heap leach relocation 306
uranophane 195–6, 202–3

vinyl chloride (VC) 35, 44, 86, 95
volatile organic compounds (VOCs) 44, 86

Walkley–Black method 142
washing or sweeping phase 309
well-based systems 23
Wiesbaden, PRB project in 68–9
Wismut mines, flooding strategies 212–18,
        309
WSPEST 243

X-ray photoelectron spectroscopy (XPS) 90
o-xylene 131
xylenes, biodegradation 131–2

zeolites 178–83, 304
zero-valent iron (ZVI) 6, 26, 35, 85, 91, 277
    uranium removal from water using 13, 44,
        203–4
zeta potential 105
zinc, chelate-induced phytoextraction of
        146–51
zinc speciation and plant uptake 144–5